The Natural and the Artefactual

The Natural and the Artefactual

The Implications of Deep Science and Deep Technology for Environmental Philosophy

Keekok Lee

LEXINGTON BOOKS
Lanham • Boulder • New York • Oxford

LEXINGTON BOOKS

Published in the United States of America
by Lexington Books
4720 Boston Way, Lanham, Maryland 20706

12 Hid's Copse Road
Cumnor Hill, Oxford OX2 9JJ, England

British Library Cataloguing in Publication Information Available

Library of Congress Cataloging-in-Publication Data

Lee, Keekok, 1938-
The natural and the artefactual : the implications of deep science and deep technology for environmental philosophy / Keekok Lee.
p. cm.
Includes bibliographical references and index.
ISBN 0-7391-0061-0 (cloth : alk paper)
1. Environmentalism—Philosophy. 2. Nature—Effect of human beings on. I. Title.
GE40.L44 1999
304.2'8—dc21 99-20325
CIP

Printed in the United States of America

∞™ The paper used in this publication meets the minimum requirements of American National Standard for Information Sciences—Permanence of Paper for Printed Library Materials, ANSI/NISO Z39.48–1992.

For Bill
A True Lover of Philosophy

Contents

Acknowledgments xi
Introduction 1
Notes, 9
Chapter 1 Worldviews: Modern and Pre-modern 11
Modernity, 11
The Old Philosophy and the Old Science, 12
The New Science and Its Method, 15
The New Philosophy, 22
Goals of the New Science, 28
Teleology, Its Forms, and Their Fortunes, 32
Notes, 41

Chapter 2 The Natural, the Artefactual, and the Technological 49
Extrinsic/Imposed Teleology and the Artefactual, 49
Modern Science and the History of Technology: Transforming the Natural to Become the Artefactual, 54
Modern Technology, the Philosophy of Technology, and the Philosophy of Science, 62
'Deep' Theories and Their Power of Control, 69
Notes, 73

Chapter 3 Independence, Human Design, and Artefacticity 81
The Natural: Different Senses of 'Nature', 82
The End of Nature? 86
Nature's Independence, 90
The Garden, 93
Biotic Artefacts and Their Residual *Tele*, 95
Notes, 97

Chapter 4 Techhnology: Threats to the Natural 107
Extant Technology and the Less Radical Threat to the Natural, 107
Biotechnology and Its Radical Threat to Biotic Nature, 112
Future Technology and the Radical Threat to the Natural, 114
Homo Faber, the Humanization of Nature, and the Naturalization of Humanity, 123
Homo Faber, Artefacts, and the Language of Machines, 135
Notes, 149

Chapter 5 Ontology and Axiology 159
Resisting Humean Projectivism, 161
Interests and the Intentional Stance, 166
Abiotic Nature and Intrinsic Value, 172
Intrinsic Value, Trajectories, and Independent Value, 177
Ontological Elimination, Dualism, and Dyadism, 180
Ecosystem Health and the Human/Nonhuman Ontological Dyadism, 184
Environmental Ethics and Environmental Philosophy: Axiology and Ontology, 189
Anthropogenic and Nonanthropogenic, 194
Narcissism and *Homo Faber*, 201
Notes, 203

Conclusion 223
Note, 228

Appendix 1 Nature as Work of Art 229
Notes, 237

Appendix 2 Is Nature a Mere Social Construct? 241
Notes, 249

Bibliography 253
Index 271
About the Author 287

Acknowledgments

I wish to thank Holmes Rolston for his suggestion, at an earlier stage, to prune a vastly larger manuscript to become a more stream-lined book. I am similarly grateful to John Nolt for his numerous suggestions to strengthen the arguments in some of the chapters as well as improve the overall organization of the book. I am convinced that the resulting monograph has benefited enormously from their interventions. However, I bear sole responsibility for the work as it stands in spite of the excellent advice I have been so generously offered. And finally but most certainly not least, I wish to express my gratitude to Kim Bevington for her patient help, given beyond the call of duty and under trying circumstances; without her generosity of time and effort, many of the technical problems faced in producing the final copy for publication would not have been solved. However, others, who also gave me invaluable technical assistance, are too numerous to mention individually; but to each and everyone of them, too, I wish to say a big thank you.

Introduction

Pollution in the last few decades has been perceived as the central disvalue generated by our industrial civilization. Furthermore, our concern about pollution, naturally enough, focuses on the damage it does to our home planet, Earth, which is unique among the planets of our solar system in possessing life. At the same time, we, *Homo sapiens*, though late comers in the history of natural evolution, have, since our emergence, humanized to an excessive extent Earth's landscape. Our modern industrial civilization is but a continuation of this process of humanization; the main difference between past and modern impacts upon the environment is traced, by and large, to a kind of degradation caused by more modern technology which can be said to be ecologically insensitive.

The emphasis on pollution produced by an ecologically insensitive technology affecting Earth's biota has meant that, on the whole, theorists have ignored several related issues, such as the dynamic character of (modern) technology itself and the kind of impact such changing technology could produce on the environment. One of the aims of this book is to redress the balance by attending to more recently established technologies such as biotechnology and to techonologies of the rising future, such as (molecular) nanotechnology. It will be argued that these more radical and powerful technologies induced by 'deep' scientific discoveries are much less polluting than those responsible for causing the environmental degradation we have been witnessing. However, their drawback lies in their capability to systematically transform naturally-occurring beings (whether biotic or abiotic) to become artefactual ones.

The philosophical implications underlying this technological transformation need to be laid bare, showing how the pre-modern worldview constructed out of its philosophy supporting its science was supplanted by the modern worldview, in turn constructed out of its philosophy supporting its own science, and how the technologies of the late twentieth century and the rising ones of the twenty-first century are but the fruits of the Modern Project of Science/Technology promised since its inception. The Modern Project was (and is) conceived within the metaphysics of Scientific Naturalism;[1] its ideological goal was (is) the control and manipulation of nature to fulfill human goals and purposes, the primary ones being the improvement of material well-being on the one hand and the achievement of self-realization on the other, in new and ingenious ways rendered possible by such technological innovations. From this viewpoint, it becomes clear that *homo faber* and its preoccupation with fabrication is the fundamental category of modernity.

That fundamental category involves two very different philosophical standpoints by which humankind's relationship with nature may be appraised; until now, the emphasis has been on the axiological dimension. In environmental ethics, since its emergence about twenty years ago, a debate has centered on the crucial issue of whether humans alone are intrinsically valuable and, therefore, morally considerable. The modern worldview characteristically denies moral considerability to nonhuman nature, thereby rendering it of only instrumental value to humankind. The philosophical challenge to it consists of various and different attempts to establish that parts of nonhuman nature on Earth (or indeed Earth's biosphere itself) may be said to warrant moral consideration. But it is time to turn to the ontological dimension. To mark this shift in focus, it would be more appropriate to refer to the new discipline in philosophy as environmental philosophy.

To appreciate this dimension one needs to highlight the distinction between the artefactual and the natural. The former is the material embodiment of human intentionality—an analysis in terms of Aristotle's causes shows that all four causes, since late modernity, may be assigned to human agency.[2] The latter, *ex hypothesi*, has nothing to do with human agency in any of its four causes. This shows that the artefactual and the natural belong to two very different ontological categories—one has come into existence and continues to exist only because of human purpose and design while the other has come into existence and continues to exist independently of human purpose and design. In the terminology of this book, the artefactual embodies extrinsic/imposed teleology while the natural (at least in the form of individual living organisms) embodies intrinsic/immanent teleology. However, the more radical and powerful technologies of the late twentieth and the twenty-first centuries are capable of producing artefacts with an ever increasing degree of artefacticity. The threat then posed by modern *homo faber* is the systematic elimination of the natural, both at the empirical and the ontological levels, thereby generating a narcissistic civilization.

In this context, it is, therefore, appropriate to remind ourselves that beyond Earth, nature, out there,[3] exists as yet unhumanized. But there is a strong collective urge, not merely to study and understand that nature, but also ultimately to exploit it, and furthermore, even to transform parts of it into *ersatz* Earth, eventually making it fit for human habitation. That nature, as far as we know, has (had) no life on it. These aspirations raise a crucial problem which environmental philosophy ought to address itself, namely, whether abiotic nature on its own could be said to be morally considerable and the grounds for its moral considerability. If no grounds could be found, then nature beyond Earth is ripe for total human control and manipulation subject to no moral but only technological and/or economic constraints. The shift to ontology in grounding moral considerability will, it is argued, free environmental philosophy from being Earthbound in the millennium about to dawn.

In slightly greater detail, the aims of this book may be summarized as follows:

1. To show how modern science and its technology, in controlling and manipulating (both biotic and abiotic) nature, transform it to become the artefactual. It also establishes that there are degrees of 'artefacticity' depending on the degree of control and precision with which science and technology manipulate nature. An extant technology such as biotechnology already threatens to imperil the existence of biotic natural kinds. Furthermore, technologies of the rising future, such as molecular nanotechnology, in synergistic combination with biotechnology and microcomputer technology could intensify this tendency to eliminate natural kinds, both biotic and abiotic as well as their natural processes of evolution or change.
2. To consider the implications of the above for environmental philosophy, and in so doing, to point out the inadequacy of the extant accounts about intrinsic value in nature. By and large (with some honorable exceptions), these concentrate on arguing that the biotic has intrinsic value but assume that the undeniable contingent link between the abiotic and the biotic on Earth would take care of the abiotic itself. But the proposed terraformation of Mars (and even of Earth's moon only very recently) shows the urgent need to develop a much more comprehensive environmental philosophy which is not merely Earthbound but can include the abiotic in its own right.
3. The book also raises a central inadequacy of today's approaches in environmental philosophy and movements. They concentrate predominantly on the undesirable polluting aspects of extant technologies on human and nonhuman life, and advocate the introduction of more ecologically sensitive technology (including this author's own earlier writing). If this were the most important remit of environmental philosophy, then one would have to admit that nature-replacing technologies (extant and in the rising future) could be the ultimate 'green' technologies as their proponents are minded to maintain, in spite of their more guarded remarks about the environmental risks that may be incurred in running such technologies.[4] Such technologies would also achieve what is seemingly impossible, as they promise to make possible a world of superabundance, not only for the few, but for all, without straining and stressing the biosphere as a sink for industrial waste. But this book argues that environmental philosophy should not merely concern itself with the virtuous goal of avoiding pollution risks to life, be that human or nonhuman.

It should also be concerned with the threat that such radically powerful technologies could render nature, both biotic and abiotic, redundant. A totally artefactual world customized to human tastes could, in principle, be designed and manufactured. When one can create artefactual kinds (from what Aristotle calls 'first matter,' or from today's analogue, what we call atoms and molecules of familiar elements like carbon, nitrogen, hydrogen, etc.) which in other relevant respects are indistinguishable from natural kinds (what Aristotle calls 'second matter'), natural kinds are in danger of being superseded. The ontological category of the artefactual would replace that of the natural. The upholding of the latter as a category worth preserving constitutes, for this book, the most fundamental task in environmental philosophy. Under this perspective, the worrying thing about modern technology in the long run may not be that it threatens life on Earth as we know it to be because of its polluting effects, but that it could ultimately humanize all of nature. Nature, as 'the Other,' would be eliminated.

4. In other words, the ontological category of the natural would have to be delineated and defended against that of the artefactual, and some account of 'intrinsic' value would have to be mounted which can encompass the former. The book argues for the need to maintain distinctions such as that between human/nonhuman, culture/nature, the artefactual/the natural. In other words, ontological dyadism is required, though not dualism, to combat the transformation of the natural to become the artefactual. The book also argues that the primary attribute of naturally-occurring entities is an ontological one, namely, that of independence as an ontological value. Such an attribute is to be distinguished from secondary attributes like intricacy, complexity, interests-bearing, sentience, rationality, etc., which are said to provide the grounds for assigning their bearers intrinsic value. In this sense, ontology precedes axiology.
5. In the course of constructing such an account, some of Aristotle's insights embodied in the notion of *telos* as well as about the character of artefacts will be reformulated and made to cast light on the ontological distinction between the natural and the artefactual.
6. The book also shows that the philosophical basis for the transformation of the natural into the artefactual has been laid down as early as the beginnings of modern science itself in the seventeenth century, but that it will be the technology of the late twentieth and twenty-first centuries (such as biotechnology and molecular nanotechnology) which will truly make good that philosophical promise. That basis, among other themes, includes Humean projectivism and the metaphysics of Scientific Naturalism.
7. The crucial role played by modern science and its technology appears to bear out the centrality of the conception of *homo faber* and its elevation of the life of production/fabrication to be the highest human good in our present industrial culture, which is unique in the history of human civilization itself.

The book contains five chapters and two appendixes.

Chapter 1 outlines the concept of modernity as it emerged in Western Europe roughly from the seventeenth century onward as well as its philosophical framework within which modern science and its new method and goals may be accommodated and supported. It emphasizes two main themes:

(a) That science and technology are expected to play a key role in controlling and manipulating nature as human wishes dictate. To achieve that, nature was mathematized and quantified as well as analyzed into its smallest component parts, ultimately at the level of atoms. The contributions of Descartes, Galileo, Kepler and Newton are clearly significant in formulating this mechanistic and reductionist program.

(b) The new science is sustained by the new philosophy—indeed, the two mutually interlock to produce the metaphysics and the epistemology of Scientific Naturalism. The new philosophy replaces the sensuous organismic with the mechanistic worldview. Of Aristotle's four causes, the material and the efficient, alone, are retained, thereby creating ontological space only for inert matter as the object of study for science, and epistemologically speaking, recognizing only the observable, the measurable and the quantifiable as constituting knowledge. Furthermore, the initial elimination of the final and the formal causes from the study of natural entities and phenonema prepares the way ultimately for the introduction of human designs and purposes in the project of transforming the natural into the artefactual *via* science and technology. In other words, human agency, substituting for natural processes, becomes the predominant efficient cause. As the last half of the twentieth century shows, even the material cause is reduced to the ontological minimum, as nature is being manipulated increasingly at the level of atoms and molecules, not at the level of natural kinds, by human agency. By driving out Aristotle's thesis of intrinsic/immanent teleology (based on the *telos* of organisms), by rejecting the ancient Greek ideal of the *bios theoretikos* as the highest human aspiration, modernity distorts Aristotle's thesis of external teleology (the hierarchy of natural beings with humans at the top) to sustain its own aggressive form of strong anthropocentrism by elevating **the new teleology** above all else, namely, an extrinsic teleology imposed on matter (and energy) by modern science and technology. It banishes values from the natural world, permitting their re-introduction only under human aegis in two ways: (a) the familiar Humean route which holds that values are entirely humanly generated and bestowed upon the world, and (b) increasingly, the technological route of transforming the natural to become the artefactual.

Chapter 2 focuses on the philosophical foundations for the technological transformation of the natural to become the artefactual which involve not merely the philosophy, but also the history, of technology in order to bring out, on the one hand, the relationship between deep science and deep technology, and on the other, that between science-induced technology and the greater degree of artefacticity. The following themes, in particular, will be emphasized:

(a) The relationship between the **new teleology**, in particular the thesis of extrinsic/imposed teleology, and technology will be explored in order to highlight the distinction—crucial to the fundamental theme pursued by the book—between the natural and the artefactual.
(b) The history of technology in Western Europe shows that unlike the case of modern science, the radical cleavage between modern and pre-modern technology did not occur until the mid nineteenth century in spite of the programmatic pronouncements on its behalf in the seventeenth century. But since then, each new science, 'deeper' than its predecessor, induces its own more powerful technology in transforming the natural into the artefactual.
(c) It shows too that both biotic and abiotic nature may be transformed into artefacts.
(d) The relationship between the philosophy of technology and the philosophy of science will also be explored. The book argues that the epistemological goals of scientific theorizing and of technological practice may not be identical—the former is interested in 'truth,' in the formulation of universal regularities, the latter, in efficient solutions to practical problems, in the formulation of rules prescribing procedures to achieve practical ends. However, it remains true that the most efficacious rules turn out, in general, to be those grounded in fundamental laws.

Chapter 3 sets out primarily to

(a) Examine the notion of 'nature,' identifying certain crucially different senses of the term as well as establishing the pivotal significance of one of these senses to environmental philosophy, namely, the natural as the foil to the artefactual. It will also be argued that this sense is inextricably bound up with the notion of independence and independent value.
(b) Clarify the ambiguities surrounding the notion of 'human impact upon the environment,' distinguishing unwanted side-effects of technological impact from the transformation of the natural to become the artefactual. The artefactual is a deliberate intentional structure, designed and created by humans through technology, and is not simply the accumulation of the unintended consequences of using technological means. In this context, McKibben's claim about the end of nature is critically assessed.

Chapter 4 explores the following:

(a) It distinguishes between the main threat posed to nature by most of extant technology and that posed by technologies which have recently been put in place or will come on stream in the near future. The former is, by and large, polluting and, therefore, ecologically insensitive—it has given rise to problems like acid rain, global warming, and contamination of soil, water and air. But it will be superseded by much less polluting technologies which, however, turn out to be nature-replacing rather than nature-saving. If pollution is perceived to be the fundamental disvalue, what good reason is there, then, for entertaining doubts about such ecologically sensitive technologies? The book argues that the resistance is an ontological one, against the supersession of nature as 'the Other.'

(b) The chapter also argues that the notion of *homo faber* is central to modernity. Under it, humanity is primarily celebrated in terms of the intelligence displayed in tool-making and in the manufacture of artefacts. Its associated theme of the humanization of nature is another way of saying that the natural is both ontologically and morally void (in keeping with the metaphysics of Scientific Naturalism) unless endowed with being and value by humans in their endeavor to transform it to become the artefactual *via* human labor and technology. Furthermore, fabrication permeates the whole of human existence in the sense that whatever we do—eat, make love, climb mountains—our activities are mediated increasingly by highly technologized artefacts. *Homo ludens* rides on the back of *homo faber*; so does the kingdom of freedom if the kingdom of necessity is to be left behind.

(c) The chapter further explores the language of machines and that of artefacts to show how deeply the notion of *homo faber* has penetrated all theoretical discourse, even theoretical biology itself. The writings of Maturana and Varela, two distinguished theoretical biologists, are examined to show that their conception of individual organisms as machines is a reflection of this urge on the part of *homo faber* to transform even biotic entities to become artefactual ones.

Chapter 5 seeks to establish the following:

(a) As the categories of the natural and the artefactual are ontologically distinct, the defense of nature in axiological terms alone is not sufficient. It must be linked to one in ontological terms.

(b) The book rejects dualism but argues for the retention of vital distinctions, such as between the natural and the artefactual, in the form of ontological dyadism. Curiously, the logic of *homo faber* and fabrication leads to the same conclusion as that strand of postmodernist thinking in environmental philosophy itself as represented, for instance, by Callicott and Rothenburg, namely, the elimination of nature as 'the ontological Other.'

(c) Biotic nature is crucially different from abiotic nature in the following respect: while it is possible to ground values in biotic nature within a quasi-Aristotelian perspective, it is not possible to do likewise for abiotic nature. An alternative framework has to be explored—while abiotic nature has no *telos* and appears incapable of realizing itself or initiating behavior to maintain or retain its own identity and integrity, it and its underlying processes, nevertheless, have a trajectory of their own and in that sense, an autonomy which is totally independent of human agency and its intentions. As such, it is ontologically distinct from human intentions and their end products which are their translations into action. To eliminate abiotic nature and its processes, as much as eliminating biotic nature and its processes, amounts to an ontological elimination of 'the Other.'

(d) Two senses of intrinsic value should be distinguished. Human beings are intrinsically valuable 'in themselves' in virtue of the unique type of consciousness they possess. Biotic beings are intrinsically valuable 'for themselves' in virtue of the fact that they strive to maintain their own functioning integrity. But biotic and abiotic beings and the natural processes underpinning them are valuable 'by themselves'[5] in virtue of the fact that they have come into existence and will continue to exist independent of humankind. They had existed and will exist in a world without humankind. Their independence in this fundamental sense constitutes their ontological value.

(e) Contrary to Humean projectivism, the chapter argues that human consciousness is not the source of all values. Neither is it the sole locus of intrinsic value. However, human consciousness does add another dimension to the matter of values in the world. Nature, biotic and abiotic, embodies and displays 'mutely-enacted' values. But human consciousness renders 'recognized-articulated' values possible. A world without human consciousness is, therefore, a world without recognized-articulated values but it is, nevertheless, a world with mutely-enacted values.

(f) Ontological difference, rather than axiological similarity *simpliciter*, is fundamental to environmental philosophy.

(g) The elimination by *homo faber* of 'the ontological Other' shows that such a civilization is a profoundly narcissistic one.

The two appendixes deal with issues which, though not directly relevant to the main arguments of the book, nevertheless, may require brief attention,[6] as failure to confront and challenge them could lead to a misunderstanding about their possible significance in undermining the thesis which the book attempts to defend.

Appendix 1 argues that it is a mistake to see nature as work of art, as art is artefactual, whereas nature is not. Art has no existence and meaning apart from the intention and design of its human creator. In contrast, nature exists indepedently of human intention and design.

Appendix 2 looks at the thesis of nature as social construct in some of its various interprctations and shows that it does not, under any of the versions identified, succeed in undermining the claim that nature has independent existence and value apart from whatever account any (human) group or individual may give of it.

If the arguments set out above are plausible and defensible, then the book would have succeeded in shifting the focus in environmental philosophy in at least three major ways: from the axiological, relying on secondary characteristics displayed by entities, like complexity or creativity, to ground moral considerability, to the ontological based on the primary characteristic of independence displayed by naturally-occurring entities; from an earthbound environmental philosophy to one whose remit includes other planets of our solar system (if not beyond), making it intelligible to say that abiotic nature is morally considerable; from pollution being perceived as the fundamental disvalue produced by our industrial civilization to the ontological (and empirical) loss of independent value brought about by the systematic transformation of the natural to become the artefactual *via* increasingly more radical and powerful technologies (which are relatively non-polluting) induced by the discoveries of 'deep' scientific theories.

Notes

1. In spite of quantum physics and its associated characterization of reality, a considerable amount, if not the bulk of science practiced today, is conducted within the Newtonian-*cum*-Euclidean framework.

2. Up to roughly the mid nineteenth century—see Chapter 2—the material causes of artefacts were, by and large, naturally given.

3. In Chapter 3, different senses of 'nature' will be identified and delineated. In certain key senses, the term does refer to the entire universe, not merely our solar system or even our galaxy. One of the central claims of this book is that deep technologies, generated by deep science, make it possible for humankind to transform nature into human artefacts. However, the deep technologies discussed are only those which either have emerged recently (like biotechnology) or will emerge in the not so distant future (like molecular nanotechnology). So, perhaps, it would be wise to restrict 'nature, out there' to only our solar system in this context of discussion, as the technology of terraformation conceived today would only enable us to terraform, at best, some of the other planets in our solar system.

4. See Gregg Easterbrook, *A Moment on the Earth*, whose subtitle is *The Coming Age of Environmental Optimism* (London: Penguin Books Ltd., 1996). Easterbrook labels his position 'ecorealism.' Among the several theses which make up ecorealism is indeed the

view that the environmental crisis as it has been perceived in the last quarter of this century—based on pollution as the fundamental disvalue—would have become passé come the twenty-first century, at least in the affluent, developed economies in the world.

5. For a gloss on the terminology—'in themselves,' 'for themselves' and 'by themselves'—see note in Chapter 5.

6. The issue dealt with in Appendix 2, namely, whether the notion of nature is a mere social construct, is, of course, immensely vast and complex. However, to ignore it altogether on the grounds that to do justice to all its aspects and their ramifications is, clearly, beyond the remit of this appendix, seems too easy an option. This short examination, admittedly, is but a compromise.

Chapter One

Worldviews: Modern and Pre-modern

Modern civilization is scientific and technological with profound impact upon the natural environment as well as profound implications for human relationship to nature. But such a civilization is built within a philosophical framework which also differs profoundly from that which preceded it.[1] This chapter outlines the two contrasting philosophies and worldviews, modern and pre-modern, and in turn their respective conceptions of what a 'proper' scientific explanation demands in terms of their respective metaphysics and methodologies.

Modernity

Modernity as a concept is European in origin but has since acquired global coinage. It may be understood and defined in many ways. Here, only two aspects will be singled out for comment—its philosophy and its science, which are inextricably linked. The new philosophy and the new science will be contrasted with the 'old' philosophy and the 'old' science which are also inextricably linked. From the philosophical perspective, it is defensible and plausible to define modernity in terms of the overthrow of the Aristotelian worldview and its organismic teleological philosophy by the philosophy of mechanism as well as of positivism which began to take shape in the development of Western philosophical thought, more or less systematically, from the seventeenth century onward.[2] Thomas Aquinas, by effecting a synthesis between Catholic theology and

Aristotle's philosophy in the form of Thomist Aristotelianism, had previously provided the dominant intellectual and philosophical foundation not only for Roman Catholicism but also for European cosmology from the thirteenth century onward.[3]

Moreover, Thomism had provided, at the same time, the philosophical framework within which late medieval science was conducted—Aristotelian physics required Aristotelian philosophy to back it up. Galileo was a leading pioneer of the new science. However, he made no systematic effort to formulate a new philosophy to underpin the new science, although he had things to say which have bearing on the new scientific method[4] and the new philosophy. Indeed, he himself, in 1610, had asked to be appointed not merely Chief Mathematician but also Philosopher to the Grand Duke of Florence. Moreover, his new science and its method embodied many of the tenets which eventually made up the core of the new philosophy. For instance, his distinction between primary and secondary qualities was incorporated, as we shall see, later by Locke in the latter's systematic construction of empiricism which forms a key component of positivism itself.[5] Furthermore, in defending Copernicanism, Galileo did not so much consider the Church as such to be his real opponent as the philosophers in the universities teaching and upholding Aristotelianism.[6] In other words, both Galileo the scientist and the Aristotelian philosophers recognized that the new science and its method posed a challenge not only to the old science but to the old philosophy as well.

The Old Philosophy and the Old Science

From Aquinas onward , Aristotle was generally regarded as the Philosopher with a capital P and in Dante's words, "the master of them that know." However, it is very important to remember that the Aristotelianism (Hobbes called it "Aristotelity" and attacked it) which became the dominant European worldview during the late Middle Ages and right up to the seventeenth century bore resemblance, but was not faithful in all ways, to the philosophy of Aristotle himself. Somewhat simplistically put, Aristotelianism may be said to be a distortion of Aristotle's philosophy in two main ways:

1. Aristotle was interested in synthesis as the goal of science although his own thought never actually achieved it.[7] In this attempt, he engaged in extensive empirical research, especially in biology.[8] Such research might take many generations of science and scientists. Thomism ignored this, taking for granted that it was already completed.
2. Aristotle did not share in the Thomistic goal of personal salvation, although it is true that he considered rational activity, like engaging in philosophical thinking, to be nearly divine. But Aristotelians, of one sort or other, used philosophy to justify God's ways to man. Aristotle's own philosophy might

involve a reference to the divine but that has nothing to do with an authoritative theology that guarantees human salvation either in this world or the next.

Although Aquinas himself was aware of a fuller range of Aristotle's works than his predecessors, it remains true that he made use of the master's thoughts for his own purpose, to effect a grand synthesis between Catholic theology with a history of Augustinism in the Church and the Church's exposure to Platonism.[9] What Thomism has done is to co-opt some of Aristotle's theses, his categories and his logic to lay down a cosmological framework within which the universe is conceived to be finite, hierarchically ordered as well as purposefully ordained by God. Its center is the earth; its outermost boundary is the fixed stars. Everything in the space between has its proper place. In a nutshell, it is this conception of the universe which Copernican heliocentrism, as championed by Galileo, challenged and ultimately demolished.

The new astronomy thus threatened not merely the old Ptolemaic astronomy but also the old cosmology and philosophy which underpinned it. It challenged it in a truly radical manner, not because it downgraded man but because, as would soon be made clear, it replaced qualitative differences with quantitative ones. Collingwood has pointed out that it is "both philosophically foolish and historically false"[10] to construe its significance as dethroning man and his planet from the center of the universe and diminishing their importance in the cosmic scheme of things. After all, Boethius in his *De Consolatione Philosophiae*, considered to be the most widely read book of the Middle Ages, had long acknowledged the cosmic insignificance of man and earth.[11] Copernicus and Galileo were, therefore, not likely to risk the charge of heresy for merely affirming a commonplace belief. For the same reason, when Copernicus's posthumous publisher wrote that heliocentrism for the purpose of studying and calculating the orbits of the planets would be a convenient device, his remark need not be interpreted as one of mere caution and timidity, as if he admitted that the established view was true and correct. On the contrary, those who championed heliocentrism were aware that it would amount to a revolution in cosmology. Collingwood holds that it amounted to denying that the universe has a center, and by so doing, thoroughly undermined the view, inherited from Greek thought, of spherical world-organism—the earth in the center, surrounded by water, air, then fire, and for Aristotle, the *quinta essentia* which constituted the outermost layer. But with no center, differentiated parts would not be required either. This then left the way open to the rival paradigm that the universe is homogeneous as far as the kind of matter it is made out of is concerned. The stars do not possess a divine substance. The laws of terrrestrial physics, like the law of gravitation, apply to both the sublunary regions as well as celestial space. Terrestrial and celestial physics could become one. Nature is not made of heterogeneous substances differing in quality, as Aristotelians taught. There is one qualitatively uniform substance. The differences it displays are differences of quantity and of geometrical structure. This, as we shall see, enables Galileo and others to develop and bring to maturity the mechanistic worldview *via* the new scientific method.[12]

But the Aristotelian philosophical worldview was not merely entwined inextricably with Ptolemaic astronomy but also laid down the terms in which terrestrial science was conducted. In outline, its main points are:

1. The human intellect is concerned to explain why things are as they are (why a dog is a dog and not a cat), why they could not be otherwise (why a dog does not become a cat), and why it is best that they be as they are. A satisfactory explanation consists of (a) giving the causes of things and (b) comprehending the purpose behind events in nature.[13]
2. There are four causes in terms of which a thing or event would be satisfactorily explained—material, efficient, formal and final. Take a statue. It is made of marble (material cause); it is made by the sculptor using certain tools (efficient cause); it is in the shape of a horse (the formal cause which guides the sculptor as he chips away at the marble); it is to commemorate the fabulous exploits of Bucephalus, the horse of Alexander the Great (final cause). Also to understand why something is what it is, it is necessary to understand it in terms of the four causes.
3. Nature (as already mentioned) is composed ultimately of four elements —earth, water, air and fire. Each of these may have two qualities attributed to them in terms of the pairs hot/cold, wet/dry—earth is cold and dry; water, cold and wet; air, hot and wet; fire, hot and dry. Each of these elements has a natural home assigned to them; for earth and water it is the ground, for fire and air it is above the ground. Moreover, things which pertain to earth and water possess the natural tendency of gravity enabling them to strive to return to their natural abode if they were removed from it—if a stone were thrown up into the air, it would eventually fall to the ground. Similarly, things pertaining to air and fire possess the natural tendency of levity. Explanations in terms of these natural properties are called teleological. (See the last section of this chapter for a further discussion of teleology.)
4. Causes of things are not ascertained by means of the senses, although their effects could be so perceived. Rules then have to be established by which one could determine causes through their effects.
5. As things perceived in nature both differ as well as are similar to one another in numerous aspects, one needs to distinguish accidental from non-accidental properties. The latter constitutes the inner essences of things which serve to distinguish them from one another. Definitions capture these essences.
6. Aristotle distinguished between forms of knowledge, each with its respective source and goal—*techne* and *episteme*. *Techne* is concerned with how to do things based on an awareness of or reasoning about the proper nature of the thing to be made or the activity undertaken. *Episteme* is knowledge whose

source was the understanding of things through their causes. *Phronesis* is practical, that is, moral and political, knowledge.

The New Science and Its Method

Although Galileo (1564-1642) bears the distinction of bringing to maturity the new revolutionary mechanistic paradigm and worldview in his scientific method, others also paved the way. Among these is Kepler (1571-1630), a contemporary. Kepler, building on the work of William Gilbert on magnetism (published in 1600), formulated the principle of inertia, that bodies tend to remain stationary wherever they might be. This challenged the Aristotelian conception of natural movements, as noted in thesis 3 in the section above. Furthermore, Kepler explained gravitation in terms of mutual affection which draws a body toward neighboring bodies— the stone falls to the ground because the ground attracts it. In the same way, the tides change because the moon attracts the water. To these innovations, he added a third—that, in physics, the word *anima* be replaced by *vis*. The former embodies the conception of a vital force or energy capable of producing qualitative changes, the latter that of a mechanical energy or force, which is itself quantitative, bringing about quantitative changes. This semantic change paves the way for rendering modern science objective and value-free. In 1595, Kepler had written:

> [A]s the eye was created for color, the ear for tone, so was the intellect of humans created for the understanding not of just anything whatsoever but of quantities. It grasps a matter so much the more correctly the closer it approaches pure quantities as its source. But the further something diverges from them, that much more do darkness and error appear. It is the nature of our intellect to bring to the study of divine matters concepts which are built upon the category of quantity; if it is deprived of these concepts, then it can define only by pure negations.[14]

In other words, Kepler was really referring to the mathematization of nature entailing not only a change in scientific method, but also a profound change in worldview, from an organic to a mechanistic one. (Kepler held that the human intellect was created by God to apprehend the world in terms of mass and number. But his theological belief simply underpinned his mechanistic worldview, rather than challenged it.[15]) Galileo is credited with having formulated the new paradigm even more forcefully, especially in that famous remark of his about the "book of nature":

> Philosophy[16] is written in this grand book—I mean the universe—which stands continually open to our gaze, but it cannot be understood unless one first learns to comprehend the language and interpret the characters in which it is written. It is written in the language of mathematics, and its characters are triangles, circles, and other geometrical figures, without which it is humanly impossible to understand a single word of it; without these, one is wandering about in a dark labyrinth.[17]

Galileo saw mathematics and mathematical measurement as indispensable to the study and understanding of nature as well as a tool to make it disclose its secrets to us. The marriage of mathematics to physics constituted a radical departure from Plato's and Aristotle's views of the relationship between the two subjects. Plato disparaged the physical world as being transient and subject to decay—true knowledge is about objects which are immutable and eternal. Pure mathematical ideas seem to qualify for such a status. Hence Plato thought the forms alone worth studying. Aristotle, on the contrary, inferred from the very abstract character of mathematical procedure that mathematics could have nothing to offer to physics, as the latter is concerned with the study of matter and its motion, which mathematics precisely ignores.

For Galileo, mathematics enables one to make calculations which could then be tested to see if they fit observation. If they do not, this should not be construed that either calculations are irrelevant (Aristotle) or that observation is not required (Plato). A bad fit could signal that the scientists have left something out of account and that they should go back to redo their homework. For Galileo, observations and measurements yield scientific facts, and if these conflict with existing philosophical beliefs, it is (orthodox) philosophy and not science that should give way.

Galileo distinguished between philosophizing and studying philosophy, which is analogous to the distinction between drawing from nature and copying pictures. It is good practice for an artist to start by copying other artists' drawings of objects. But if one were never to draw from nature, one could never become a perfect artist.

> In the same way a man will never become a philosopher by worrying forever about the writings of other men, without ever raising his own eyes to nature's works in the attempt to recognize there the truths already known and to investigate some of the infinite number that remain to be discovered. This, I say, will never make a man a philosopher, but only a student of other philosophers and an expert in their works. I do not believe that you would esteem as a good painter a man who had made so great a study of the drawings and canvases of all painters that he could promptly identify the style of each one, even if he could also imitate them.[18]

Careful observation of the moon's surface through the telescope has shown that it is not smooth, but has craters and mountains; dismissing such evidence in the name of Aristotelianism would amount to a mere dogmatic appeal to authority. This would be neither good philosophizing nor practicing good science but sterile mouthing of the philosophy of others. Galileo was hostile to the Aristotelians precisely because he was against the dogmatism they displayed.

Mathematical physics soon established itself as the queen of the new sciences. But whether Galileo and others anticipated or intended its implication is immaterial; its success definitively ushered in the mechanistic worldview. The new science and its method imply a new philosophy and cosmology. The mathematization of nature meant that a new 'reality' emerged, one based on abstraction, isolation, measurement and quantification. Galileo gave expression to it in *The Assayer*, a work which may be read as his attempt to formulate a philosophy of science, albeit not in a systematic manner. The gist of it may be found in yet another passage from it:

> Now I say that whenever I conceive any material or corporeal substance, I immediately feel the need to think of it as bounded, and as having this or that shape; as being large or small in relation to other things, and in some specific place at any given time; as being in motion or at rest; as touching or not touching some other body; and as being one in number, or few, or many. From these conditions I cannot separate such a substance by any stretch of my imagination. But that it must be white or red, bitter or sweet, noisy or silent, and of sweet or foul odor, my mind does not feel compelled to bring in as necessary accompaniments. Without the senses as our guides, reason or imagination unaided would probably never arrive at qualities like these. Hence I think that tastes, odors, colors, and so on are no more than mere names so far as the object in which we place them is concerned, and that they reside only in the consciousness. Hence if the living creature were removed, all these qualities would be wiped away and annihilated. But since we have imposed upon them special names, distinct from those of the other and real qualities mentioned previously, we wish to believe that they really exist as actually different from those.[19]

From the above, one may infer the following theses:[20]

1. What are real and reside in (material) substances are what Locke later called "the primary qualities," namely, shapes, numbers and motions. These alone would be sufficient to excite in us tastes, odors, sounds and colors—what Locke calls "the secondary qualities." Galileo wrote: "I think that if ears, tongues and noses were removed, shapes and numbers and motions would remain, but not odors or tastes or sounds."[21]
2. The elimination of secondary qualities, that is, of qualitative differences between things, is required because what is real and intelligible in nature is what is measurable and quantifiable. The ontology implied consists precisely of holding that what is real is what is measurable and quantifiable, and only what is measurable and quantifiable is real.
3. The elimination of secondary qualities permits the reduction of a complex whole (with its sensuous qualities) to the relatively simple matrix of what could be weighed, measured and counted.
4. Not only are the secondary qualities derivative and dependent upon the primary ones but they are also totally mind-dependent and hence are mere appearances with no objective existence whatsoever. Galileo called them "mere names" to which there are no referents in the 'objective real' world, but

at best refer to mental phenomena residing in living, sensible (human) beings.

5. This means that the natural world studied by the new science is necessarily a world of pure quantity from which living and sensible beings have been excluded. In other words, it is a dead, inert nature that is being studied.
6. As such, the new science concentrates on the overt, the outer, the public, the impersonal, capturing their quantifiable features in laws of nature which are meant to be universal in scope. By the same token, it ignores or downgrades immediate experience (which is of secondary qualities), the qualitative, the covert, the inner, the private, the personal or the particular.
7. Its epistemology consists of holding that what is knowable is what is measurable and quantifiable and only what is measurable and quantifiable is knowable.
8. Humans, the student of nature, stand outside nature. As Collingwood puts it:

> Nature, so regarded, stands on the one hand over against its creator, God, and on the other over against its knower, man. Both God and man are regarded by Galileo as transcending nature; and rightly, because if nature consists of mere quantity its apparent qualitative aspects must be conferred upon it from outside, namely by the human mind as transcending it; while if it is conceived no longer as a living organism but as inert matter, it cannot be regarded as self-creative but must have a cause other than itself.[22]

9. The scientists become instruments for recording and analyzing the real and the knowable. Apart from the processes of thinking which involve their intellect-ual/logical capabilities, their sensory and emotional reactions are neutralized or eliminated both in the design of the experiment and the analysis of its result. Scientific data are emotion and value-free. Science becomes the most rational, if not the only, form of rational activity.

This amounts to an outline of what is also sometimes called the metaphysics of Scientific Naturalism. On this mechanistic view of scientific method and of nature, the behavior of natural entities, their processes of change and maintaining dynamic stability are understood as regularities or uniformities, as mere movements which are the result of the impact of one body on another body, the attraction of one body toward another body, or the repulsion of one body by another body. Hume's analysis of the notion of cause later in the eighteenth century articulates this conception most forcefully.[23] Regularities—phenomena of kind A followed by phenomena of kind B—replaced tendencies which are the result of effort on the part of the beings which are studied. Why do plants lean towards light? Because the plants, in order to grow and develop in a way they are capable (or have the potential), require light and so strive to reach it. The new science and its philosophy render this kind of explanation both redundant and unintelligible but would instead sanction a regularity type of explanation —whenever plants are found to grow well, they have leaned towards light or that they would not grow well, if at all, in the absence of light.[24]

Indeed, to modern philosophy, the entire Aristotelian conceptual apparatus in terms of 'wants,' 'desires,' 'striving to fulfill,' etc. , is suspect and must be rejected. It is condemned as "teleological," as it conceived changes and processes in nature to be directed or dictated by goals or ends which did not yet exist but which would be ultimately realized.[25] As mentioned earlier, for the Aristotelians (and Aristotle), a full and proper explanation has to be in terms of the four causes. But to the new science and the new philosophy, two of them—the final and the formal—smack of the teleolgical. Only the material and the efficient causes which lend themselves to measurement and quantification are retained. To explain why a coastline is indented in the way it is, one needs only to refer to the kind of rock or rocks the coast is made of, the strength and direction of the waves, the force with which the waves hit the shores, the temperature of the water, the direction and strength of the prevailing winds, etc.

Final and formal causes are suspect because they appear to be tied up with essences. Essences are grasped through reason and given by definitions, according to Aristotelianism. Why does fire rise? Because it is of its essence or in its nature to do so. But to Galileo and those who professed the new science and its philosophy, these are mere words, signifying and referring to nothing in reality. For them, only result obtained through calculation and measurement is to count as scientific knowledge. But as essences are not amenable to such treatment, they do not form part of the province of science.

And even worse, in the hands of the Aristotelians, final causes even led to anthropomorphism. For instance, Galileo, as much as the Aristotelians, noticed that a falling object like a stone, falls faster and faster in its downward journey. The Aristotelian physicists (or Thomists) would explain the phenomenon thus: a stone belongs to the element, earth, whose natural home is at ground level, the surface of the planet. Suppose you had been away for a long time from your loved ones. As you got nearer and nearer your destination on the return journey, you would get more and more excited and walk or ride your horse faster and faster. Similarly, a stone would fall faster and faster as it approached nearer and nearer its natural abode—the impetus being the joy of getting there.

Galileo would regard such anthropomorphism to be singularly unhelpful. He preferred to observe and measure the rate of fall and to determine the law of acceleration in precise mathematical terms. The 'why' is of no concern to the scientist. Only the fact that the object fell in the way it did which could be measured is of significance.

The new method and the mechanistic worldview ushered in by Galileo and others is necessarily empirical (using mathematics not merely as a tool, but also thereby mathematizing nature), anti-metaphysical[26] (no essences in terms of hidden entities and mechanisms), anti-teleological (no final and formal causes, only material and efficient ones). As an earlier quotation from Galileo shows, he complained that Aristotelian doctrines were upheld as dogmas at the expense of empirical evidence. This was because Aristotelianism believed in *episteme*, knowledge arrived at by means which, to Galileo and others, were obscurantist and "metaphysical." The spirit of modernity consists precisely in repudiating all such superstitions and "idols" upheld by traditional authorities of one kind or other. As far as Galileo himself was concerned, the Church, as an authority, was fine provided it confined itself to matters purely of faith and did not meddle with matters of science which were outside its jurisdiction. The only authority in matters of science which he would acknowledge was the authority of those who practiced the new scientific method implicitly backed up by the new philosophy and not that of Aristotelian science and its philosophy. In other words, a new epistemological authority replaced the old.

That replacement is reflected, curiously, in the very emergence of the word "epistemology" in modern Western philosophy, to denote that branch of philosophy also referred to as "the theory of knowledge," dealing with scientific and other types of knowledge claims. But philosophy conducts no experiments or makes observations and measurements, although it tries to render intelligible the findings yielded by experiments, observations and measurements. The etymological root of "epistemology" is the word "episteme." *Episteme*, as we have seen, is knowledge, the understanding of things uncovered by the use of reason through grappling with their causes and essences, which is contrasted with *techne*, delineating the domain of know-how. But for Galileo and other moderns, scientific knowledge is not *episteme*, but what is yielded by measurement, quantification, calculation, observation (what is grounded in sensory experience, in the broad sense of the term). Modern scientific knowledge is, therefore, more like *techne*, as the latter, too, operates within the realm of experience (but only in this very restricted aspect—see below for the fundamental differences between the two).

From *techne* is derived the word "technology." At least in the English language, "technology" was given its modern usage by the last half of the seventeenth century, appearing for the first time in a dictionary in 1706 referring to "a Description of arts, especially the Mechanical." Later, it was extended to the industrial arts and practical arts in general. Building on this, Jacob Bigelow in the preface to his book *Elements of Technology* (1831) used the word to refer to "the principles, processes, and nomenclatures of the more conspicuous arts, particularly those which involve applications of science." And in the German language, Christian Wolff, in his *Preliminary Discourse on Philosophy in General*, as early as 1728, used it to refer to "the science of the arts and of the works of art" or to the use of physics to give "the reason of things which occur through art."[27] Contemporary usage clearly reflects this historical lineage. Today when we speak of science and technology, we roughly mean the distinction between theory and its application in ultimately manufacturing products.[28]

Even more interesting, "technology" is derived not only from *techne* but also from its being conjoined with *logos*. But as Mitcham reminds us, strangely enough, when, for the first time in Greek thought, the two words *techne* and *logos* were joined together into a single word by Aristotle, he used it in the context of rhetoric, so that the Greek term "technology" meant "the study of grammar or rhetoric" and the term "technologist" referred to the grammarian or rhetorician.[29] What could account for the dramatic change in denotation of the words between Aristotle's use of the terms and their modern usage? Mitcham points to a deep ontological divide between the moderns and Aristotle (and Aristotelians) in their conception of nature and the world around them.[30]

According to Aristotle (and those who followed his worldview), *techne* involved *logos*, but *logos* had nothing to do with mathematical or quantitative concepts or reasoning—carpentry and flute-playing involved great *techne*. What could be grasped by *techne* through *logos* was merely the form or the "whatness" of the thing that was being made or done. But the matter itself out of which the thing was made and the actual processes of making it fell outside of *logos*. In producing an artefact such as a table, the form is the idea in the head of the artisan:

> [I]ts union with matter is, as it were, at the mercy of matter and its specific receptivity. Form cannot be forced into or imposed upon matter; an artisan must let the matter guide the way it will receive form. The ultimate decision in action rests not with reason but with sensation, aisthesis. ... Indeed, on one occasion [*Physics* I, 9; 192a18] Aristotle goes so far as to describe the coming together of form and matter, the becoming of an entity, as dependent on a 'desire' or "reaching out" on the part of matter for form.[31]

Aristotle understood matter to be embodied in particulars. Hence the place of *logos*—the logical universal—was necessarily limited in knowledge about particulars. Knowledge of particulars is acquired essentially through imitation, practice and experience—one does not become a builder by reading manuals on building, but by building and thereby coming to know intimately the properties and propensities of the stone that one is building with.

This limitation of the role of *logos* in the case of *techne* would also explain why Aristotle produced a *logos* of the *techne* of persuasion, because language is a rarefied medium and is not material in the way that blocks of stone are material particulars. Here one can lay down a systematic discourse about the means and processes involved in the art of persuasion—a recognition that words, even when divorced from reason, are a powerful means to get an audience to accept certain ends or do certain things. There is a logic of means—a set of general rules and devices—irrespective of the ends to which the means may be put, which could be laid down and learned. Although Aristotle did claim that one could produce similar discourses about every other art, it remains true he never did, except in the case of the *Rhetoric*, so much so, that the term "technology," as noted earlier, comes to mean no more and no less than simply the study of grammar or rhetoric. Aristotle as good as conceded that as far as *techne* in general is concerned, apart from grasping form, there is no *logos* of the activity involved *qua* activity.

In contrast, modern technology is precisely predicated on the assumption that there is a process of production which has nothing to do with the particular forms of things. Aristotle and Aristotelians regarded matter as taking on forms, and held that there was a desire on the part of matter to unfold itself in accordance with the forms the particulars involved. But with modernity as it emerged under the influence of Galileo, Descartes, Newton and others, matter, as we have seen, becomes inert, dead matter. According to the Cartesian view it is mere extension, which is devoid of form, potentiality or *telos*. Being deprived of any desire or aspiration of its own, it opens the way for what today is called strong anthropocentrism, which regards humans as the sole source and locus of intrinsic value and nature as being of only instrumental value to humans.

As matter is uniformly inert, there can, then, be a general process of production which consists ultimately of the rearrangement of the elements of such matter to serve human ends. So technology in modern terms is the study of the manipulation of nature.[32] From the manipulation of words it becomes the manipulation of matter. Such a drastic change in meaning reflects the revolution in worldview, from the Aristotelian (ultimately, ancient Greek) paradigm of living, organic matter to that of mechanism and its conception of dead, inert matter. Therein lies the (or at least one very significant) passage from the Middle Ages to modernity.

The New Philosophy

The last section has dealt with some aspects of the new philosophy behind the new science and its method. To complete that outline, one must look at the contribution of Hobbes (1588-1679) who, for the first time, articulated systematically the new philosophy to go with the new science. This is the philosophy of empiricism-cum-positivism.[33] We have seen that Galileo had initiated its ontological and epistemological core by introducing the new conception of inert matter, the primary qualities of which alone are real as they are quantifiable and measurable. The new systematic philosophy accepted the Galilean initiative and built on it. Unfortunately, it is not possible in this brief section to do justice to a philosophy which has existed for nearly four centuries and which has developed during that time to yield rich and variegated complexities.[34] Moreover, it is a philosophy which has held sway since the seventeenth century and in spite of the pronouncement of its official demise is still extremely influential because it sets the very agenda which other philosophies feel bound to challenge.

The themes singled out for some attention include the following:

1. The ontological and epistemological core.[35]
2. The law of intellectual development.
3. The setting of the aims and the goals of science in the light of 1 and 2.

As already observed, the new philosophy is empiricist in outlook. To prevent misunderstanding, perhaps, one should briefly distinguish between "empirical" and "empiricist." Aristotelian science clearly relied on empirical observation, as must all science. But the new philosophy goes beyond merely using observation and indeed even measurement.[36] It lays down that the world as ascertained by the senses is the only world we can come to know; it professes an empiricist epistemology. Furthermore, it holds that anything not grounded in sense experience is not real but is metaphysical; it professes an empiricist ontology. Whatever is known by the senses is real and nothing is real unless known through sensory experience. In this sense of "metaphysics," the word no longer means what Aristotle had meant—to Aristotle, metaphysics was the study of being-as-such[37] to be distinguished from the study of particular beings which formed the study of the different and separate sciences. The metaphysical realm on the new understanding came to be identified with what is beyond sensory experience and hence cannot be real. Pronouncements about such a domain would only amount to empty words, if not outright unintelligibility.

In *Leviathan*, Chapter 1, Hobbes wrote: "Concerning the thoughts of man ... the original of them all, is that which we call SENSE, for there is no conception in a man's mind, which hath not at first, totally, or by parts, been begotten upon the organs of sense. The rest are derived from that original."[38] This prompts a commentator, McNeilly, to say: "That reads, of course, like the slogan for an Empiricist Manifesto. It would not be completely misleading indeed, to describe Hobbes as a sort of empiricist."[39]

A central preoccupation of the new positivist philosophy is the search for criteria in terms of which scientific knowledge may be identified and defined and which would in turn serve to distinguish such knowledge from the non-scientific or pseudo-scientific. This was Hobbes' enterprise. But to understand it better, it is necessary to leap nearly two centuries forward to take a brief look at Comte (1798-1857), as he is officially credited with having not only coined the term "positivisme" but also articulated it for the first time in his monumental work entitled *Cours de Philosophie Positive*, the first volume of which was published in 1830. In it, he announced that he had discovered a fundamental law which governs the development of the human mind, both at the level of the species and the individual. "This law is that each of our principal conceptions, each branch of our knowedge, passes successively through three different theoretical states: the theological or fictitious, the metaphysical or abstract, and the scientific or positive."[40]

The first and most primitive, the theological stage, is concerned with the search for first and final causes. The human mind tries to look for the hidden nature of things by posing the question, why things happen in the way they do, and answering it by postulating divine or supernatural beings in man's own likeness. There is a storm at sea. Why? This is because Neptune, the god of the sea, is angry. In Greek mythology, the cycle of the seasons is explained in terms of the loves and lives of the gods and goddesses, of the comings and goings of Persephone from

Earth to the underworld where Pluto lives. Polytheism itself progresses to monotheism, so that by the time of Aquinas, God is the first cause which guarantees the existence of everything else in the world.

Comte himself was not entirely scornful of this mode of explanation. He saw it as an embryonic form of knowledge which stimulated and often led to fruitful developments. For instance, astronomy owed much to astrology—the belief that the motions of the stars and the state of human beings are intimately linked led to accurate measurements and, hence, predictions about the motions of the heavenly bodies.

The second or metaphysical stage is still concerned to ask the question 'why,' but instead of populating the universe with supernatural agencies modeled on human beings, it postulated secular but hidden constructs like 'forces,' 'powers,' 'natures' or 'essences.' Why does water flow downhill? Because it is in its nature to do so. Why does opium send one to sleep? Because of its 'dormative virtue.' But such explanations are really verbal or tautologous. Opium sends one to sleep because it sends one to sleep, for 'dormative virtue' is just a high-sounding name to refer to the observable fact that it does put people to sleep.

It is only when a branch of intellectual study reaches the third or positive stage, that its status as scientific is secured. The positive mode of thought is the scientific mode and is characterized as follows:

> [T]he human mind recognising the impossibility of attaining to absolute concepts, gives up the search for the origin and destiny of the universe and the inner causes of phenomena, and confines itself to the discovery, through reason and observation combined, of the actual laws that govern the succession and similarity of phenomena. The explanation of the facts, now reduced to its real terms, consists in the establishment of a link between various particular phenomena and a few general facts, which diminish in number with the progress of science.[41]

and:

> Is the nature of positive philosophy to regard all phenomena as subject to invariable natural <u>laws</u>, the discovery of which, and their reduction to the least possible number, is the aim and end of all our efforts, while causes, either first or final, are considered to be absolutely inaccessible, and the search for them meaningless? ... Every one knows that in positive explanation, even when it is most perfect, we do not pretend to expound the generative <u>causes</u> of phenomena, as that would be merely to put the difficulty one stage further back, but rather to analyse the circumstances in which the phenomena are produced, and to link them to one another by the relations of succession and similarity.[42]

From the above, one can see that Comte too shared Hobbes' empiricism. But like him and other early positivists, Comte failed explicitly to distinguish between two theses within empiricism, although they were implying both—that (i) the source of knowledge is sensory experience, (ii) what counts as knowledge must be testable in terms of sensory experience. That is why, according to Comte, first and final causes must be abandoned as well as the search for the "generative causes of phenomena," because these are occult, hidden and not accessible to observation.

It may be argued that Hobbes, in substance, if not in name, anticipated Comte's law of human intellectual development. But it may be apt here, before proceeding further, to point out that although both Hobbes and Comte were against the theological mode of explanation, they were not necessarily against theology. Indeed, both gave it a prominent role in their system of thought, provided theology, religion and its institutions were purged of so-called superstitious elements.

Hobbes had no use for the theological mode of explanation simply because he had resolved to transform Galileo's science of motion and the law of inertia to become the new philosophy of materialism and mechanism. The storm at sea is to be explained not in terms of Neptune's anger but of the motion of particles and bodies. It is clear that for Hobbes, the aim of science is to demonstrate the causes of things and that the most universal cause is motion. But note that his causes are not the obscurantist ones which Comte rejected. Instead, they are quite un-mysterious. For instance, to find out the cause of a particular thing, like a square, one must first of all by analysis resolve it into a plane bounded by four straight lines of equal lengths and four right angles. In this way, one arrives at things common to all matter—line, plane, angle, and so on. These Hobbes called universals. And if one finds the cause of these, one will then be able to combine them all to form the cause of a square. Or to take another example, gold. By resolution, one gets to ideas of solidity, visibility, heaviness and so on, which are more universal than gold. These in turn could be resolved further until one gets to the most universal cause, motion.

As everything else is to be explained in terms of motion, motion itself cannot be understood to have any other cause besides motion. All change is the result of motion. Hobbes held that in principle everything could be explained in terms of motion and, furthermore, that all the sciences (even including the science of law, of politics, of the passions or morals) could in the end be reduced to the science of motion. In this, Hobbes went beyond most positivists in maintaining both the unity of method thesis as well as the unity of science thesis. All positivists subscribe to the former but only some to the latter. Moreover, for Hobbes, there was a close connection between the two—"For nature worketh by motion; the ways and degrees whereof cannot be known, without the knowledge of the proportions and properties of lines and figures."[43] Geometry was the paradigmatic science not only because its method was none other than the scientific method but also because its subject matter dealt with motion and nothing else—geometry, for Hobbes, was the effect produced by a body that was moving.

In other words, one could say that Hobbes provided a new metaphysics for the new science. In this context, however, the term 'metaphysics' is to be equated with 'ontology' and not to be identified with the so-called metaphysical mode of explanation which both Hobbes and Comte denigrated. For Hobbes, unlike some twentieth-century positivists, science and metaphysics were not mutually exclusive. Science required a metaphysics or what he called a *philosophia prima*. Of course, not any metaphysics would do. He objected to the scholastic variety. It must be a metaphysics appropriate to the scientific, positive mode of thought, namely, motion, materialism and mechanism.

Hobbes distinctly identified the metaphysical mode of explanation with Aristotle and Aristotelianism or what he called "Aristotelity." He could find nothing good to say in its favor. He wrote:

> To conclude, there is nothing so absurd, that the old philosophers, as Cicero saith (who was one of them) have not some of them maintained. And I believe that scarce anything can be more absurdly said in natural philosophy, than that which is now called Aristotle's *Metaphysics*; nor more repugnant to government, than much of that he hath said in his *Politics*; nor more ignorantly, than a great part of his *Ethics*.[44]

Aristotelity led to absurdities primarily because it failed the Hobbsian theory of truth:

> [T]ruth consisteth in the right ordering of names in our affirmations ... in the right definition of names lies the first use of speech; which is the acquisition of science: and in wrong, or no definitions, lies the first abuse; from which proceed all false and senseless tenets; which make those men that take their instruction from the authority of books, and not from their own meditation, to be as much below the condition of ignorant men, as men endued with true science are above it. ... Words are wise men's counters, they do but reckon by them; but they are the money of fools, that value them by the authority of an Aristotle, a Cicero, or a Thomas, or any other doctor whatsoever, if but a man.[45]

"The right ordering of names" should, however, as already pointed out, not be construed as evidence that Hobbes was an extreme nominalist—he was a nominalist with regard to universals but a 'realist' with regard to 'accidents' or properties. This meant that nominalism played only a limited role within his empiricist framework.

When the procedure of generating true propositions by ordering names correctly is violated, the words used are at best empty sounds and at worst meaningless noises. Their utterance can only lead to the obfuscation of thought, and cause mischief and chaos of all kinds, intellectual, social and political. And this was precisely what Aristotelians in the Universities were doing, in Hobbes' opinion. He condemned their doctrine of abstract essences and substantial forms—"From these metaphysics, which are mingled with the Scripture to make School divinity, we are told, there be in the world certain essences separated from bodies, which they call abstract essences, and substantial forms."[46] But what do terms like 'entity,' 'essence,' 'essential,' 'essentiality' signify? According to Hobbes, they are:

> no names of things; but of signs, by which we make known, that we conceive the consequence of one name or attribute to another: as when we say, *a man is a living body*, we mean not that *the man* is one thing, the *living body* another, and the *is*, or *being* a third; but that the *man*, and the *living body*, is the same thing; because the consequence, *if he be a man, he is a living body*, is a true consequence, signified by that word *is*. Therefore, *to be a body, to walk, to be speaking, to live, to see*, and the like infinitives; also *corporeity, walking, speaking, life, sight*, and the like, that signify just the same, are the names of *nothing*.[47]

Just as Comte exposed the emptiness of the metaphysical mode of explanation, so did Hobbes with regard to Aristotelian physics. To explain why some bodies fall to the ground, you say, because they are heavy, but to say that they are heavy is to say no more than that they fall to the ground.

> Then for *physics* ... they render none at all, but empty words. If you desire to know why some kind of bodies sink naturally downwards toward the earth, and others go naturally from it; the Schools will tell you out of Aritstotle, that the bodies that sink downwards, are *heavy*; and that this heaviness is it that causes them to descend. But if you ask what they mean by *heaviness*, they will define it to be an endeavour to go to the centre of the earth. So that the cause why things sink downward, is an endeavour to be below: which is as much as to say, that bodies descend, or ascend, because they do. Or they will tell you the centre of the earth is the place of rest, and conservation for heavy things; and therefore they endeavour to be there: as if stones and metals had a desire, or could discern the place they would be at, as man does; or loved rest, as man does not; or that a piece of glass were less safe in the window, than falling into the street.[48]

The new philosophy as formulated by Hobbes, being materialistic and mechanistic, is not only against the metaphysical mode but also aggressively against the teleological mode of explanation. Like Galileo as well as later positivists, he had no use for all the four Aristotelian causes—the formal and final were discarded. Bodies which are real and exist and about which we can have knowledge are material and motion is the efficient cause of all changes in such bodies.

In the light of the discussion above, it is not implausible to claim that Hobbes systematically constructed a new philosophy to go with the new science, a philosophy whose core tenets are shared by a philosophy which later came to be called positivism. Hobbes anticipated Comte's law of intellectual development. Like him, he was against the theological and metaphysical modes of explanation. This hostility, for both, is a consequence of their empiricist ontology and epistemology which also ruled out of court the teleological mode of explanation in terms of final and formal causes. Hobbes, in reinforcing Galileo's attempt, fiercely formulated the new worldview of materialism and mechanism, thereby abandoning the Aristotelian organic cosmos. Focusing on Hobbes' contribution to the construction of modernity in this way should, however, not be taken to mean that other thinkers in the seventeenth century did not also make the effort—for instance, Descartes most certainly did. But while Descartes is invariably acknowledged, Hobbes is often, if not invariably, ignored in spite of the fact that he systematically articulated for the first time the philosophy of positivism.

Goals of the New Science

Let us next move to the goals of the new science in the light of the new philosophy. Today, we commonly identify three such goals—prediction, explanation and control. We have mentioned earlier that the positivist methodology and philosophy of science upholds the unity of method thesis and, moreover, that the logic of predicting/explaining an event as well as testing a theory or hypothesis is symmetrical. On this conception, the ability of science to make predictions is crucial. To predict a phenomenon is to invoke a law (a regularity or uniformity of sequence) which licenses the prediction; in turn a law is tested in terms of the prediction it licenses. Prediction then is the lynch pin of an epistemology which decrees that the scope of knowledge is delimited by the sensory given. A system of thought which does not issue in testable predictions cannot count as knowledge. This entails a positivist exclusion from the scientific domain any theory which is incapable of issuing in testable (hence precise) predictions—on this view, for instance, geology is in danger of not qualifying to be a science since its principles or laws permit explanations but not predictions precise enough to be testable.

Apart from the crucial role of prediction in epistemological terms, the possibility of prediction is also linked to the possibility of control. Comte, certainly, held this to be so. If one can successfully make predictions with the help of laws, then one can either (i) take steps to avoid the event predicted if it is considered to be undesirable (the weak sense of control), or (ii) alter or modify the circumstances so that certain desired results could be brought about and other undesired ones prevented from arising (the strong sense). Astronomical knowledge enables one, for instance, to predict an eclipse of the sun at a certain place and on a certain date; then one can arrange to be there to observe it if its observation can be used to further some other task, like Eddington's expedition in 1919 to test Einstein's theory. Alternatively, if an eclipse of the sun is considered to have undesirable effects—suppose observing one causes cancer of the eye—then one could take appropriate avoiding action.[49]

The second possibility allows one to interfere more directly with the workings of nature. According to the laws established about plant growth, a certain degree of warmth, and not merely exposure to light, encourages plant growth. If one wishes to encourage growth, then one ought to put the plants in a warm place.

For Comte, the possibility or the lack of direct intervention depends on the type of phenomenon studied—astronomical phenomena are too large in scale and too far away for us to influence, whereas physiological phenomena are not.[50] However, the possibility of control in both the weak and the strong senses provides the link between science and what Comte called "art," or between science and technology, as we would put it today. In this way, the new science has always been connected up with utility (for humans)—a theme that Francis Bacon had made familiar.

The weak form of control is not the real goal. It is *faute de mieux* and at best a prelude to the aspiration of controlling nature in the strong form. Being able to predict the onset of drought or rain is clearly better than not being able to do so at all. But it would be better if scientific theoretical understanding of meteorological phenomena ultimately enables one either to generate rain (when drought is undesired) or to hold rain at bay (when dry weather is desired).

To Bacon's voice on this matter, Descartes also added his:

> [A]s soon as I had acquired some general notions in physics and had noticed, as I began to test them in various particular problems, where they could lead and how much they differ from the principles used up to now, I believed that I could not keep them secret without sinning gravely against the law which obliges us to do all in our power to secure the general welfare of mankind. For they opened my eyes to the possibility of gaining knowledge which would be very useful in life, and of discovering a practical philosophy which might replace the speculative philosophy taught in the schools. Through this philosophy we could know the power and the action of fire, water, air, the stars, the heavens and all the other bodies in our environment, as distinctly as we know the various crafts of our artisans; and we could use this knowledge—as the artisans use theirs—for all the purposes for which it is appropriate, and thus make ourselves, as it were, the lords and masters of nature. This is desirable not only for the invention of innumerable devices which would facilitate our enjoyment of the fruits of the earth and all the goods we find there, but also, and most importantly, for the maintenance of health, which is undoubtedly the chief good and the foundation of all the other goods in this life. ... We might free ourselves from innumerable diseases, both of the body and of the mind, and perhaps even from the infirmity of old age, if we had sufficient knowledge of their causes and of all the remedies that nature has provided.[51]

Hobbes distinguished science from prudence. The latter is based on extrapolating future experience from that of the past. It rests, in other words, on inductive logic.[52] Hobbes knew of Bacon's inductivism and had used it in his earlier writings but after he set eyes on a copy of Euclid's geometry, he repudiated it. Science, for Hobbes then, relied on deductive logic, as exemplified in geometry, his paradigmatic science. (To avoid a possible misunderstanding, one ought, perhaps, to point out here that in repudiating the inductive methodology, he was not necessarily repudiating the empirical. He could not do so, as he was an empiricist as well as a materialist. When he maintained that science was not prudence based on the accumulation of sense experience, but the imposition of a method upon the deliverances of sense experience, he merely intended to say that sense and memory on their own could not yield science and also that science must be mathematically exact and systematic, which prudence is not.) But within such a methodological framework allied with the ontology of materialism and mechanism, Hobbes was as eager as Bacon to pursue the goal of controlling nature *via* science.

His attempt may be grasped by looking at another distinction he made, namely, between science, on the one hand, as reasoning from cause to effect and from effect to cause and, on the other, as knowledge which allows us to generate an effect from the knowledge of its cause. The former means we can explain a phenomenon when we have discovered its cause. The latter permits us to go beyond mere explanation as we can actually produce the phenomenon by bringing about the conditions that are said to be its cause. When we come across ashes in a certain location, we can reason, using the method of resolution, from the palpable effect, that is, the ashes back to the cause, fire, and in explaining the phenomenon of fire itself, one works back to yet another cause until one gets to the most universal cause, motion. In principle, from the most universal cause, by the method of composition, one could deduce all phenomena like fire, ashes, etc. But this still does not mean we know how to bring about fire in order to bring about ashes. To do that we need to be able to assemble all the relevant causal conditions to see if we could then generate the phenomenon in question.

For Hobbes then, certainty of scientific knowledge was demonstrated in two ways: (a) when scientific truths are validly derived using deductive logic (like theorems in geometry) from a set of "apt imposing of names" (that is, basic terms and axioms), (b) when we know how to produce the effect by producing those conditions which our theoretical knowledge of its cause gives. The first type of certainty is involved with logical certainty. The 'certainty' of the second is what enables us to control nature.

Like all positivists, Hobbes was fundamentally interested in understanding and generating order in the world, not by means of the theological or metaphysical mode of thought, but by the application of scientific method to the study of both natural and social phenomena as well as moral and legal ones. The theological mode did secure order in the world, or at least Europe, for a long time, by relying on supernatural entities, superstitious beliefs and practices. The metaphysical mode, usually in conjunction with the former, was but another attempt to use unreason to procure order in human thought and behavior. The superiority of the positive mode over the other two lies precisely in its use of reason and science to achieve order. In other words, positivism, while disagreeing with the earlier modes about the means to achieve order, nevertheless, is in agreement with them on the end they all aim at attaining. That is why it may be distinguished from them by calling it the scientistic philosophy of order. The order which both the physical and social worlds exhibit is not God-decreed;[53] rather, it is the outcome of an attempt by the human intellect to grapple with the complexities of physical and social life using the methodology of positivism, that is, of science.

Order in the study of natural phenomena takes the form of systematically structuring sense experience into a coherent interconnected body of knowledge so that knowledge about one phenomenon could ultimately be understood by being derived from knowledge about others within it.[54] Not only does such an axiomatic structure allow explanation, prediction and theory testing to take place, as we have seen, but it also enables us in the end to control nature (in the strong sense earlier identified). And this bears out the Baconian dictum that "knowledge is power."

In the light of the above, it would be fair to conclude that built into the new scientific method and its accompanying philosophy from the seventeenth century onward is the aspiration to control and manipulate (and in that way to dominate) nature. Bacon, Descartes and Hobbes all unhesitatingly declared it to be so. It does not look as if the ideal of knowledge for its own sake, what Einstein called "the holy curiosity of inquiry," ever existed in its neat purity at the inception of modernity (or at any time, later, for that matter). The philosophical as well as the ideological requirements of the new philosophy ensure that science as technology and science as theoretical knowledge go hand in hand. While humans had used and controlled nature in the past, modern science makes it possible for them, more systematically than ever before, to control (to exploit) nature.

This new opportunity for manipulating nature has prompted several radically different responses. The majority holds that the exploitation of nature redounds to the good of all humans. Some argue that the possibility of exploiting nature would displace the exploitation of men by fellow men only when capitalism has been superseded, and envisage, thereafter, a cornucopia for all humans. Others hold that the exploitation of nature is yet another means to sustain the exploitation by some humans of others (whether capitalism is dislodged or not) and that the exploitation of nature and of humans must together be overcome. Yet others recognize even the possibility of exploiting certain humans while emancipating nature from exploitation. Those who subscribe to Adam Smith's "invisible hand" argument represent the first (which is the dominant) attitude. Marx stands for the second, utopian socialists for the third, and the so-called eco-fascists for the fourth.

The crucially built-in goal of controlling nature in modern science has taken on another dimension in the last thirty years or so with the establishment of molecular genetics as a theoretical discipline and its accompanying technology called biotechnology, a form of genetic engineering. Apart from this significant actual development, promises of more spectacular ones are already on offer, such as (molecular) nanotechnology. The next chapter will argue that this involves a deeper kind of control of nature than the earlier types of scientific theories and their associated technologies were capable of. This, surprisingly, in turn permits the dramatic reimposition of teleology upon the world, the restoration of formal and final causes, two of the four Aristotelian causes which modernity from the seventeenth century onward has cast into the outer darkness. But before the next chapter can go on to demonstrate this, one needs first to clarify the notion of teleology, to distinguish between its different forms and, in turn, to relate these to the four causes. Furthermore, one needs to draw the distinction between what may be called old teleology on the one hand and new teleology on the other; the former embodies pre-modernity and a more passive form of anthropocentrism while the latter, modernity and a correspondingly more aggressive form of anthropocentrism.

Teleology, Its Forms, and Their Fortunes

The relegation of the formal and final causes to the realm of the superstitious or the 'metaphysical' is often regarded as constituting the definitive break from the medieval worldview or the hallmark of modernity. As already commented upon, the theological and metaphysical modes of explanation are considered to be redundant.

But to say that the rejection of the formal and final causes is synonymous with the rejection of the teleological worldview may be too simplistic and, hence, misleading. [55] To begin with, the rejection of the former was accompanied at the same time with the establishment of strong anthropocentrism, the claims that only humans have intrinsic value, and nonhuman naturally-occurring beings, therefore, have only instrumental value for humans. But this view is held to be remarkably similar to that of Aristotle (and Aristotelians). It is indeed true that for Aristotle, in his hierarchy of beings, humans are higher than animals because they possess reason to a greater degree. Furthermore, he also believed that the purpose of beings further down the hierarchy of rationality is to serve those higher up:

> [W]e may infer that, after the birth of animals, plants exist for their sake, and that the other animals exist for the sake of man, the tame for use and food, the wild, if not at all, at least the greater part of them, for food, and for the provision of clothing and various instruments. Now if nature makes nothing incomplete, and nothing in vain, the inference must be that she has made all animals for the sake of man.[56]

Kant said something quite similar, that "so far as animals are concerned, we have no direct duties. Animals are not self-conscious and are there merely as a means to an end. That end is man. We can ask, 'Why do animals exist?' but to ask, 'Why does man exist?' is a meaningless question."[57] So it is held that Aristotle as well as the moderns share the view that other nonhuman naturally-occurring beings are for the benefit and use of man. This, however, amounts to a form of teleology. There might or might not be a God who created the world especially for humans, but it appears that an unbroken dominant tradition runs through the history of Western thought from Aristotle *via* the Aristotelians to modernity, that the nonhuman natural world exists for the sake of humans.[58] And yet it is also commonly claimed that modernity broke with the medieval worldview precisely by its rejection of teleology. Perhaps what has led to the confusion is the ambiguity within the notion of teleology itself.

In the first instance, one may have to distinguish between two possible theses, **external teleology** on the one hand and **intrinsic/immanent teleology** on the other.[59] The latter is what may be involved with formal and final causes (which will be examined later). The former is about perceived hierarchy in the world ordered in terms of certain criteria or attributes, with the superior beings at the top, and the related belief that those further down exist to sustain and maintain those above them. Aristotle chose rationality as the appropriate attribute to order his hierarchy,[60] Kant, self-consciousness, and Descartes, linguistic capacity or soul. This set of related characteristics unsurprisingly, according to their critics—like Routley and Routley[61]—enthrones humans at the summit of the pyramid. From this perspective, external teleology and anthropocentrism go hand in hand. This conjunct holds true in modernity no less than it did in the medieval cosmology. But while modernity requires the rejection of intrinsic/immanent teleology, medieval cosmology did not. The difference may be traced to the goal of controlling nature built into the methodology and ideology of modern science. Its ontology of materialism and mechanism, pioneered by Galileo and philosophically systematized by Hobbes, Descartes and others, which renders matter inert and dead, entails the rejection of intrinsic/immanent teleology, at the same time extolling anthropocentrism. If matter were truly inert, then clearly, humans need not be constrained by its *telos* and may, therefore, do what they please with it entirely to suit human ends and purposes, including re-fashioning and re-modeling it. While the predecessors of modernity simply held nature, as they found it, by and large, to be ordained for the use of man, their successors in the modern era go one beyond, and consciously aspire through their scientific method to control (dead) nature by molding it in accordance with their own ends. For instance, as already observed, the scientific developments of the last twenty years, in particular, molecular genetics and its accompanying genetic engineering,[62] as well as nanotechnology on the horizon, embody the ultimate truiumph of this aspiration. One may conclude that the conception of nature as dead matter is what constitutes the definitive break between modernity and its medieval past in European thought.

The notion of intrinsic/immanent teleology is most clearly expressed by Aristotle, especially through his formal and final causes. One needs then to consider Aristotle's philosophy of nature. The pre-Socratics had held that there was a rational explanation for the things that existed and the events that happened. But Plato taught that such matters belonged to the world of Appearances. This sensory world was illusory, delusory and fleeting. About it, at best one could only have beliefs. Behind it stood the world of Reality, of immutable forms of which alone there could be knowledge. Aristotle, in spite of the Platonic indictment of the world of Appearance, tried to return to the pre-Socratic enterprise which regarded the study of natural, sensory phenomena (that is, natural philosophy) as a truly 'scientific quest,' as fit objects of the quest for knowledge.

Aristotle studied and investigated nature in numerous ways. He was astronomer, physicist, biologist, etc. The biologist, in particular, has to confront change, a matter which Plato could not take seriously, having relegated it to the world of mere Appearance. Immutable permanence, the defining characteristic of Platonic forms, obviously, has no application to biological phenomena. Instead, they seem to display permanence as well as change—the beech in the field is covered with leaves in the summer, yet it is the same tree even when it sheds them in the autumn. No biologist can ignore this kind of change; nor can the philosopher who studies the subject ignore it. The world is neither in total Heraclitian flux nor is it in absolute Parmenidean stasis.

Perhaps the best way of appreciating such an orientation in Aristotle's philosophy is through his understanding of *φύσις*, usually translated as 'nature' in English.[63] Aristotle, who devoted one book of his *Metaphysics* to investigating its meaning, contended that the word originally meant "the genesis of growing things—the meaning which would be suggested if one were to pronounce the *υ* of *φύσις* long" (*Metaphysics* 5, 4, 1014b 17-18).[64] If so, it is probably derived from φύω, which has a long *υ* in most of its forms. This word in turn connotes begetting and growth. The original word *φύσις*, then, would mean the process of growing. But it was also applied in early Greek thought to the origin of growing things. This leads to postulating 'nature' as the principle of all movement and rest. Aristotle consistently held that "nature is the principle and cause of motion and rest in those things, and those things only in which she inheres primarily as distinct from incidentally."[65]

The pre-Socratics were engaged in finding "the primary material out of which any natural object is made" (*Metaphysics* 5, 4, 1014b27-28). Their search for the 'nature' of things was a search for an active principle of movement and becoming, of the source out of which all things were formed. Some thought it was fire, others, earth, air, water or combinations of them.

Aristotle went beyond the pre-Socratic enterprise by taking into account the evidence of the senses. He investigated into the source of things or substances which are presented in sensory experience. For instance, a tree is not merely one of the four elements that makes up the tree. When the tree dies and decays, the tree, essentially and substantially, would be no more. In other words, the substance or essence of a tree is not given reductionistically by citing the appropriate element or combination of elements which underlies the make-up of the tree. The tree has a 'nature' which is a primary and *per se* principle of motions and characteristics proper and peculiar to the tree. Here the 'nature' refers to the form which gives the tree its substantial essence, quite different from the sense of 'nature' as understood by the pre-Socratics in terms only of the four elements.

Aristotle maintained that they had failed to distinguish between first matter and second matter. First matter may be said to be primary as ultimate matter, in the absolute sense. For instance the ultimate matter of bronze is probably water. But Aristotle's natural philosophy focused instead on second matter which consists of primary matter and the form or essence, and is the end of all becoming. By so doing, he was able to recognize the reality of substances that could change, which the pre-Socratics could not or which Plato had simply denied.

From this, Aristotle concluded that the primary and proper meaning of 'nature' is the essence of things which have in themselves (*qua* themselves) a principle of motion which is present in them either potentially or actually. By invoking this distinction between potentiality or potency and actuality or act, he was able to explain substantial change. The former refers to the ultimate, radical possibility of change inherent in matter; the latter, to the manifestation of that potency. This actuality, he called 'form,' what makes a thing to be what it is by definition. However, this form was given an extended meaning when used in connection even with entities which display no movement or behavior—for example, one talks of the form, the nature or essence of a triangle. But this is a derivative of the primary meaning of 'nature' as a spontaneous source of characteristic movement and rest. It is this extended meaning which allows one to say that every thing is a kind of essence and every essence is a nature.

> Primarily for Aristotle the term 'nature' has two closely linked senses. In the sense of 'form,' it is active, dynamic, and spontaneous. It explains what is specific, characteristic, and formally recognizable in different natures we observe in the world. But there is another sense of *φύσις*, as when we apply it to what is potentially so, or material. In other words, ... there are not only 'formal' principles of motion toward a term and the possession of that term, but also the real 'ability' (material capacity) to have that form, as the seed and the ovum are only potentially a man before they are united in the embryo. That 'ability' or 'potentiality' can also be called 'natural' when it specifically belongs to the thing under consideration.[66]

The scholastics in the Middle Ages developed these two meanings of the word 'nature' in terms of an active and a passive principle. Aquinas maintained that natural bodies possessed an active or formal principle when they spontaneously moved. But when they must be moved, they displayed the passive principle, which was matter. But only observation and experience could tell whether bodies moved spontaneously or were moved by an external force. However, they did not query the fundamental framework of Aristotle's philosophy of nature. They fully accepted his definition of nature as a principle of motion and rest and like him, they considered nature as a principle which could be studied and understood rationally in a scientific way.

Nature, for Aristotle, not merely manifested change but also change in certain clearly defined ways. In other words, change involved development which in turn implied a *nisus*, a direction and a goal toward which the changes were orientated. The acorn, in germinating and pushing its way through the soil, may be said to be striving to become an oak, the whole process of which tends toward the form of the mature plant. If this were so, the formal and final causes (in natural bodies) become identical. In the case of certain natural bodies, though not in all, the formal and final causes themselves are also identical with the efficient cause. "Aristotle ... conceives the notion of a final cause which not only directs but also excites or awakens the energy which it controls, by arousing in the appropriate object a nisus towards its own realization in bodily form."[67] The efficient cause, then, is an immaterial cause. Furthermore, as already mentioned, Aristotle, by focusing on second matter, held that in nature, form and matter go hand in hand, as it is form incorporated in matter which is subject to the processes of change or becoming. In other words, as far as Aristotle was concerned, in nature, the four causes are often inseparable (not identical) and may be isolated only for the purpose of intellectual clarification.

One can see from this brief account of Aristotle's philosophy of nature that the role played by the formal and final causes is crucial, rendering the theory a teleological one. Such an approach is, undoubtedly, incompatible with the materialism and mechanism of modern science. To quote Collingwood again:

> The conception of development is fatal to materialism. According to a materialistic metaphysics, that is to say, a metaphysics according to which bodily existence is the only kind of existence, whatever works or produces results must be a body: in other words, there can be no immaterial causes. But development implies an immaterial cause. If a seed is really developing into a plant, and not merely changing into it by pure chance owing to the random impact of suitable particles of matter from outside, this development is controlled by something not material, namely, the form of the plant, and of that specific plant ... as the formal cause of the full-grown plant and the final cause of the process by which the seed grows into it.[68]

Moreover, and equally crucial, it is also incompatible with the aspiration of controlling nature by re-fashioning or re-modeling it. Naturally-occurring beings possessing their own *telos* would develop in ways independently of the will and wishes of humans. To make room for such an aspiration, (biotic) nature's intrinsic/immanent teleology must be denied.

In order further to highlight the differences between the modern worldview on the one hand and the pre-modern worldview as informed by Aristotle and the spirit of his philosophy on the other, one must go back to his four causes, and in so doing, identify yet another thesis of teleology, namely, **extrinsic/imposed teleology**, in addition to the two others already distinguished as external and intrinsic/immanent teleology. The identification of this third thesis is done in the context of relating his elucidation of art and artefacts to his four causes.

Aristotle was obviously influenced by the notion of art and artefact in arriving at the four causes. This is because with artefacts, the four causes are clearly distinct. That is why in illustrating them, Aristotle found himself liberally using examples from the realm of art and the artefactual.[69] They can be identified separately, by reference to four distinct and different sources—in the case of a building, stone is the material cause, the architect, the formal cause, the mason who built it, the efficient cause, the owner with a particular purpose in mind who commissioned the design and the building of it, the final cause. Unlike the oak tree springing forth from the acorn, a building does not come into existence in the same way, under its own steam, so to speak. In the case of the Taj Mahal, Shah Jahan conceived the idea of commemorating his wife by building a monument of a certain type. He next talked about his idea with his architect who translated his conception into a plan. He then hired builders to execute that plan, and to realize it as the building we know and admire today. This is not to deny that, in practice, one person could embody the roles of commissioner, architect and builder. The division of labor which splits them simply reinforces the key point which is being established, namely, that in artefacts, but not in naturally-occurring objects—at least in the case of those which are biological entities—it is in principle, as well as in practice, possible to separate out and identify each of the four causes in terms of four distinct and different sources.

With natural biological entities, we have seen that Aristotle argued for the fusion of the four causes in the existence of any one of them. Matter is always second matter (rather than merely first matter). Second matter is always informed by the form of the being which shapes its development—hence, the formal and the final causes are identical. Furthermore, the final cause is also the immaterial efficient cause. The germinating acorn struggles through the soil to reach sunlight while it grows roots to get nutrients from the surrounding soil so that it would eventually become a mature oak tree. Second matter, first as acorn, then as oak tree, grows and develops under its own internal impetus, according to its own program. To account for an oak, one has to refer to the intimate fusion of the four causes in the entity itself. Distinguishing between them is only made for the purpose of intellectual clarification.

Humans, when they engage in creating artefacts, do so for an end or purpose, be it a utilitarian end such as providing shelter, a personal, sentimental end of preserving or celebrating the memories of some dead beloved, an aesthetic end of realizing a particular conception of beauty, a moral end of teaching the difference between good or evil. In other words, artefacts, necessarily, embody human ends. Accomplishing these ends necessarily requires actions or the deliberate initiation of certain processes. The activities involved are directed activities, with a point and a pointedness to them, so to speak. In other words, they are (human) intentional structures.[70]

This leads Aristotle to the analogy of art and nature. In nature, too, just as in art—art being an artefact—many activities are directed activities. A bird builds a nest. The nest is not a human artefact, but like it, it displays 'guidedness,' although whether one goes so far as to call it an avian artefact remains more controversial in this context.[71] Just as the house a human builds is end-oriented involving end-directed activities, so too does the nest the swallow builds.

Aristotle postulated that something comes about in one of two ways—either through chance/coincidence, or for an end. To him, it is obvious that neither the activities of nest building nor the nest itself happen through chance. So they must be for an end.

> If then, it is agreed that things are either the result of coincidence or for an end, and these cannot be the result of coincidence or spontaneity, it follows that they must be for an end; and that such things are all due to nature even the champions of the theory which is before us would agree. Therefore action for an end is present in things which come to be and are by nature.
>
> Further, where a series has a completion, all the preceding steps are for the sake of that. Now surely as in intelligent action, so in nature; and as in nature, so it is in each action, if nothing interferes. Now intelligent action is for the sake of an end; therefore the nature of things also is so. Thus if a house, e.g., had been a thing made by nature; it would have been made in the same way as it is now by art; and if things made by nature were made also by art, they would come to be in the same way as by nature. ... If, therefore, artificial products are for the sake of an end, so clearly also are natural products. ... If then it is both by nature and for an end that the swallow makes its nest and the spider its web, and plants grow leaves for the sake of the fruit and send their roots down (not up) for the sake of nourishment, it is plain that this kind of cause is operative in things which come to be and are by nature.[72]

At this point, it would be wise to dispose of an obvious red herring. It is a mistake to accuse Aristotle himself of committing anthropomorphism. The end-directedness or purposiveness that he was talking about is not meant to be the conscious, intentional variety which is displayed in the human creation of artefacts. He did not say that the plant knew what it was doing and planned it in the same way the sculptor did when he chipped away at the block of stone according to a blueprint of some kind he carried in his head, or according to external instructions he had received from the commissioner of the artefact as well as the conventional requirements of the activity of sculpting objects. It is not unintelligible to say that the plant strives, albeit in the absence of consciousness, to get to the source of sunlight so that it could grow properly. Conscious intentions do not exhaust the notion of purposiveness and end-directedness.[73] Aristotle's final cause does not in general entail conscious intentionality, although in the context of creating human artefacts, it does refer to conscious intentions and purposes. But Aristotle also commented that conscious intentions and purposes could be attributed even in contexts where the agent could not be observed to be deliberating. "It is absurd to suppose that purpose is not present because we do not observe the agent deliberating. Art does not deliberate."[74] From this he concluded that there could be purpose in nature even though there is no deliberation. "If the ship-building art were in the wood, it would produce the same results by nature. If, therefore, purpose is present in art, it is present also in nature."[75]

The point to grasp is that Aristotle was simply drawing an analogy between art and nature. An analogy between two things precisely entails that they are not similar in all respects. While purposiveness as end-directedness is what they have in common, conscious intentionality is a point of disanalogy between humans with their particular type of consciousness on the one hand, and naturally-occurring biotic beings on the other.

There is yet another point of disanalogy, already referred to. Living organisms embody intrinsic/immanent teleology. Indeed, intrinsic/immanent teleology is used even to define 'nature' or 'the natural,' as we have seen. "[T]those things are natural which, by a continuous movement originated from an internal principle, arrive at some completion ."[76] In contrast, 'the artificial' are "those things which, by a movement originated from an external source, arrive at some end." In other words, art/artefact is a form of 'the artificial.' Art embodies what may be called an extrinsic/imposed teleology. The end comes from outside the thing itself and is imposed upon it by an outside source, namely, human agency. The Taj Mahal came into existence, and took the shape and character it did, because of the intention entertained by Shah Jahan, in the light of which, he set in motion certain actions leading to its ultimate realization.

It is now time to pause and clarify the relation of this third identified form of teleology called **extrinsic/imposed** to the two forms earlier contrasted with each other, namely, **external** and **intrinsic/immanent** teleology. However, before this can be done, one needs first to refine the thesis of external teleology itself, in terms of two subvarieties which may be distinguished—the unmodified from the simplified. The former involves a thorough-going hierarchy of beings, with the more lowly, like plants, serving those above them, namely, animals, while plants and animals in turn serve the ends of humans. This worldview, represented in ancient Greek philosophy by Aristotle, leads to the notion of the Great Chain of Beings in medieval thought. Modernity, however, rejects this more finely differentiated account in favor of a simplified one, which simply holds that all naturally-occurring beings have only instrumental value for humans who alone possess intrinsic value.

But paradoxically, differentiated external teleology, at least on Aristotle's account, goes with what may be called a more passive form of anthropocentrism and instrumentalism, as intrinsic/immanent teleology takes precedence over it. That is to say, for Aristotle, naturally-occurring living beings primarily exist as ends for themselves, unfolding their *tele*, and only secondarily, serve the ends and purposes of humans.[77] The revised account, on the other hand, involves the rejection of intrinsic/immanent teleology while holding simplified external teleology. The rejection of intrinsic/immanent teleology, it has been argued, is a requirement of modern science, its methodology, its ontology of materialism and mechanism as well as its goal, both epistemological and ideological, of controlling nature. *Pari passu*, its adherence to simplified external teleology is married to an aggressive form of anthropocentrism and instrumentalism. The logic of such strident anthropocentrism involves not merely simplified external teleology but also extrinsic/imposed teleology. This means that while, on the one hand for Aristotle, it makes perfect sense to combine intrinsic/immanent teleology with differentiated external teleology to produce a less aggressive form of anthropocentrism, on the other, it is coherent for the moderns to reject intrinsic/immanent teleology, adopt extrinsic/imposed teleology, and in so doing, transform simplified external teleology into a more aggressive form of anthropocentrism.

The various theses of teleology, an elucidation of them in terms of Aristotle's four causes as well as their relationships to one another may be further clarified by distinguishing between what may be called **Old Teleology** (which expresses the pre-modern worldview) on the one hand and **New Teleology** (which expresses the modern worldview) on the other as follows:

Old Teleology

1. Intrinsic/immanent teleology—formal, final as well as material and efficient causes are inseparable in the case of many naturally-occurring beings, such as biological ones.
2. External teleology, but in its differentiated form, involves a complex hierarchy of naturally-occurring beings, such that those lower down serve the ends and purposes of those above them, with humans right at the top.
3. For Aristotle, intrinsic/immanent teleology precedes external teleology—first and foremost, plants and animals exist, grow and develop under their respective *tele*, and only secondarily do they exist for the use of humans.
4. Hence a more passive anthropocentrism and instrumentalism could emerge.[78]

New Teleology

1. The retention of only material and efficient causes while rejecting final and formal causes amounts to a rejection of intrinsic/immanent teleology.
2. External teleology is held in a simplified form with a more stream-lined hierarchy—all naturally-occurring beings have only instrumental value for humans who alone possess intrinsic value. Such a form leads to imposed/extrinsic teleology as its logical extreme.
3. Rejection of intrinsic/immanent teleology is required to make room for an ontology of materialism and mechanism based only on material and efficient causes. Furthermore, the latter is understood to be external to the entity itself.
4. Hence an aggressive anthropocentrism and instrumentalism could emerge, underpinned by modern science, its method and its ideological goal of controlling nature through its technology.

Notes

1. For a recent systematic account of the history of Western philosophy from the earliest Greek to contemporary times about the relationship between being and value, see Frederick Ferré, *Being and Value: Toward a Constructive Postmodern Metaphysics* (Albany: State University of New York Press, 1996).

2. The philosophy of mechanism may be taken to embrace any one or all of the following theses:

1. The (modern) science of mechanical motion as established by Galileo, Newton and others.
2. Reality is nature mathematized and quantified.
3. The natural world is actually a machine to be studied and understood by sciences like the science of mechanistic motion using mathematics for the formulation of their laws.
4. The fundamental stuff of the universe is matter in motion (for instance, Hobbes' materialism) to which everything else in the universe may ontologically be reduced.
5. The laws of motion or of physics are the most fundamental laws to which other laws about natural phenomena are to be ultimately reduced.
6. The universe is a deterministic one—given the laws of nature (discovered under 1, 2 and 3 above) and the positions and velocities of the particles of matter at time t, one could predict precisely their positions and velocities at time t_1 (Laplacean determinism).

In this chapter, the mechanistic paradigm and worldview are discussed but primarily in the context of 1 and 2 above. Chapter 4 section entitled **Homo Faber, Artefacts and the Language of Machines** will look at 3 in particular.

For a fuller discussion of the emergence of positivism, especially with regard to Hobbes and Bacon, see Keekok Lee, *The Positivist Science of Law* (Hants: Gower Publishing Company Limited, 1989), 35-66. Positivism itself plays a significant part in formulating the philosophy of mechanism. But another very important contributor is, of course, Descartes, whose dualism between the material and the spiritual/mental makes it possible for science to treat all things in the universe (except humans, in so far as they have minds and souls) as mere objects, subject only to the laws of mathematics and physics.

The respective accounts of modern philosophy and science given here are not intended to imply that they arose absolutely *de novo* from nowhere and, therefore, to deny that precursory work in medieval philosophy and science based on developments of Aristotelianism might not have paved the way—see, for example, Edward Grant, *The Foundations of Modern Science* (Cambridge and New York: Cambridge University Press, 1997). (Mentioning a date, such as the beginning of the seventeenth century, as the starting point of modernity is in one sense necessarily arbitrary, though unavoidably so.)

3. For qualifications, see next section.

4. As with the work of all great thinkers, Galileo's writings are subject to different assessments. One interpretation even claims that there is nothing really new about his science. For an account of some of these interpretations, see Dudley Shapere, *Galileo: A Philosophical Study* (Chicago and London: Chicago University Press, 1974). Shapere in his own evaluation also differs from that of Stillman Drake, *Galileo* (Oxford: Oxford University Press, 1980). For a selection of Galileo's writings, see Stillman Drake,

Discoveries and Opinions of Galileo (New York: Doubleday Anchor Books, 1957). See also Stillman Drake and C. D. O'Malley, *The Controversy on the Comets of 1618* (Philadelphia: University of Pennsylvania Press, 1960).

5. Positivism is often coupled with empiricism as in the term 'empiricism-cum-positivism'.

6. For one such interpretation of his inquisition, see Stillman Drake, *Galileo* (Oxford: Oxford University Press, 1980). St. Augustine, an extremely important early Church Father, was of the opinion that religion and science belonged to separate domains; a heretic, after all, could be more knowledgeable about astronomy than a Christian. Galileo accepted this separation between matters of faith and of science. The Church was the legitimate authority over the former but would be inappropriate for arbitrating issues which could be settled by experience, observation and measurement alone. Moreover, Thomist philosophy also distinguished between faith and knowledge—faith belonged to Sacred Theology while knowledge fell under the domain of philosophy.

7. For a recent short assessment, see Jonathan Barnes, "Life and Work," in *Cambridge Companion to Aristotle*, Jonathan Barnes, ed. (Cambridge: Cambridge University Press, 1995).

8. For an assessment of Aristotle's empirical research in his biology, see G. E. R. Lloyd, "Empirical Research in Aristotle's Biology," in *Philosophical Issues in Aristotle's Biology*, Allan Gotthelf and James G. Lennox, eds. (Cambridge & New York: Cambridge University Press, 1987). For detailed discussions of his philosophy of biology, see Anthony Preus, *Science and Philosophy* (New York: Georg Olms Verlag Hildesheim, 1975); *Philosophical Issues in Aristotle's Biology*, Gotthelf and Lennox, eds. For a shorter more recent assessment of his science and his philosophy of science including his biological thought, see R. J. Hankinson, "Philosophy of Science," and "Science" in *The Cambridge Companion to Aristotle*, Barnes, ed.

9. See Warner Wick, "Aristotelianism," in *The Encyclopedia of Philosophy*, vol. 1, Paul Edwards, ed. (London: Collier-Macmillan, 1967), 148-51.

10. R. G. Collingwood, *The Idea of Nature* (Oxford: Clarendon Press, 1945), 96.

11. The crucial passage cited below makes a point which, according to Collingwood, must be well known to nearly every educated person in Western Christendom:

> Thou has learnt from astronomical proofs that the whole earth compared with the universe is no greater than a point, that is, compared with the sphere of the heavens, it may be thought of as having no size at all. Then of this corner, it is only one-quarter that, according to Ptolemy, is habitable to living things. Take away from this quarter the seas, marshes, and other desert places, and the space left for man hardly even deserves the name of infinitesimal (Book ii, Prosa vii). (Collingwood, *The Idea of Nature*, 97)

12. We shall see later in the book that the scientific research program based on the metaphysics of such a worldview has by no means run out of steam toward the end of the twentieth century; indeed, it promises a revolutionary technology of manufacture, namely, molecular nanotechnology, sometime in the next century.

13. The points made in this paragraph are 'Aristotelian' rather than Aristotle's.

14. Max Caspar, *Kepler* (London and New York: Aberlard-Schuman, 1959), 62.

15. Caspar, a biographer, has written:

> Nothing in the world was creatd by God without a plan; this was Kepler's

principal axiom. His undertaking was no less than to discover this plan of creation, to think the thoughts of God all over again, because he was convinced that "just like a human architect, God has approached the foundation of the world according to order and rule and so measured out everything that one might suppose that architecture did not take Nature as a model but rather that God had looked upon the manner of building of the coming human." (*Kepler*, 62)

16. The term 'philosophy' is here used in the sense of 'natural philosophy' to refer to the study of what we call physics today. The ancient Scottish Universities still use the term 'professor of natural philosophy' to refer to the professor of physics.

17. Stillman Drake, *Discoveries and Opinions of Galileo* (New York: Doubleday Anchor Books, 1957), 237-38. For a dissenting voice on the overrated significance of this passage, see James Maclachlan, "Drake v. the Philosophers," in *Nature, Experiment, and the Sciences: Essays on Galileo and the History of Science in Honour of Stillman Drake*, Trevor H. Levere and William R. Shea, eds. (Dordrecht/Boston/London: Kluwer Academic Publishers, 1990).

18. Drake, *Discoveries*, 225. (What Galileo wrote in his notes on Lagall's book.)

19. Drake, *Discoveries*, 274.

20. For a related account of the new science, its methodology and its metaphysics, see Richard S. Westfall, "The Scientific Revolution of the Seventeenth Century," in *The Concept of Nature: The Herbert Spencer Lectures*, John Torrance, ed. (Oxford: Clarendon Press, 1992). In the new order, nature was quantified, mechanized, perceived to be other and secularized.

21. Drake, *Discoveries*, 276-77.

22. Collingwood, *The Idea of Nature*, 102-3.

23. For an account of Hume's analysis within positivist methodology, see Lee, *The Positivist Science of Law*, 67-71.

24. Such a low-level regularity may in turn be derived from a higher-level regularity, so that ultimately a science consists of a hierarchical or pyramidal structure of regularities—see Lee, *The Positivist Science of Law*.

25. However, one should not overlook that for Aristotle "for the sake of which" did not imply want, desire or ends separate from the processes themselves.

26. This term will be explained in the next section, which looks primarily at Hobbes.

27. On the above points, this author follows Carl Mitcham in his "Philosophy and History of Technology," in *The History and Philosophy of Technology*, George Bugliarello and Dean B. Doner, eds. (Urbana and London: University of Illinois Press, 1979), 184-85.The dictionary referred to is the John Kersey's edition of Edward Phillips' *The New World of English Words*, published in 1706.

28. Having said this, one must still point out that in mankind's history of tool- making and tool-using, there are different types of technology, only the latest stage of which is quite so systematically science-induced. This point will be looked at in detail in Chapter 2.

29. Mitcham, "Philosophy and the History of Technology."

30. The account that follows summarizes Mitcham's arguments (without misrepresenting them, one hopes).

31. Mitcham, "Philosophy and the History of Technology," 178.

32. For a more detailed discussion on the relation between modern science and modern technology, see in particular Chapter 2 section entitled, **Modern Technology, the Philosophy of Technology and the Philosophy of Science**. There, two approaches will be mentioned—the more orthodox holds that pure science (at least in what Chapter 2 calls

Phase II) paves the way for technology as applied science, and the less familiar, the Heideggerian inversion, namely, that pure science is really theoretical technology. The line of thought developed in this book acknowledges the Heideggerian inversion at the level of the metaphysics and the ideology of Scientific Naturalism, but not necessarily at the more concrete level of theory formulation, theory testing and theory application as technology. But these points will be pursued in Chapter 2 itself.

33. For a detailed account and assessment, see Keekok Lee, *The Legal-Rational State: A Comparison of Hobbes, Bentham and Kelsen* (Hants: Gower Publishing Company Limited, 1990), 11-106.

Hobbes' contribution is frequently, if not invariably, overlooked because:

1. He had the misfortune to write before Newton, so that after Newton, physics became the "queen of the sciences" or the paradigmatic science. For Hobbes, the most mature science (except possibly, astronomy) was geometry.
2. The geometry he celebrated as the paradigmatic science was Euclidean geometry, which since the late nineteenth and early twentieth centuries has been shown by logicians and mathematicians to be only one geometry among others. The plurality of geometries means that geometrical truths are no longer considered to be both certain and informative and, hence, no longer satisfy what used to be the philosophical Holy Grail, the twin desiderata of being informative and certain. As Einstein commented, insofar as they are informative, they are uncertain and insofar as they are certain, they are uninformative or tautologous.
3. He did not coin the term "positivism," the honor going to Comte, almost two centuries later.

34. For details, see, for instance, Lezek Kolakowski, *Positivist Philosophy* (UK: Penguin Books Ltd., 1972); Peter Halfpenny, *Positivism and Sociology: Explaining Social Life* (London: Allen and Unwin, 1982); Lee, *The Positivist Science of Law* and *The Legal-Rational State.*

35. Although one of its most notable features is its scientific methodology, this would not be covered in this brief excursion. All forms of positivism—be it the Baconian/Benthamite (the so-called inductivist), the Popperian (the hypothetico-deductive, based on physics as the paradigmatic science), the Vaihingerian (the conventionalist) or the Hobbesian (deductive but based on Euclidean geometry as the paradigmatic science) variety, the nineteenth century Comtean or the twentieth century logical positivist form—uphold the unity of method. This consists of maintaining that (i) there is a symmetry in the logic of prediction, explanation, theory-testing, as well as of justification, (ii) to predict, explain an event, etc., means to derive a statement containing the prediction/explanation from a conjunction of two other statements—namely, one which is nomological or law-like (which can serve as the hypothesis under test in a context of theory-testing) and the other, statements about observable particulars, often called statements of initial conditions. (See Lee, *The Positivist Science* and *The Legal-Rational State.*)

36. As pointed out by Dijksterhuis: "The Aristotelian-Thomistic theory of knowledge is markedly 'sensationalist': all our knowledge is due to the experience gained during our present lives with the aid of the senses, and there is no room for any knowledge inborn or deriving from an anterior existence" (E. J. Dijksterhuis, *The Mechanization of the World Picture* Oxford: The Clarendon Press, 1961, 132). Moreover, Albertus Magnus (another important Dominican thinker who contributed to the so-called Thomistic synthesis) also emphasized the role of experimentation, under all possible circumstances, in gaining

scientific knowledge. However, in spite of these 'modern' characteristics, it would not be right to ignore the overall medieval context—social and philosophical—in which they were embedded.

37. It could also mean the study of unchanging being, of a transcendent supersensible being of which God would be an example.

See Jonathan Barnes, "Metaphysics" (in *The Cambridge Companion to Aristotle*, Barnes, ed.) for a short account of Aristotle's notion of metaphysics.

38. Thomas Hobbes, *Leviathan* (New York: Collier Books, 1962), 21.

39. F. S. McNeilly, *The Anatomy of Leviathan* (London: Macmillan, 1968), 77.

There is, undoubtedly, a nominalist strain in Hobbes' philosophy but Hobbes never allowed it to get out of hand so as to render his empiricism redundant—see Lee, *The Legal-Rational State*, 13-18.

But note, too, that nominalism had already appeared in the works of William of Ockham and John Duns Scotus, two, among others, of notable Schoolmen. In retrospect, one could say that this philosophy of language constitutes one of the numerous preparations for the emergence of the new science and the new philosophy to back it.

Indeed, the ideas of Nicholas of Autrecourt approximate so closely to that of modern skepticism that he has been called the medieval Hume. He held that:

> A conclusion is really reliable only when it was already implied in the premises, i.e. when, to use a modern term, it amounts to a tautology. All conclusions based on this principle possess the same degree of certainty—in mathematics, for instance, all proved propositions, no matter how far they are removed from the axioms. It is on the ground of this conception that Nicholas of Autrecourt denies that the existence of any thing can ever be inferred with certainty from the existence of any other thing. Nor can it ever be proved that one thing is more perfect than another or that one thing can be the end or the cause of another. Thus it can never be verified with certainty that outside ourselves there are material substances which cause our sensations. It is no more possible to infer the existence of substances—*a fortiori* of substances of a mental or abstract character, i.e. those not perceptible to the senses—from accidents such as qualities and relations than it is possible to deduce the cause from the effect. (Dijksterhuis, *The Mechanization of the World Picture*, 169)

In 1346, he was forced by the University of Paris to recant several of his skeptical theses. He claimed that he himself did not hold any of them to be true but merely defended them in a disputative spirit. He was happy to retract them and even burnt his writings with his own hands in public. Fortunately, his works survived in spite of the conflagration.

40. *The Essential Comte*, S. Andreski, ed. (London: Croom Helm, 1974), 20.

41. *The Essential Comte*, 20.

42. *The Essential Comte*, 124.

43. Hobbes, *Leviathan*, 481.

44. Hobbes, *Leviathan*, 481.

45. Hobbes, *Leviathan*, 36-37.

46. Hobbes, *Leviathan*, 483.

47. Hobbes, *Leviathan*, 484-85.

48. Hobbes, *Leviathan*, 487.

49. If it is objected that such a possibility does not qualify as weak control, then only the possibility of control in the strong sense remains as the goal of science.

50. Comte could not possibly have anticipated late twentieth-century projects like terraformation.

51. Descartes, "Discourse on the Method," in *The Philosophical Writings of Descartes*, vol. 1 (Cambridge: Cambridge University Press, 1992), 142-43. It is evident that Descartes held a view which Richard Routley, in "Is There A Need for a New, an Environmental Ethic?" in *Proceedings of XVth World Congress of Philosophy* (Bucharest: World Congress of Philosophy, 1973), called "human chauvinism," which embodies the following theses: (a) The Sole Value Assumption (humans are the sole locus of intrinsic value); (b) The Greater Value Assumption (in a conflict of interests between humans and nonhumans, those of humans always have priority); (c) Nonhumans only have instrumental value for humans (who may manipulate and control them without constraint with the help of their science and their technology).

52. Inductive logic does have a role to play in positivism in that variety, which may be called the inductivist wing, as exemplified in Bentham's and Mill's writings. According to such positivists, inductive logic accounts for the provenance or genesis of the theory/hypothesis. Other positivists belonging to the hypothetico-deductive variety (like Popper) distinguish between the contexts of discovery and justification—for them, the former is not governed by any logic and may be an entirely irrational or non-rational matter. Yet others belonging to the scientific conventionalist variety (like Vaihinger) maintain that a theory may be regarded as a fiction. (See Lee, *The Positivist Science of Law*, 77-101; *The Legal-Rational State*, 20-44, 198-204.)

53. With regard to the social world, it must be pointed out that positivists are not ideologues of order *per se* but of order and reform. They are not revolutionaries; neither are they the upholders of the *status quo* at all cost. Indeed, they all agree on the need for rational change (that is, change in accordance with scientific methodology), which is managed change through reform. They aim to avoid on the one hand, the Scylla of fossilization and the Charybdis of chaos and disorder on the other. Indeed war and disorder worried Hobbes greatly in the seventeenth century just as the excesses of the French Revolution worried Bentham in the nineteenth century and the Bolshevik Revolution and the First World War the logical positivists in the twentieth century. (See Lee, *The Positivist Science of Law*, 164-200.)

54. As far as order in the normative political, moral and legal domains is concerned, it is obvious that such order is entirely human-made. In these, there is no analogue to prediction in the domain of natural phenomena by which one could test a theory or hypothesis. That is why the logic of prediction or explanation or theory testing is transmuted to become the logic of justification—see Lee, *The Positivist Science of Law*, 132-63.

55. For discussions on Aristotle's notion of final cause, see Wolfgang Kullman, "Different Concepts of the Final Cause in Aristotle," in *Aristotle on Nature and Living Things*, Allan Gotthelf, ed. (Pennsylvania and Bristol: Mathesis Publications Inc. and Bristol Classical Press, 1985) and Allan Gotthelf, "Aristotle's Conception of Final Causality," in *Philosophical Issues in Aristotle's Biology*, Gotthelf and Lennox, eds.

56. Aristotle, *Politics* (1, 8, 1256b16-22), in *The Basic Works of Aristotle*, Richard McKeon, ed. (New York: Random House, 1941).

57. Immanuel Kant, *Lectures on Ethics* (New York and Evanston: Harper and Row, Publishers, 1963), 239.

58. Aquinas wrote: "[A]nimals are ordered to man's use in the natural course of things, according to divine providence. Consequently, man uses them without any injustice, either by killing them or by employing them in any other way. For this reason, God said to Noe: 'As the green herbs, I have delivered all flesh to you'" (*Summa Contra Gentiles* Notre Dame: Notre Dame University, 1975, 119).

59. Soon another form, to be called extrinsic/imposed teleology, will be introduced in this section; the relationship between this and the two identified here will then be further explored.

60. However, unlike modern philosophers, Aristotle did not hold that humans alone were rational or intelligent, only that they were the most rational and the most intelligent—see Larry Arnhart, "Aristotle, Chimpanzees and Other Political Animals," *Biologie et Vie Sociale* 29 (1990): 477-557.

61. Richard Routley and Val Routley, "Against the Inevitability of Human Chauvinism," in *Ethics and Problems of the 21st Century*, K. E. Goodpaster and K. M. Sayre, eds. (Notre Dame and London: University of Notre Dame Press, 1979).

62. In the opinion of some economists, the combination of biotechnology and information technology is expected to fuel the next Kondratieff/Schumpeter long wave of economic growth—for instance, see Andrew Tylecote, "Ecology, Technology and the Next Long Wave Upswing," in *Environment, Technology and Economic Growth: The Challenge to Sustainable Development*, Andrew Tylecote and Jan van der Straaten, eds. (Cheltenham, UK and Northampton, MA, USA: Edward Elgar, 1997). But for a critical view about the ascendancy of biotechnology, see Jeremy Rifkin, *The Biotech Century* (London: Gollancz, 1998).

63. This account leans on James A. Weisheipl, O. P. "Aristotle's Concept of Nature: Avicenna and Aquinas," in *Approaches to Nature in the Middle Ages*, Lawrence D. Roberts, ed. (New York: State University of New York at Binghamton, 1982).

64. However, in the opinion of Sir David Ross (a distinguished twentieth-century British scholar of Aristotle), the word never had the meaning of 'birth' or 'growth.' Collingwood (*The Idea of Nature*, 81-82) endorsed this.

65. Weisheipl, "Aristotle's Concept of Nature," 139-40.

To elucidate what Aristotle could mean, Weisheipl gives the following example:

> [I]f a statue falls to the ground and breaks, it does so not because it is a statue by Michelangelo, but because it has weight and is made of breakable material. We are not concerned about what may be incidental to the point under discussion, but are concerned only with a specific natural motion and its *per se* principle or source, leaving aside everything that may be irrelevant to that motion and termination, such as, Who made it?, Why? or any monetary value. ("Aristotle's Concept of Nature," 140)

66. Weisheipl, "Aristotle's Concept of Nature," 145.

67. Collingwood, *The Idea of Nature*, 84.

68. Collingwood, *The Idea of Nature*, 83-84.

69. See Mitcham, "Philosophy and the History of Technology," 176.

70. For a more formally-spelled out definition, the following is a handy one to use:

> By an "artifact" I mean here an object which has been intentionally made or produced for a certain purpose. According to this characterisation, an artifact necessarily has a maker or an author, or several authors, who are responsible for

> its existence. ... Artifacts are products of intentional making. Human activities produce innumerable new objects which are entirely unintentional (or unintended); such objects and materials are not artifacts in the strict sense of the word. When a person intends to make an object, the content of the intention is not the object itself, but rather some description of an object; the agent intends to make an object of a certain kind or type. Thus what I want to suggest is that artifacts in the strict sense can be distinguished from other products of human activity in the same way as acts are distinguished from other movements of the body; a movement is an action only if it is intentional under some description (Davidson, 1980, pp. 43-61), and I take an object to be an artifact in the strict sense of the word only if it is intentionally produced by an agent under some description of the object. The intention "ties" to the object a number of concepts or predicates which define its intended properties. These properties constitute the intended character of the object. I shall denote the intended character of an object o by "IC(o)."
>
> Thus an object o is a proper artifact only if it satisfies the following Dependence Condition:
>
> ... The existence and some of the properties of o depend on an agent's (or author's) intention to make an object of kind IC(o). (Rosto Hilpinen, "Belief Systems as Artifacts," *Monist* 28 (1995): 138-39)

Note, however, that Hilpinen's definition of 'artefact' is much wider than that used by this book, which stipulates that the human intentionality be embodied in a material medium. On this account, unlike Hilpinen's, belief systems cannot be said to be artefacts.

71. Aristotle himself did not call it an artefact. To do so would be controversial in the light of the distinction made in this book between 'the natural' and 'the artefactual,' where the latter is defined in terms of human agency as the efficient, final and formal causes, and the former's four causes in terms of the absence of human agency. (Arguments, however, will be provided later in Chapters 3 and 5 for maintaining a distinction between humans and their artefacts on the one hand, and nonhumans including the products of some of their activities on the other.)

72. Aristotle, *Physics*, II. 8. 199a2-30.

73. There is further discussion on this point in Chapter 5, in the section entitled **Interests and the Intentional Stance**.

74. Aristotle, *Physics*, II. 8. 199b27-28.

75. Aristotle, *Physics*, II. 8. 199b28-30.

76. Aristotle, *Physics*, II. 8. 199b15-16.

77. This point will be taken up again in Chapter 5.

78. There are also other reasons which contribute to this, but these will be explored in Chapter 4. For instance, in ancient Greek philosophy, the supremacy of *bios theoretikos* provides additional constraint against an aggressive form of anthropocentrism toward nature.

Chapter Two

The Natural, the Artefactual, and the Technological

As shown in the last chapter, the main elements of the new worldview and its philosophical framework within which modern science with its new method and goals was also emerging, were put in place as early as the seventeenth century. In turn, this lays the ground for the eventual systematic transformation of the naturally-occuring to become the artefactual. However, technologically speaking, the ultimate logic of this transformation is not reached until the last fifty years or so. This chapter looks in greater detail at the philosophical foundation for the technological transformation itself—the relationship between (a) the philosophy of science and the philosophy technology, (b) the philosophy of technology and history of technology, and (c) the 'depth' of theoretical scientific discoveries, the power of the technologics they may generate and the level of artefacticity embodied in the products of such technologies. And it begins by exploring the relation between the third thesis of teleology identified in the last chapter, namely, extrinsic/imposed teleology on the one hand, and the notions of the artefactual and degrees of artefacticity on the other.

Extrinsic/Imposed Teleology and the Artefactual

The last chapter has shown that the purest embodiment of extrinsic/imposed teleology is art/artefact. (For the purpose of this section, talk of artefact covers talk of art, as art is a form of artefact. But later, further complexities to the discussion about the matter will be introduced.) Aristotle locates a crucial difference between the natural and the artefactual in terms of the efficient cause. Naturally-occurring beings,[1] like individual organisms, not only have within themselves the source of movement, change or rest, but they also possess the active potential of changing other things in certain ways, as well as the passive potential of being changed by other beings in certain ways. In other words, the entity carries within it its own efficient cause—remember, though, that for Aristotle, in a naturally-occurring being, like an individual organism, all four causes are fused. In an artefactual entity, the efficient cause (as much as the formal and final causes) comes from without.

While agreeing with this essential characterization respectively of the natural and the artefactual, however, for the purpose of laying bare the philosophical transformation of the natural into the artefactual, this may not be enough. It may be necessary to look in greater detail at the artefact from the point of view of its material cause. This is because, while the meaning of 'artefact' could be clearly delineated as any entity or object which does not exist in nature (without human intervention), but is created by humans, according to human designs to fulfill human purposes, the range of second matter, which humans can use in producing artefacts, is varied and may, moreover, be itself the product of technological innovations.

The complexities may be approached by, first of all, dealing with a common association which people have between an artefact on the one hand and an item of exbiotic or abiotic nature on the other. A wooden chair conjures up the exbiotic, a stone statue, the abiotic. This, of course, is not what Aristotle himself said or implied. But he used, in the main, examples of what today we call abiotic artefacts in discussing the four causes. Perhaps this could have led others unthinkingly to confine artefacts entirely to the exbiotic or the abiotic. However, this would be a mistake. The matter out of which artefacts could be created may, of course, also be biotic.[2] Domesticated flora and fauna qualify to be such artefacts. The claim that they are would only appear counter-intuitive if one were mesmerised by stone statues or wooden chairs as the paradigm of an artefact. A black tulip or a chihuahua is no less a human creation than a statue or a chair. Their formal, final and efficient causes lie in humans and with humans. Humans can and do make artefacts not only out of exbiotic and abiotic nature but also of living organisms.[3] There is no conceptual difficulty involved. The constraint is at best a technological one.

And, when technology permits, humans have also been known to create substances which do not exist in nature. For instance, plutonium is one such substance, plastic is another. In particular, organic and inorganic chemistry and their accompanying technologies generate and produce such artefacts. However, the creation and existence of this kind of artefacts increase the possibility of conceptual confusion. They are referred to as 'artificial' in the sense of 'synthetic.'

Yet there is also another sense of 'synthetic' which occurs in the example of synthetic rubber. *Ex hypothesi* it is not natural rubber—the latter comes from a tree but the former has been created in a laboratory out of certain chemicals and chemical processes.[4] This leads to the sense of 'artificial' as imitative—it apes the qualities of natural rubber. But yet it is not imitative and *ersatz* in quite the same way as a silk rose can be said to be artificial, imitative, *ersatz*, trying to pass off as the real thing in terms of attributes and appearance. For a silk rose, no matter how like the real rose in appearance, is not a 'real' flower and could never behave like a real flower, unlike synthetic rubber, which in nearly all other ways, behave like natural rubber. What Aristotle says in a slightly different context may be borrowed to make the point—"If a bed were to sprout, not a bed would come up but a tree."[5] If a silk rose were to sprout, not a rose would come up but silken threads, perhaps. The characteristics which make one mistake it for the real thing are entirely based on cursory observation of them *via* usually visual, though sometimes also olfactory, cues. At a distance, the silk rose might look no different from a real rose. One's nose might even detect a whiff of fragrance emanating from it. But usually, closer inspection would unmask the deception. A more systematic investigation into its nature would give all the evidence one needs to determine its artificiality. What is artificial in this sense does not and cannot come to possess those characteristics possessed by its real counterpart, an entity which belongs to a certain natural kind.[6] In contrast, a black tulip is a real flower and, therefore, clearly not 'artificial' in that sense. It is real because it grows, matures, changes, decays, is the result of the photosynthetic activity of the tulip plant, like any other ordinary tulip, whereas a silk rose does not grow or mature, is not the result of photosynthetic activity although it may change and ultimately decay. But all the same, a black tulip is not a naturally evolved entity, and would not be found growing, but for human design and intervention.

The discussion above indicates that at least two senses of 'artificial' should be distinguished: firstly, whatever object or state of affairs is the product of deliberate human creation is 'artificial$_1$'which is synonymous simply with 'artefactual' and secondly, in the sense of 'imitative' or '*ersatz*'—call this 'artificial$_2$.' A silk rose is an artefact but is artificial in both senses, whereas a black tulip is an artefact but is 'artificial' only in the first sense. This book is by and large not concerned with 'artificial$_2$' artefacts but with artefacts which are 'artificial$_1$.' Artefacts in the first sense of 'artificial' can be biotic, abiotic or exbiotic. The black tulip is an instance of the biotic and plastic of the abiotic variety. (For short this category will simply be referred to from now on as artefacts. If and when objects of which the silk rose is the paradigm are meant, they will be referred to as 'artificial$_2$' artefacts.)

As hinted earlier and to be developed in greater detail later, a sub-division in the category of artefacts is called for, namely, between those which are the products of a science and technology which is less 'deep'[7] and those which are the products of a 'deeper' type of scientific theory and its accompanying technology. With biotic artefacts, an example of the former would be the result of advances made by the science of Mendelian genetics and its accompanying distinctive technology of hybridization.[8] An instance of the latter would be the product of the more recent establishment of the science of molecular genetics as well as that of molecular biology and their accompanying technology of molecular genetic engineering or biotechnology.[9] With abiotic artefacts, an example of the former would be plastics and an example of the latter would be, say, a new type of diamondoid material to be constructed *de novo* from carbon atoms which molecular nanotechnology promises to be able to do in the not-too-distant future.[10]

One then detects a gradation in the 'artefacticity,' as it were, in an artefact.[11] In the case of biotic artefacts, in ascending order of artefacticity, the following levels may be identified:

1. Produced through the trial and error method of traditional selection and breeding by farmers.
2. Produced through the combined contribution of Mendelian genetics and hybridization technology.
3. (a) Produced through the combined contribution of molecular genetics and genetic engineering; (b) to be produced through the combined contribution of molecular genetic engineering, molecular nanotechnology as well as microcomputer technology and the various sciences which underpin them.

The ascending order of artefacticity in the case of abiotic artefacts may be displayed as follows:

1. A granite statue, for instance, may be produced by a technology of blasting, chipping, using tools like hammer, wedge, etc. But the material is granite, a natural kind, found in nature and is not, therefore, itself an artefact.
2. A plastic statue, on the other hand, is produced by a different technology informed by the sciences of chemistry, geology, etc. The material, plastic, is not found in nature but is itself an artefact, although it is derived from a natural material, namely, oil.
3. A statue (to be made in the future) out of a material which is created *via* molecular nanotechnology—unlike plastic, this kind of material is not derived from a naturally-occurring material like oil or sand, but constructed *de novo* from atoms and molecules of one of the elements like carbon.

The above shows that the degree of artefacticity of an artefact depends on two related things:

1. The matter or stuff (the material cause) of which the object is made—whether it is found in nature, like wood or stone, whether it is derived from something found in nature, like plastic from oil, or whether it is constructed *de novo*, using atoms and molecules provided by nature in abundance. In other words,

it depends on whether the material is a natural kind (what Aristotle calls second matter), whether it is derived from a natural kind, or whether it is directly synthesized by humans as an artefactual kind from an analogue of Aristotle's first matter, or what we call atoms and molecules of the elements.

2. The degree of artefacticity is, therefore, at its highest when humans succeed in designing the artefact from scratch, from atoms themselves. Such artefacts may then be said to be the total product of human design and technology. But this highest level of artefacticity is only achieved by deep science generating the appropriate technology to go with it.

In the light of the above, it is not inappropriate to hold that transgenic organisms are artefacts possessing a degree of artefacticity analogous to that of plastic toys.[12] The fact that one is biotic and the other abiotic does not affect the issue. Transgenic organisms are constructed out of existing natural kinds and so are derived from them. Similarly, plastic toys are constructed out of an existing natural kind and so are derived from it. Transgenic organisms have lost their own intrinsic/immanent *tele* while plastic toys have none of their own to start with. The *tele* of such organisms have been designed and imposed on them by humans. Humans are their creators. They have become creatures to serve human ends in much the same way as humans have created plastic tables to further other human goals. That they are alive, that is, breathe, ingest nutrients, reproduce, is, from this perspective, totally irrelevant, as these capabilities of theirs have been captured by humans so that as 'designer organisms' they no longer live out their own *tele*.

To repeat the point but putting it slightly differently: not only are the formal, final, and efficient causes of transgenic organisms identified in human terms, even their material cause is the handiwork of humans. Humans have put together bits of genetic material in order to create them according to their own design. *Ex hypothesi*, such genetic materials would not have come together without deliberate human intervention. The technology of molecular genetic engineering (or biotechnology) backed up by the theoretical sciences of molecular genetics and biology permits an extremely high order of precision and control in such intervention.

In contrast, organisms which are the products of Mendelian genetics and its accompanying hybridization technology may be said to possess relatively a lower degree of artefacticity. Although their formal, final and efficient causes may be the work of humans, at least humans have not designed their genetic matter in the way molecular genetic engineering and manipulation have succeeded in doing. Hybridization as such could and, indeed, does occur without human intervention, that is, naturally, whereas transgenic organisms by their very nature could not have come about without deliberate human design and execution. Mendelian genetics merely gave rise to a technology which improves the efficiency of hybridization under the traditional technology of domestic and field breeding. It leaves less to chance than the method of trial and error it replaces, but from the point of view of precise human control, it also leaves much to be desired.

Because the control is not so precise, there is still something, even though, diminished, left to the intrinsic *tele* of such organisms. At least it could be said that even if certain traits have been bred out of their genotype, it still remains true that they retain something of what their forebears had before humans intervened. They may be artefacts but they are not the embodiment of an extrinsic/imposed teleology to the same extent as transgenic organisms appear to be. Their minimal intrinsic *tele* are preserved insofar as Mendelian genetics and its hybridization technology operate within the theoretical framework of the inheritance of genetic material. In radical contrast, molecular genetics and molecular genetic engineering, at a stroke, replace such inheritance by direct human manipulation and rearrangement of genetic material across organisms and species. This is why transgenic organisms may be said to possess a much higher degree of artefacticity compared with organisms produced through the traditional methods of breeding—their efficient, final and formal causes are humanly inspired, imposed and executed, while their material cause is technologically derived from other existing natural kinds.

Modern Science and the History of Technology: Transforming the Natural to Become the Artefactual

Technology, as we shall see, has a very much longer history than modern science. We need then to clarify the relationship between the two by distinguishing several phases of development within the history of technology itself, and then to argue it is only in its relatively recent phase, dating from the mid-nineteenth century onward, that technology has been science-induced, thereby rendering it immensely more powerful than before in its attempts to transform the natural to become the artefactual.

But first a preliminary comment is in order about the referential scope of technology, whether pre-modern or modern.[13] Technology, obviously, involves skills, but for the purpose of this discussion, it should not be so diffusely understood as to be synonymous with any skills whatsoever. The skills must be practical ones. This would then exclude the so-called non-utilitarian arts which are done 'for their own sakes'[14] rather than to achieve the over-arching utilitarian end of securing material gains, either directly or indirectly.[15] That is why the term "technology" is said to coincide with "the applied arts" or "the industrial or practical arts."

If the technological involves skills, then it is an expression of calculated intelligence of the kind often referred to as means-end rationality.[16] Given a certain end deemed to be desirable, one sets out systematically—even trial and error is a systematic procedure[17]—to find an efficient means for achieving it. One might stumble upon a solution by accident, but having stumbled upon it, its adoption to solve the problem would, and could no longer, be said to be accidental.

In the "industrial arts," such intelligence and skills are embodied in material objects of one kind or other. In other words, they go beyond being 'in the head,' and must be expressed in a material medium. On this account, mathematical calculations *per se* may involve techniques but are not technologies, while an abacus or an electronic calculator are technological objects. This proposed definition of the technological excludes techniques.[18] Techniques, too, involve efficient means toward ends, but because they are not necessarily embodied in a material medium, they do not count as technological processes or objects, although they may in turn give rise to such processes and objects. For instance, the division of labor—the breaking down of a task into simpler components—is a powerful technique for increasing productivity in many contexts, but it, itself, does not amount to a technology even though it has given rise to the conveyor belt technology of mass production.

The distinction between technique and technology on the one hand, and that between the non-utilitarain arts and the 'industrial or practical arts' on the other, may be justified by reference to the distinction between human doing and human making.[19] If one were to take the performing arts as the paradigmatic representative of the non-utilitarian arts, such arts primarily involve techniques, and only secondarily, technologies. For instance, singers (folk or operatic) or actors rely on their respective techniques of breathing, of voice projection. But some singers and some actors also rely on certain technological products, such as the microphone. Techniques are ways of doing—singing, acting and dancing are primarily ways of doing, rather than ways of making things. But technologies are precisely ways of making things outright as objects, and through the objects made, to cause other objects to be made and certain happenings to occur. For instance, steam technology involves the steam engine. The steam engine, as the steam pump, made deep mining possible, and as the steam locomotive, it made the steam train possible.

Techniques, no matter how intimately entwined with technological objects, are detachable from them. Writing as a technique was long entwined with the paper, brush or pen as technological products. But with the invention of the typewriter as a new technological object in the early part of this century, writing to a great extent became detached from paper and pen, and became entwined with the typewriter. And today, the wordprocessor has, by and large, superseded the mechanical or electric typewriter. The typewriter is an instance of the machine-with-rigid-but-movable-parts type of technology. The wordprocessor is a different kind of machine and is the outcome of the electronic technology.

But the above argument may be said to be weakened if the fine, instead of the performing arts, were cited as paradigmatic forms of non-utilitarian arts. In the case of the fine arts, it is less easy to divorce the techniques of, say, painting, from the material media of canvas, paint and brush, issuing forth finally in a portrait. However, in spite of this admittedly more intimate relationship between technique and material medium, nevertheless, it remains true that painting a portrait falls more into the domain of human doing than human making. The portrait, indeed, is an artefact, as it is an entity which would not have come into existence without direct human agency with its determination to execute intention into action. But all the same, it is not itself a technological artefact, although its execution may, and does, involve particular technological objects like canvas.

It follows, then, that the domain of artefactual entities is wider than that of technological material objects—the technological is part of the artefactual. Artefactual entities include (i) portraits, (ii) microphones, and (iii) songs. In other words, as the terms have been defined, whereas all technologies are artefactual in nature, not all artefactual entities are necessarily technological material objects. Someone just singing may use certain techniques but does not rely on any technological object. The song and the singing may be said to belong to the domain of the artefactual in the sense that they are the embodiment of human intentions expressed through an endosomatic organ, that is, the human voice. But if the singer were doing karaoke, the performance clearly is inextricably entwined with that particular ensemble of technological objects, which may collectively be referred to as the karaoke machine.

With that clarification about the scope of technology out of the way, let us now give a brief account about technology in general and its history. It must straightaway be admitted that technology is not peculiar to modernity.[20] Technology had always existed since the first adze made by our stone age ancestors. It should not be understood as merely co-terminus with our contemporary variety rooted in modern science.

To do justice to all historical forms of technology and to provide a comprehensive framework for a philosophical analysis of technology, Mitcham proposes a tentative schema,[21] suggesting four modes of the manifestation of technology in terms of (a) technological knowledge, (b) technological volition,[22] (c) technological activities (making and using), (d) technological objects (or artefacts). However, this book cannot hope to be a contribution to this larger and more ambitious project that Mitcham has delineated. Its aims are far more modest and confined. With regard to (b), it takes the view (without further discussion) that while pre-modern technology primarily involves the will to survive or to satisfy basic biological needs of one kind or another, technology in modernity is primarily about the will to control and manipulate nature with the goal of improving the material well-being of humans in particular[23] and/or advancing human ends including those of freedom and self-realization. With regard to (b) and (c), Chapter 4 argues that, in modern thought, the essence of the human species, so to speak, is captured by the notion of *homo faber* and its related notion of fabrication, and that through the activity of fabrication, *homo faber* seeks material affluence on the one hand and freedom and self-realization on the other. But in the main, it explores the relation between (a) and (d), to show how modern technology ties in with the program of modern science, that increasingly since roughly the mid nineteenth century, scientific theories induce and inform technology, that the 'deeper' the theory, the more powerful the technology generated. The exploration is intended to cast light on the implications of modern science and technology for environmental philosophy in general, and in particular, to argue for the ontological distinction between the natural and the artefactual, as well as to defend the former category (that is, the natural) from the threat of elimination posed by modern technology and science.

Scholars of (European) technological civilization have suggested dividing it up into various phases. For instance, Mumford proposes a three-fold division (the edges of which are meant to be overlapping) in terms of the type of energy and characteristic materials used—the eotechnic phase, which is a water-wind-and-wood complex, the paleotechnic phase, which is a steam-coal-and-iron complex, and the neotechnic phase, which is an electricity-and-alloy (as well as synthetic compounds) complex.[24] The first, for him, stretches roughly from 1000 AD to 1750, the second, from 1750 to the 1850s, and the third, from the 1850s to the present.

Mumford's classification is heuristically enlightening in general but, perhaps, less helpful from the standpoint of this book. So a different division is proposed here, not based so much on the conjoint variables of energy and material, but on whether the technology is craft- or science-based, and in the case of the latter, on the relationship between the technology and the kind of science it might or might not rely on. The suggested classification in the context of European technological history is as follows[25] (bear in mind, though, that the boundaries between them are not meant to be neat and tidy, but overlapping):

Phase I Relatively automous craft-based (though not necessarily guild-based) technology.
- A Roughly equivalent to Mumford's eotechnic phase.
- B Roughly equivalent to Mumford's paleotechnic phase.

Phase II Science-theory-led technology.
- A Roughly equivalent to Mumford's neotechnic phase but ending by the 1940s.
- B Roughly from the 1940s to the present.

Note that this division fails to superimpose neatly upon that which obtains in the history of science. There the radical cleavage, as we have seen, is between pre-modern science (up to the seventeenth century) and the rise of modern science (from the seventeenth century onward). Phase IA falls clearly into the pre-modern scientific era, but Phase IB (roughly up to the 1830s) falls clearly into the modern scientific period. In other words, the major cleavage has been drawn between the kind of technology which is theory led and inspired, in contrast to that which is relatively autonomous of scientific theorizing itself. Although Phase IB, in terms of temporal location, coincided with the rise of modern science, the technology it represented was, nevertheless, by and large, not a spin-off of theoretical advances.

On the contrary, during this period, it often happened that technology inspired theoretical research rather than that theoretical advances led the way to new technologies. For instance, this relationship of technology preceding theory is true in the case of the invention of the steam engine, which first appeared in the form of the steam pump, as a response to the demands of the coal mining industry to mine seams at deeper levels where flooding occurred. It later made railway transportation possible as the steam locomotive, and replaced sailing ships on the high seas in the form of the steamer. Attempts to improve its efficiency eventually led to the establishment of the abstract, fundamental science of thermodynamics. Sadi Carnot, a French army officer and engineer, set out to understand how the steam engine works, hoping thereby to improve its efficiency. The English had invented the machine, enabling perfidious Albion to be superior both in war and industry. He studied the phenomenon of heat with the goal of recapturing that superiority for France. He found an intrinsic inefficiency in the conversion of heat to work. The steam engine works because parts of it are very hot and other parts very cold. Heat moves from the hot to the cold and in so doing, work is performed. But when the parts reach the same temperature, that is to say, a state of equilibrium, no further work can be performed. A difference in temperature between parts of the system—a difference in energy concentration—must obtain for work to occur. He also discovered that as energy moves from a higher to a lower level, less energy is available for work on the next round. In a waterfall, the water can be used to drive a wheel, but it is no longer available to perform any further work once it reaches the pool at the bottom.

Later, famous scientists like Joule, Kelvin, Clausius and Boltzmann added to Carnot's efforts. Boltzmann's contribution consists of linking the behavior of matter at the macro level to the behavior of matter at the atomic level, thus providing a unifying theory of vast scope to explain the nature of change in the world. It is apt to quote Atkins here:

> The aims adopted and the attitudes struck by Carnot and by Boltzmann epitomize thermodynamics. Carnot traveled toward thermodynamics from the direction of the engine, then the symbol of industrialized society: his aim was to improve its efficiency. Boltzmann traveled to thermodynamics from the atom, the symbol of emerging scientific fundamentalism: his aim was to increase our comprehension of the world at the deepest levels then conceived. Thermodynamics still has both aspects, and reflects complementary aims, attitudes, and applications. It grew out of the coarse machinery: yet it has been refined to an instrument of great delicacy. It spans the whole range of human enterprise, covering the organization and deployment of both resources and ideas about the nature of change in the world around us. Few contributions to human understanding are richer than this child of the steam engine and the atom.[26]

Even more remarkably, during Phase IB, technological discoveries, which formed the very basis of the Industrial Revolution, were made by people who knew no science, had no formal education and, indeed, in some cases, could not even read or write. The most famous of these apprentices and craft-based mechanics is George Stephenson. Later in life, when he became famous and rich, he was only partially successful in overcoming his illiteracy. What is now called the Davy Lamp—the safety lamp for miners which first appeared in 1815—was also an invention by Stephenson. But because of his humble background, illiteracy and ignorance of physics and chemistry, Humphrey Davy—fellow and later president of the Royal Society on whom a baronetcy was eventually conferred—could not credit Stephenson as a fellow inventor. Instead, he accused him of having stolen his idea. Davy died in 1829 convinced that Stephenson had cheated in spite of all the evidence to the contrary. Eventually, it took a House of Commons Committee in 1833 to vindicate Stephenson.[27]

This incident illustrates not merely the more humble origins of the inventors of many remarkable technological discoveries, but also the class-based difference at the time—at least in Britain—between, on the one hand, technology and the practical (those who work with their hands) and, on the other, science and the theoretical (those who work with their brains). The ancient Universities of Britain, then, did not want to know either science or technology. The Royal Society was established to cater in the main for (pure) science, and was supported and patronized by gentlemen and members of the Establishment. Technology, instead, belonged to the Mechanical Societies which grew up in the eighteenth century in the cities of Britain, and was nurtured and supported by the combined zeal of entrepreneurs, industrialists, engineers, unlettered and untutored mechanics—in other words, of people who dirtied their hands in one way or other with industry and manufacturing.

From this point of view, it is not unreasonable to argue that Phase IB and Phase IA, in spite of differences between them, share the essential similarity of being craft-based and relatively autonomous of explicit scientific/theoretical input. Phase IA includes inventors like Leonardo da Vinci (1452-1519), but in spite of the ingenuity of his many inventions, he is not celebrated in history for his contribution to science, but as a Renaissance genius in the design and execution of artefacts, belonging to both the fine and practical arts. He considered himself to be a "man without letters." Other giants of the period, like Galileo, were hired by rulers, for instance, to improve their weapons of war which, in turn, led them to so-called pure scientific research and to establish new sciences. Yet others like Sir Isaac Newton (1642-1727), a Cambridge mathematician and physicist, often hailed as the greatest scientist of all times, concentrated on the theorizing and did not dabble at all in technological inventions (although he dabbled a lot in alchemy, so much so that John Maynard Keynes was moved to say that "Cambridge's greatest son" was "not the first of the age of reason" but "the last of the magicians"). In other words, Phase IA and B both displayed a split between science and technology—either science was pursued relatively autonomously of technology or that technology led the way to scientific theorizing. The causal direction the other way around, of theory inducing technology, by and large, did not occur until much later on.

One difference between the two stages of Phase I worth commenting on is this: IA is, on the whole, as Mumford points out, an era of creative syncretism.[28] Western Europe collected unto itself the technological innovations of other civilizations, adapted and built upon them. To mention just a few—the watermills, already in place in the earlier part of the Christian era, could be traced back to the waterwheel of the Egyptians who used it to raise water. The windmill had probably come from Persia in the eighth century. Gunpowder, the magnetic needle and paper came from China, the last two *via* the Arabs. Europe by 1000 AD was ready to receive these and other discoveries (such as algebra from India, again *via* the Arabs). Glass technology (known as far back as the Egyptians), improved and developed, laid the foundation for the development of astronomy by Galileo *via* the telescope (invented by a Dutch optician, Johann Lippersheim in 1605), and of bacteriology by Leeuwenhoek in the mid seventeenth century made possible by the invention of the compound microscope (by another Dutch optician, Zacharias Jansen in 1590).

Mumford sums up succinctly what he has called the paleotechnic phase and what is called Phase IB as follows:

> The detailed history of the steam engine, the railroad, the textile mill, the iron ship, could be written without more than passing reference to the scientific work of the period. For these devices were made possible largely by the method of empirical practice, by trial and selection: many lives were lost by the explosion of steam-boilers before the safety-valve was generally adopted. And though all these inventions would have been better for science, they came into existence, for the most part, without its direct aid. It was the practical men in the mines, the factories, the machine shops and the clockmakers' shops and the locksmiths' shops or the curious amateurs with a turn for manipulating materials and imagining new processes, who made them possible.[29]

While the focus of Phase IA was largely Western Europe (but leaving out England), in contrast, England became the main center of Phase IB. But even here it must be admitted that the steam engine, the symbol of the Second Industrial Revolution, could be said to owe its conception to Hero of Alexandria, the translations of whose works in the sixteenth century had made people turn to the steam engine as a possible source of power and energy. As Mumford points out, the relative backwardness of England, ironically, made it more ready to welcome and push through the developments associated with Phase IB.[30]

As we have seen, the history of science and the history of technology in modern Western Europe, at one level of understanding, are not neatly harnessed in tandem. In Phase I, technology stood relatively autonomous of theoretical/scientific input. Phase II shows a marked difference—the major technological innovations are theory led or induced. With regard to Phase IIA, on the theoretical side, by 1850, most of the fundamental scientific discoveries had already been made. Regarding electro-magnetism, Faraday, in 1831, found that a conductor cutting the lines of force of a magnet created a difference in potential. This, together with the work done by Volta, Galvani, Oersted, Ohm, Ampere and Henry, provided the theoretical foundation for the conversion and distribution of energy as well as for such significant inventions like the electric cell, the storage cell, the dynamo, the motor, the electric lamp. During the last quarter of the nineteenth century, these were spectacularly translated into industrial terms in the form of the electric power station, the telephone, the radio telegraph. Augmenting these were the phonograph, the moving picture, the steam turbine, the airplane.

That was on the physics front. On the chemistry front, it was the isolation of benzine by Faraday in the 1830s (and later the use of naphtha) which made the industrial use of rubber possible. Advances in organic chemistry permitted the industrial utilization of coal beyond using it as a direct source of energy. From one ton of coal, one could get 1,500 pounds of coke, 111,360 cubic feet of gas, 12 gallons of tar, 25 pounds of ammonium phosphate and 4 gallons of light oils. From coal tar itself, the chemist produced new medicines, dyes, resins, perfumes. Metallurgy also took revolutionary steps forward—however, aluminum, discovered by Oersted as early as 1825, had to await the arrival of electricity, as the cheap source of energy before its commercial exploitation became feasible in the last decade of the century. Rare metals were incorporated into the industrial processes—for example, selenium, whose electrical resistance varies inversely with the intensity of light, was used in automatic counting devices and electric door-openers. To quote Mumford again, he has aptly written as follows:

> In (this) phase, the main initiative comes, not from the ingenious inventor, but from the scientist who establishes the general law: the invention is a derivative product. It was Henry who in essentials invented the telegraph, not Morse; it was Faraday who invented the dynamo, not Siemens; it was Oersted who invented the electric motor, not Jacobi; it was Clerk-Maxwell and Hertz who invented the radio telegraph, not Marconi and De Forest. The translation of the scientific knowledge into practical instruments was a mere incident in the process of invention. While distinguished individual inventors like Edison, Baekeland and Sperry remained, the new inventive genius worked on the materials provided by science.
>
> Out of this habit grew a new phenomenon: deliberate and systematic invention. Here was a new material: problem—find a new use for it. Or here was a necessary utility: problem—find the theoretic formula which would permit it to be produced. The ocean cable was finally laid only when Lord Kelvin had contributed the necessary scientific analysis of the problem it presented: the thrust of the propeller shaft on the steamer was finally taken up without clumsy and expensive mechanical devices, only when Michell worked out the behaviour of viscous fluids: long distance telephony was made possible only by systematic research by Pupin and others in the Bell Laboratories on the several elements in the problem. Isolated inspiration and empirical fumbling came to count less and less in invention.[31]

In other words, it was only roughly from 1850 onward that modern society began to reap the material benefits promised by modern science, its method, its philosophy and its ideological goal of controlling nature. This promise took more than two centuries to materialize when the paths of pure (theoretical) science and technology no longer diverged, acting, by and large, independently of each other, but began to be harnessed to work as joint forces. However, at least on one level of understanding, the team may be said to be led by pure science, the senior partner, while technology follows. (Yet at a deeper level, this may be an oversimplification—for qualifications, see section below.) In Phase I when each was relatively autonomous, technology, sometimes, led the way to theoretical advance—witness the relationship between the steam engine and the fundamental science of thermodynamics. However, under the new settlement, technology has lost that causal initiative and now becomes, much more so than before, the executive arm, so to speak, of pure science.

Modern Technology, the Philosophy of Technology, and the Philosophy of Science

In the light of the relatively recent specific partnership between modern science and technology noted above, a terminological issue, about how the word "technology" ought to be used, springs immediately to mind. However, to sort this out satisfactorily, one has to delve into the philosophy of technology and the philosophy of science as well as the relationship between the two.

Initially, one would be tempted just simply to distinguish between Phase I and Phase II by proposing that the word "technology" be confined only to the former, and that some other term, such as "applied science," be used in connection with the latter. It follows from this proposed usage that (a) the relationship between science and technology in Phase I is a contingent one, and (b) the relationship between science and applied science in Phase II is more than contingent. However, this possible way of defining terms may not find favor, as it produces too much of a discontinuity in the history of humans in their attempts to modify nature in order to secure their own ends, be it survival, improvement of material well-being or whatever. The new technology is but a form of technology in the long history of that subject. It would be less misleading and distorting in recognizing it as such. So it would be clearer to say that science and technology are really two separate, though related, forms of activities, and that the very intimate relationship that has grown up between the two since roughly 1850 is, nevertheless, a contingent one, in spite of the avowed aim of modern science to produce a technology which can control nature in a thoroughly systematic manner, guided by theoretical understanding rather than crude empirical happenstance.

To prevent misunderstanding of what has just been said, one needs to return to one of the main points raised in the last chapter. There, it was argued that (a) modern science from its first beginnings was backed up by the new philosophy, in particular by its metaphysics of Scientific Naturalism, and (b) its ideological goal was the advancement of material well-being *via* its technology to control and manipulate nature. These two theses may be said to constitute the Modern Project of Science and Technology. The ideological goal to control and manipulate nature renders the Modern Project *au fond* a technologically oriented one. Under the Modern Project, modern science may be said to be really theoretical technology, a view associated with, for instance, Heidegger and Jonas. From this standpoint, the science and the technology appear to be inextricably linked—the linkage is more than an accidental one. As such, it is more than merely contingent. It is, then, not surprising that such Science should eventually spawn successful Technology, even though the Modern Project itself took over two hundred years, since its inception, 'to deliver the goods,' so to speak.

To quote Mitcham:

> For Heidegger what lies behind or beneath modern technology as a revealing that sets up and challenges the world is what he calls Ge-stell.
>
> Ge-stell names, to use Kantian language, the transcendental precondition of modern technology. ... "Ge-stell" refers to the gathering together of the setting-up that sets up human beings, that is, challenges them, to reveal reality, by the mode of ordering, as Bestand" or resource. ... "Ge-stell refers to the mode of revealing that rules in the essence of modern technology and is not itself anything technological." ... Not only does Ge-stell "set-up" and "challenge" the world ... it also sets upon and challenges human beings to set upon and challenge the world. ... "The essence of modern technology starts human beings upon the way of that revealing through which reality everywhere, more or less distinctly, becomes resource."[32]

The same point about Heidegger is, more or less, made by Zimmerman:

> Far from being a dispassionate quest for truth, scientific methodology had become the modern version of the power-oriented salvific methodologies developed in the Middle Ages. Hence, Heidegger argued, even though modern science preceded the rise of modern technology by about two hundred years, modern science was already essentially "technological" in character, i.e., oriented toward power. ... Science ... seeks not to let the entity show itself in ways appropriate to the entity in question, but instead compels the entity to reveal those aspects of itself that are consistent with the power aims of scientific culture.[33]

Jonas has written in the same vein about Bacon's view of science:

> Theory must be so revised that it yields "designations and directions for works," even has "the invention of arts" for its very end, and thus becomes itself an art of invention. Theory it is nonetheless, as it is discovery and rational account of first causes and universal laws (forms). It thus agrees with classical theory in that it has the nature of things and the totality of nature for its object; but it is such a science of causes and laws, or a science of such causes and laws, as then makes it possible "to command nature in action." It makes this possible because from the outset it looks at nature qua acting, and achieves knowledge of nature's laws

> of action by itself engaging nature in action—that is, in experiment, and therefore on terms set by man himself. It yields directions for works because it first catches nature "at work."
>
> A science of "nature at work" is a mechanics, or a dynamics, of nature. For such a science Galileo and Descartes provided the speculative premises and the method of analysis and synthesis. Giving birth to a theory with inherently technological potential, they set on its actual course that fusion of theory and practice which Bacon was dreaming of.[34]

In the light of the above and of the points raised in the preceding section, there is, perhaps, some justification in saying that Modern Science is Theoretical Technology. But all the same, Modern Technology, nevertheless, is applied science. To see why this latter claim may be justified, one must distinguish the Modern Project itself—embedded in a certain metaphysical and ideological framework—from (a) the formulation and the testing of specific scientific theories in the history and philosophy of science, (b) the relationship, if any, between a specific theory and a related specific technology, and (c) the epistemic goals of theory formulation and theory testing on the one hand, and the testing of technological hypotheses on the other. Here, as we have seen, the linkage in the case of any one specific theory and any one specific technology throughout the modern period, in particular during Phase I, appears to be much looser than the postulated linkage between Science and Technology in the Modern Project itself. (However, in Phase II and especially IIB the intensely intimate causal relationship does obtain between certain specific theories and the specific technologies they induce and render possible.[35]) Moreover, the epistemic goals of theory formulation and testing are also perceived to be somewhat different and distinct from those of testing hypotheses in the technological domain even in Phase II.

The recognition that their epistemic goals are distinct is reflected by the fact that while the philosophy of science has a recognized and well-established agenda, an analogous philosophy of technology does not obviously exist. Indeed, while the former is an eminently respectable part of philosophical inquiry, the latter is held at arm's length with a degree of suspicion.[36] The agenda of the one may be clear, that of the other is not. Bunge feels the need to sketch an outline for that missing agenda. He writes:

> Some of the typical problems in the philosophy of technology are these: (a) Which characteristics does technological knowledge share with scientific knowledge, and which are exclusive of the former? (b) In what does the ontology of artifacts differ from that of natural objects? (c) What distinguishes a technological forecast from a scientific forecast? (d) How are rule of thumb, technological rule, and scientific law related? (e) Which philosophical principles play a heuristic, and which a blocking, role in technological research? (f) Does pragmatism account for the theoretical richness of technology? (g) What are the value systems and the ethical norms of technology? (h) What are the conceptual relations between technology and the other branches of contemporary culture?[37]

A central aim of this book is to clarify problem b identified above. This section later also addresses d. But for the moment, what will be emphasized is one crucial difference between science and technology (in both phases), namely, their respective overarching epistemological goals. But before dealing with that, it may be helpful to point out their similarities under Phase II in two essential aspects, as Bunge has done:[38]

1. Methodologically, a technological research program is no different from that of a scientific one. They include the following elements—identifying and articulating the problem, solving it with extant empirical or theoretical knowledge, and failing that, putting forward new hypotheses and ways to try to solve it, working out a solution within the new framework, testing the solution by experimentation and in the light of that, amending the hypothesis under test or even reformulating the original problem.
2. Epistemologically and ontologically, technology and pure science (at least in one conspicuous tradition in the philosophy of science)[39] share certain common assumptions: that an external world exists, that we can come to know it partially, though never totally, and that knowledge of such a world can be improved upon and increased, though again recognizing that the goal of complete and total knowledge can never be reached. In other words, they both subscribe to what may be called critical realism; technologists would realize, just as the pure scientists, that their theories cannot, literally, be pictures of reality but are symbolic oversimplified representations of a fairly abstract kind of 'the reality' that they are grappling with. (On this conception of the philosophy of science, in Phase I, technologists would have tended to be naive realists, if they had at all confronted themselves with this philosophical issue.)

However, whether under Phase I or II, it is said that the overarching epistemological goal of technology differs from that of science. Even in the latter phase, the critical realism of the technologist is subordinated to the crucial requirement that the solution works—in other words pragmatism is an overriding demand. Unlike pure scientists who often claim that in principle they are interested in knowledge for the sake of knowledge, technologists are primarily interested in scientific knowledge (if it exists) as a mere means to the end of providing a solution to the practical problem in hand. If scientific knowledge is non-existent or unhelpful, they will look elsewhere for assistance. Nor would they be unduly worried should the viable solution turn out for the moment to lack a proper complete scientific explanation.

To put it even more strongly, scientific knowledge *per se* seems neither to be a necessary nor a sufficient condition for what counts as a successful technological solution to a problem. An example that it is not the former is the success shown by the traditional methods of artificial selection in breeding plants and animals. Until the so-called rediscovery of Mendelism in 1900, there was no adequate or proper explanation to account for their success. An example that illustrates the latter is plate tectonic theory in geology and seismography, which have not so far, at least, led to a technology of forming new mountains, or of controlling the movements of the earth's crust or, indeed, even of accurate predictions of earthquakes.

Technology's goal of getting practical results also affects its relationship with the concept of truth. To quote Bunge on this point:

> Although in practice (the technologist) adopts the correspondence conception of truth as adequacy of the intellect or mind to the thing, he will care for the true data, hypotheses and theories only as long as they are conducive to the desired outcomes. He will often prefer a simple half-truth to a complex truth. He must, because he is always in a hurry to get results. Besides, any error made in neglecting some factor (or some decimal figure) is likely to be overshadowed by unpredictable disturbances his real system may undergo. Unlike the physicist, the chemist, or the biologist, he cannot protect his systems against shocks other than by building shock-absorbing mechanisms into them. For similar reasons, the technologist cannot prefer deep but involved theories when superficial ones will do. However, unless he is a pseudotechnologist, he will not shy away from complex and deep theories if they promise success. ... The technologist, in sum, will adopt a mixture of critical realism and pragmatism, varying these ingredients according to his needs. He will seem to confirm first one and then another epistemology, while actually all he intends to do is to maximize his own efficiency regardless of philosophical loyalties.[40]

The epistemological target of (pure) scientific theorizing is truth (or at least, approximation to truth) according to a dominant tradition in the philosophy of science.[41] When technology applies the findings of pure science—for instance, when a theory of flight is based on the theory of fluid dynamics—the epistemological target of such technological theories is efficiency, not truth. Indeed it may be said to adhere to the following methodological rule: only adopt as deep a scientific theory as is adequate for the problem in hand. In this sense, it is theoretically less sophisticated than pure science, although it makes up for this theoretical simplicity by being wholly opportunistic in using knowledge of any kind, from any domain (whether ordinary, older, less sophisticated or the latest sophisticated deep theory in science). For example, in constructing an optical instrument, the technologists would rely, in the main, on ray optics, a theory of optics based on what was known about light around about the middle of the seventeenth century. They would ignore wave optics except to the extent of helping them to understand why certain effects occur, such as the appearance of colors near the edge of the lens which, to them, are considered to be undesirable.

Deeper, more complex and more accurate theories may not necessarily be the most economical to use—imagine using quantum theory to predict or explain car crashes. Efficiency demands that one uses less deep theories with less operational costs, to get as much out of them with as little input as possible. From the standpoint of technology, a true scientific theory in principle can be successfully employed but in practice, technologists may have to decline its help, so long as an alternative exists which can do the job satisfactorily, but at less cost operationally and, therefore, usually, economically. The alternative may indeed even be a false theory on the whole, but so long as it possesses an element of truth which can be relied on by the technology in question, it would do fine.

To emphasize the distinction between scientific and technological knowledge, Bunge says that while the former attempts to formulate laws (about universal regularities), the latter aims at establishing rules.[42] Laws are descriptive—when conditions x, y, z obtain, A obtains. Rules, on the other hand, are prescriptive and are what may be called hypothetical imperatives—if one wishes to achieve A, then one ought to do x. Phase I technology primarily relied on prescientifc rules (rules of thumb used in arts- and crafts-based processes of production such as yeast fermentation in brewing and baking). In Phase II, technological rules are grounded in scientific laws. By this is meant that the laws must be capable of accounting for, or explaining, the efficacy of the rules. To prevent water from freezing in the car radiator in the winter, one ought to add anti-freeze to it. The rule achieving the desired end is successfully explained in terms of the differential freezing points of water and methanol or ethandiol (two commonly used anti-freeze substances), which in turn could be accounted for by further deeper theories such as the kinetic and atomic theories.

Phase I rules may be empirically very effective. But because they are not properly grounded in scientific laws, there is always the possibility that their efficacious outcome may be a mere coincidence. Suppose (in temperate climates) one adheres to the rule—do not plant in the depth of winter but in the springtime—one would indeed get a high degree of horticultural success. But one might mistakenly conclude from it that the plants grow so well because of the warmth that comes with the spring. But one would be wrong, though not totally wrong. The warmth is an important component of success, but only when it is accompanied by an increase of light in the spring and summer, which is vital to plant growth. Today the rule's efficacy is properly grounded in our theoretical understanding of the processes involved in photosynthesis and the conditions under which plant growth obtains.

The above would account for why Phase I rules, though empirically effective, provide one with less than optimal control over nature. Maybe most of the time they work but there will be cases of failure. Within the framework of technological rules, the failure cannot be explained, just observed. However, it could later be explained in terms of scientific laws when these are discovered. If so, then the laws in turn could lead to the formulation of improved, more efficacious rules (that is under Phase II) whose scope of operation may transcend that of the original rule. Using the plant growth example again, the theoretical understanding of plant physiology, chemistry, etc., enables the technologist to devise the greenhouse. Such a technological innovation makes it possible for us to overcome the constraints imposed by nature through the rythm of its seasons. Now tomatoes in northerly climates will grow the whole year round under artificially produced conditions of appropriate degrees of warmth and light. Undoubtedly in this way, the scope has enormously increased one's control over nature.

It would be helpful to summarize the arguments advanced in this section as follows:

1. Phase I technology is, by and large, autonomous of science. It flourished in cultures which lacked explicit systematic scientific theorizing of any kind. It could flourish just as readily in cultures engaged in such theoretical activities but underpinned by a metaphysics and using a methodology that differ from the modern scientific one. Such technology can be empirically efficacious and, indeed, was so historically.
2. However, Phase II technology is a much more powerful tool in manipulating nature than its Phase I counterpart. Take the treatment of hemophilia in the history of medicine. Under Phase I, the only alleviation available would have been prevention at the most elementary level, that is, for the sufferer of the condition to take steps to reduce the chances of being bruised or cut. Under early Phase II technology, hemophiliacs were given whole-blood transfusion. Further medical understanding advanced and the precise nature of the condition became known. It is now realized that there are two different forms of hemophilia: hemophilia A, in which the sufferer lacks a clotting chemical called factor 8, and hemophilia B, in which the sufferer lacks factor 9; of the two, the former is more common than the latter. In the light of this understanding, a new technology replaced whole-blood transfusion. The missing clotting chemical is injected three times a week to counter the inherited condition. The technology is more specifically targeted than the one it replaces; as a result, it is scientifically more precise. Its emergence is predicated upon advances both in theoretical knowledge and technology, allowing the clotting chemical to be either extracted from human blood plasma, or manufactured by genetically engineered organisms. This may be said to constitute the middle stage of Phase II technology.

   But today with the science of molecular genetics and its accompanying technology of genetic engineering in place, there is room to take the treatment of hemophilia to yet another stage of development *via* gene therapy. Indeed, it has just been reported that this further stage has already been taken. According to the publication *Science in China*, a team of scientists at the Institute of Genetics in Shanghai had performed it on two teenage hemophiliacs, both suffering from a lack of factor 9.[43] One of them is said, as a result of the treatment, to be now producing the clotting chemical in his own blood. If this were really so, it would be a permanent cure. Using standard gene therapy technique, the team first isolated the gene for factor 9, then inserted it into a virus. It also removed fibroblasts (cells which form connective tissue under the skin) from the two patients. The treated virus was used to infect these fibroblasts. The infected fibroblasts, now carrying the missing gene, were then injected back into the two patients. An inherited disability is now cured by gene replacement therapy.[44] This admittedly is not as radical as germ-line gene therapy which, if carried out, could in principle eradicate hemophlia by ensuring that no males would be born with the genetic disorder, not merely that males born with such an inherited condition would be permanently cured of it.

3. Although it is the case that more precise scientific theories are not necessarily always relied upon by technology, which seems to prefer the less precise and complex but still adequate alternative, such theories are, nevertheless, required to ground the efficacy of the rules, giving them the maximum epistemological support possible. Going back to the example of ray and wave optics in the construction of optical instruments, one can see why the former accounts for the instrument's overall success and the latter for its being less than totally perfect. As we have seen, while efficacious technological rules may lead to new theoretical understanding, their efficacy, on its own, is not synonymous with truth.
4. Phase II technology, although induced and led by pure scientific findings, is not entailed by them. In other words, theoretical advances and revolutions may be a necessary but not a sufficient condition for its emergence. However, to prevent misunderstanding about this claim, one has to distinguish between two contexts here: (a) pure theory providing the epistemological grounding and direction for the induced technology, and (b) a pure theory being actually used in a particular piece or type of technology. As we have seen, in context b, (i) there could be an alternative, less accurate theory the technology could rely on, or (ii) social, economic and political considerations may be hostile to the emergence of a new technology. The discussion here is only confined to i. As for context a, when a theory-induced technology does emerge, the efficacy of its technological rules is grounded in, and accounted for, by the laws of the pure theory—in this sense there is a very strong empirical as well as epistemological link between technological efficacy and scientific truth. Furthermore, they have certain concepts in common.
5. In the light of the points made above, it would be appropriate to retain the word "technology" to refer to both Phase I and Phase II technology, while, nevertheless, noting the intricate involvement of the latter with the discoveries of 'deep' science; that involvement itself, as we have seen, is to be understood within the context of relating the philosophy of technology to the philosophy of science.

'Deep' Theories and Their Power of Control

We have, so far, looked at some of the philosophical dfferences underpinning Phase I and Phase II technology. One needs to say something very briefly here about the distinction between what has been called a deeper, as opposed to a less deep, theory which underpins Phase II technology itself. 'Deep' may be understood in at least three ways:

1. A less deep theory is ultimately to be explained in terms of a deeper one. The kinetic theory is explained in terms of the atomic theory, and the latter itself is accounted for by subatomic quantum theory. Relatively speaking, the first is less deep than the second, and the second is less deep than the third. Similarly, Mendelian genetics is accounted for in terms of molecular genetics.
2. The deeper theory may also then be said to be more comprehensive in scope, explaining a wider range of data, accounting for more variables in their causal contribution to a particular phenomenon.
3. A less deep theory may contain laws about particles and their behavior at the macro level of existence and observation while a deeper theory postulates laws about particles and their behavior at the micro level of existence and observation. Newtonian macro physics may then be said to be less deep than quantum physics.

All three senses are relevant to the discussion in hand. The Modern Project of Science and Technology is built on an ontology of materialism. Ever since its inception, its central aim has been to penetrate the nature and structure of matter. As we saw in Chapter 1, macro properties of the natural world such as the so-called secondary qualities are said not to reside in the object and, therefore, are not real. Objects are constituted by their primary qualities which are real. Furthermore, matter at the macro level of existence is to be broken down analytically into its component parts at the micro level of existence. Hence the atomic theory of matter—all macro objects are made up of atoms, and molecules which are themselves combinations of atoms. Twentieth century science has gone even beyond that to the subatomic theory of matter.

On this worldview, matter then is ultimately uniform and homogeneous—their diversities in the form of different sorts of organisms, of minerals, that is, of different natural kinds, are no more than a difference in the arrangement of the primary qualities involved, of atomic particles which in turn are constituted of subatomic particles and their nuclei.

As we also saw in Chapter 1, it has been the ideological goal of The Modern Project from its very beginning in the seventeenth century to use its theoretical advances to engender powerful technologies to control nature in order to serve human ends. This promise has been made good from the mid to the late nineteenth century onward. And as its theoretical advances get deeper and deeper into the structure of matter, the theory-induced technologies get more and more powerful.

Take biology as a discipline. In the words of one well-known historian of the subject:

> Contemporary biology is characterized by several important factors. One is the

> firm belief that all biological problems can ultimately be studied on the molecular level. This view does not maintain that studies at other levels of organization, such as that of the cell, the organ, the whole organism, or the population, are of no value. In fact, there is a growing awareness among some biologists that it is equally as important to study these higher levels of organization as it is to study the lower, molecular levels. The view that reduction of a complex biological phenomenon to its simpler components (cells or molecules) is a sufficient explanation has become less prevalent among biologists in the early 1970s.[45] Nevertheless, the revolution in molecular biology in 1950s and early 1960s emphasized the importance of understanding the molecular basis of biological phenomena before trying to approach the larger, higher-level interactions.[46]

Biologists, on the whole, in the last two decades, may, indeed, have resisted strident reductionism of the kind which says that cells are mere collections of molecules, or "what is true of *E. coli* [a bacterium] is true of the elephant," a view prevalent in the 1950s and 1960s. But it remains true, they unanimously agree that molecular biology provides a deeper level of theoretical understanding than classical Mendelian genetics, leading to much more powerful technologies culminating in the creation of human-made life.

As illustrated by the treatment of hemophilia mentioned above, Phase I technology is perhaps at best only a feeble expression of the weak form of control identified in Chapter 1. But the Phase II technology involved illustrates the strong form of control at work. And each of its stages is an expression of a progressively greater degree of such control. These points may be displayed as follows using hemophilia again as an example:

1. The technological rule of Phase I yielding only weak control may be formulated thus: if unstoppable bleeding is to be avoided, the sufferer of hemophilia ought to avoid being bruised or cut. Call this TRI. The scope of TRI's efficacy is not great, in the sense that it is useless should the sufferer, unavoidably, become bruised. There are, unfortunately, many such situations arising in the lifetime of a sufferer. Its efficacy is no more impressive than its analogue in a hurricane context where one could, at best, only advise people to get out of the way of a hurricane, when the signs of its imminence are detected, there being no means of deflecting it or defusing its strength. This minimal degree of control is a reflection of the lack of theoretical understanding of the phenomenon in question (although, as noted earlier, from the epistemological point of view, theoretical understanding is only a necessary, not a sufficient, condition for the emergence of a more powerful technology).
2. The technological rule of the first stage of Phase II, which is a manifestation of the strong form of control, may be formulated thus: to prevent unstoppable bleeding, the sufferer ought to be given a blood transfusion containing normal blood. Call this TRIIa. Undoubtedly, the scope of TRIIa's efficacy is greater

than that of TRI, for it can cope, when the sufferer unavoidably had cut himself. But it is beside the point when not enough normal blood is available for transfusion.[47] The increase in control reflects the theoretical understanding that the condition is caused by an inability of the sufferer's blood to clot, owing to its lack of a certain chemical, and that it is a genetic disability, not a functional one.

3. The technological rule of the second stage of Phase II may be formulated as follows: to prevent unstoppable bleeding, the sufferer ought to be given the clotting chemical (factor 8 or 9). Call this TRIIb. The scope of TRIIb's efficacy is greater than that of TRIIa for it overcomes the scarcity in the supply of normal whole blood, especially when the clotting agent in question can be produced *via* genetically engineered organisms.[48] Also, the agent can be more conveniently introduced into the sufferer's body through injections, rather than the more cumbersome technology of full blood transfusion itself. This greater degree of control is a reflection of the more detailed theoretical understanding about the nature of blood in general, and the specific deficiency isolated in the blood of hemophiliacs.
4. The technological rule of the third stage of Phase II may be formulated as follows: to prevent unstoppable bleeding, the sufferer ought to be given gene replacement therapy. Call this TRIIc.[49] The scope of TRIIc's efficacy is greater than that of TRIIb for it renders repeated and tiresome injections of the clotting agent throughout the lifetime of the sufferer redundant. And even more tellingly, the sufferer formerly identified as a hemophiliac is transformed under such treatment into a non-hemophiliac. His status has spectacularly altered. His genetic disability has been removed once and for all (if the treatment is truly successful). This still greater degree of control reflects yet more advanced theoretical understanding of the nature of heredity *via* molecular genetics.
5. The technological rule of the fourth stage of Phase II may be formulated as follows: to prevent unstoppable bleeding in individual males from ever occurring, germ-line therapy ought to be given to the female carriers of the condition, yielding male genotypes that lack the gene to produce factor 8 or 9. Call this TRIId. The scope of TRIId's efficacy is in turn greater than that of TRIIc, for it actually tackles the problem at an earlier stage by ensuring that no males would be born hemophiliac in the first place. This ultimate degree of control is a further reflection of knowledge in molecular genetics and of the nature of hemophilia as a genetic disability.

One caveat should be entered. The correlations between the efficacy of technological rules, their corresponding degree of control on the one hand and theoretical advances in the relevant pure sciences on the other, as set out above, are not meant to reflect actual historical correlations. They are meant to make the epistemological linkage between technological rules and scientific laws, namely, to bring out more clearly the point made earlier, that laws ground the efficacy of rules. And in doing so, one is, at the same time, laying bare the philosophical foundations for the ideological goal of modern science to control nature in the strongest form possible, to make it serve human ends, be it the alleviation of pain, the promotion of material-well being or of freedom and self-realization.

Notes

1. See Andrew Brennan's way of marking the difference between so-called naturally-occurring and artefactual entities—"The Moral Standing of Natural Objects," *Environmental Ethics* 6 (1984): 35-56. He says that natural entities have no "intrinsic functions," as they have not been designed by any one.

2. Richard Sylvan introduces the notion of a **natural artefact** but he uses it to refer to "a natural object where there has been some low level restoration activity" (*Against the Main Stream: Critical Environmental Essays* Canberra: Department of Philosophy and Law, Australian National University, 1994, 69 [footnote 20]).

3. This view, though very close to that of Eric Katz, cannot, however, be said to be identical with his. It is true that we both draw the distinction between the natural and the artefactual in the same way. But he seems to have arrived at this through a series of steps in reasoning which may be reconstructed as follows:

1. He begins with the argument that "(t)he possibility of determining values for natural entities ... requires a clear distinction between artefacts—such as Frey's tractor—and living entities as teleological centres of life and activity" (Eric Katz, "Artefacts and Functions: A Note on the Value of Nature," *Environmental Values* 2 1993: 228).
2. He then points out that the distinction drawn above, between tractors on the one hand and flora/fauna on the other, has the effect of excluding "soils, waters, mountains, the atmosphere" from the domain of moral value and moral consideration.
3. To include the abiotic natural entities, he then feels obliged, following Leopold, to exclude from his environmental ethic so-called "living artefacts" such as "domesticated animals, biologically engineered species and forest plantations. These human created entities have no place in an environmental ethic since they are not natural entities" ("Artefacts and Functions," 229).
4. This leads him to the conclusion, that "(t)he crucial distinction then is not between living beings (with biolgical functions) and nonliving 'things,' but between artefacts and natural entities, considered as living or not" ("Artefacts and Functions," 229).

This author arrives at the same distinction but through a different route. Furthermore, his preoccupation is also different—it is the Leopoldian one of excluding biotic artefacts from the domain of environmental philosophy, while this book attempts to mount an ontological resistance to the transformation of the natural to become the artefactual especially as such a transformation threatens to become near total in the face of increasingly more radical and more powerful technologies based on the 'deep' basic sciences.

Another environmental philosopher who draws the distinction between the natural and the artefactual (he uses the term 'artificial) is Robert Elliot—see especially Chapter 4 in

Faking Nature: The Ethics of Environmental Restoration (London: Routledge, 1997). However, Elliot's detailed defense is different from the one pursued here; his goes down the axiological, not the ontological, route.

4. See Robert Bud, *Uses of Life: A History of Biotechnology* (Cambridge: Cambridge University Press, 1993), 37-45, for an account of the eventual emergence of the acetone-butonal process usually associated with the biochemist Chaim Weizmann (who later in his life became the first president of the state of Israel).

5. Aristotle, *Physics*, 193, 10 in *The Basic Works of Aristotle*, Richard McKeon, ed. (New York: Random House, 1941).

6. The claim that there are natural kinds or that natural kinds exist presupposes some version of metaphysical realism (though not of a naive kind). The thesis of metaphysical realism is as follows: that there is a fundamental distinction between the world as it is and whatever thoughts, beliefs, hypotheses or theories we may have about that world. The two are logically independent; sometimes, our beliefs or theories may be quite false, even utterly false, and at other times, they turn out to be true.

This book cannot begin to examine this large assumption. (For a recent defense, see John Searle, *The Construction of Social Reality* London: Allen Lane, The Penguin Press, 1995, Chapters 7, 8, in which he uses the term 'external realism' rather than 'metaphysical realism.') Nor can it begin to give an account of natural kinds which may be said to be comprehensively philosophically grounded. For a recent attempt to give such an account, see T. E. Wilkerson, *Natural Kinds* (Aldershot: Avebury Press, 1995). This one is at best very partial and limited; moreover, it is given within a context of enquiry which is different from that of Wilkerson's. It is not surprising that the two differ in certain very significant ways:

1. Given the preoccupation of this book, it has by and large ignored the analogue of Aristotle's "first matter," like atoms of carbon, as candidates for natural kinds and has concentrated on "second matter," like granite, daffodils.
2. For this account, the crucial ontological demarcation line is between nonhuman nature on the one hand and human culture on the other. For Wilkerson's, the emphasis is on intrinsic properties possessed by a natural kind which constitutes its real essence, permitting the discovery of necessary (*de re*, not *de dicto*) truths, namely, the laws of nature, such as the laws of physics and chemistry.
3. As a result, for Wilkerson, some artefactual products belong to natural kinds—polysterene, PVC, kerosene, petrol, diesel fuel, etc.—see Wilkerson, *Natural Kinds*, 37).
4. For him, too, biotic artefacts belong to natural kinds (although he does not appear to have too much to say about this category).
5. As far as biological natural kinds are concerned, he (*Natural Kinds*, 133) argues that there are far more of them than there are species.
6. As for geological or meteorological kinds, there are not really natural kinds but what he calls "real but superficial kinds," although he concedes that geologists do "excellent scientific work and produce highly respectable explanations and predictions" (*Natural Kinds*, 39).
7. He lumps together agriculture, horticulture, geology and meteorology in spite of the crucial fact (that is, for this author, not for him) that the central, if not the only aim, of agriculture and horticulture is to produce cultivars, that is, biotic artefacts, and to study them for the purpose of improving them, whereas the objects of study in geology

and meteorology are not artefacts and it is not, at least, the sole or main aim of these studies to produce artefactual geological and meteorological forms.

Three problems immediately occur to mind, arising from Wilkerson's attempt:

A. Agricultural and horticultural entities cannot be understood without a reference to human intervention and purpose—in other words, they are logically anthropogenic and anthropocentric, like exbiotic or abiotic artefacts. As such, they have relational, not merely intrinsic properties.

B. He not only contends at 5 above, that there are more biological kinds than there are species, but also denies that "species are typically natural kinds" (*Natural Kinds*, 137). His biological natural kinds are concerned exclusively with the intrinsic features of the members belonging to such kinds in terms of their genetic composition. But he admits that "the biologist's account of species rests both on considerations about intrinsic genetic properties and on considerations about the historical relations between different groups of individuals" (*Natural Kinds*, 137). To be consistent with his central account of natural kinds, he would have to say that species are not natural kinds, as these involve relational, not merely intrinsic, features. But if so, as far as the entities invoked by the natural sciences are concerned, only those which appear in the laws of physics and chemistry are natural kinds. His own "biological natural kinds" are not the objects of study of evolutionary biologists, although they could be the objects of study of molecular biologists.

C. His account of geological and meteorological kinds appears to be inconsistent. On pp. 38-39, he says that mountains, glaciers, volcanoes, seas, thunderstorms, clouds, depressions and such are not natural kinds because "they do not have real essences and do not lend themselves to scientific generalization." As scientific disciplines, their explanatory and predictive successes depend entirely on sciences that are "directly concerned with real essences". On pp. 53-57, he says that pebbles and clouds are not really natural kinds although they are "real and superficial kinds." Yet on p. 38 and pp. 51-53, he argues that mountains, glaciers, rivers, tarns, anti-cyclones, hurricanes and clouds are also not natural kinds because "they do not pick out any relevant intrinsic properties" and are R-dependent (that is, relation-dependent) kinds. The apparent inconsistency is perhaps symptomatic of the tension within his account of natural kinds caused by the criterion of intrinsic (not relational) properties chafing against that of causal powers possessed by members of a natural kind to yield scientific generalizations. The explanatory success of geology, though dependent on physics and chemistry ultimately, nevertheless, lends itself to being "a scientific discipline" and to the characterization of the entities involved as "real and superficial" though not natural kinds. On the other hand, those entities appear also to be R-dependent kinds. Can one and the same entity be a member of both an R-dependent kind and a "real and superficial" kind?

7. A later section in this chapter will explore further the notion of 'deep' theories.

8. On pre-Mendelian selective breeding techniques, see Peter J. Bowler, *The Mendelian Revolution: The Emergence of Hereditarian Concepts in Modern Science and Society* (London: The Athlone Press, 1989); Jack R. Harlan, *Crops and Man* (Madison, Wisconsin: American Society of Agronomy, 1975); Lewis Mumford, *The Myth of the Machine: Technics and Human Development* (London: Secker and Warburg, 1967); H. F. Roberts, *Plant Hybridization Before Mendel* (Princeton: Princeton University Press, 1929);

David Rindos, *The Origins of Agriculture: An Evolutionary Perspective* (Orlando and London: Academic Press, Inc., 1984); Norman W. Simmonds, *Principles of Crop Improvement* (New York: Longman, 1979). On Mendelian genetics, see Bowler, *The Mendelian Revolution*; Lois N. Magner, *A History of the Life Sciences* (New York and Basel: Marcel Dekker, Inc., 1979); Robert C. Olby, *Origins of Mendelism* (London: Constable, 1966) and "Emergence of Genetics," in *Companion to the History of Modern Science*, Robert C. Olby, Geoffrey Cantor, John Christie and Jonathon Hodge, eds. (London and New York: Routledge, 1990). On Mendelian-led biogenetic technologies, see Jack Ralph Kloppenburg Jr., *First the Seed: The Political Economy of Plant Technology, 1492-2000* (Cambridge and New York: Cambridge University Press, 1990).

9. On DNA genetics see Garland E. Allen, *Life Science in the Twentieth Century* (Cambridge: Cambridge University Press, 1979); Francis H. C. Crick, *Of Molecules and Men* (Seattle and London: University of Washington Press, 1966); H. F. Judson, *The Eighth Day of Creation: Makers of the Revolution in Biology* (New York and London: Simon and Schuster Inc., 1979); Magner, *A History of the Life Sciences*; Olby, "Emergence of Genetics," "The Revolution in Biology," in *Companion to the History of Modern Science*, eds. Olby *et al.*, and *The Path to the Double Helix* (London: Macmillan, 1974). On DNA-led biotechnology, see Jeremy Cherfas, *Man Made Life: A Genetic Engineering Primer* (Oxford: Basil Blackwell, 1982); Kloppenburg, *First the Seed*; Sheldon Krimsky, *Genetic Alchemy: The Social History of the Recombinant DNA Controversy* (Cambridge, Massachusetts and London: The MIT Press, 1982) and *Biotechnics and Society: The Rise of Industrial Genetics* (New York and London: Praeger, 1991); E. J. Sylvester and L. C. Klotz, *The Gene Age: Genetic Engineering and the Next Industrial Revolution* (New York: Charles Scribner's Sons, 1983); Edward Yoxen, *Gene Business: Who Should Control Biotechnology?* (London: Pan Books, 1983).

10. See Richard Feynman, "There's Plenty of Room at the Bottom," in *Miniaturization*, H. D. Gilbert, ed. (New York: Reinhold, 1961); Eric K. Drexler, *Engines of Creation: The Coming Era of Nanotechnology* (New York: Doubleday/Oxford: Oxford University Press, 1992) and *Nanosystems: Molecular Machinery, Manufacturing and Computation* (New York & Chichester: John Wiley and Sons, 1992); Eric K. Drexler, Chris Petersen and Gayle Pergamit, *Unbounding the Future* (London: Simon and Schuster, 1992); Edward Regis, *Nano!* (London and New York: Bantam Press, 1995); "Engineering a Small World: From Atomic Manipulation to Microfabrication," *Science* 254 (1991): 1300-304.

11. No direct correspondence is necessarily intended between the respective sub-elements in the two lists—one for biotic and the other for abiotic entities—about to be introduced. In other words, the one should not simply be superimposed on the other to yield a perfect match. The discussion which follows makes this clear.

12. Transgenic organisms are genetically modified organisms in which the genetic engineer has inserted a gene from the DNA of an organism belonging to a totally different species into the organism's genome—for instance, a human gene may be introduced into the DNA of a mouse, an animal gene into the DNA of a plant. This differs from a hybrid organism as we shall see—the hybrid organism bears genes from two organisms belonging to different varieties which can interbreed and successfully reproduce, even if, as a matter of fact, they may not do so in nature in the absence of some form of human intervention. DNA genetic engineering, unlike hybridization technology, can cross species barriers, not merely between one plant species and another, or one animal species and another, but also between a plant and an animal species. (This account presupposes the so-called biological-species concept, which though not without problems, is accepted by most biologists—for

further discussion, see Keekok Lee, "Biodiversity," in *Encyclopedia of Applied Ethics, Volume 1*, Ruth Chadwick, ed. (San Diego: Academic Press, 1998), 285-304.

13. As for the matter about the various actual usages of the term 'technology,' a discussion may be found in Carl Mitcham, *Thinking Through Technology: The Path between Engineering and Philosophy* (Chicago: Chicago University Press, 1994), 143-54. However, in the section to follow, we shall have occasion to raise a related issue, namely, about how the term ought to be used.

14. This, however, is not meant to deny that technological developments of one kind or another do not affect them—an electric guitar is not the same as a classical guitar; their musical characteristics are different.

15. The reference to indirect material gains has military technologies in mind. In one obvious sense, bows and arrows, guns and cannons, atomic and hydrogen bombs damage or destroy people and their material goods. But wars are engaged in ultimately to capture and establish power over people and material goods. Even so-called 'holy wars' are no exceptions.

16. No more will now be said about calculated intelligence, as Chapter 4 will explore the notion of *homo faber* and its construal of human intelligence in terms of instrumentation.

17. Some animals, too, are said to be capable of this method. But the trial and error method used by humans is greatly transformed by the fact that humans possess language, which can be used to make propositions which, in turn, can be used to construct arguments. The empirical and philosophical debate about animal intelligence and language is an on-going one which, however, is beyond the remit of this book to explore.

18. For a broader account of the technological which may incude techniques, see Mario Bunge, "Philosophical Inputs and Outputs of Technology," in *The History and Philosophy of Technology*, George Bugliarello and Dean B. Doner, eds. (Urbana: University of Illinois Press, 1979).

19. See Mitcham, *Thinking Through Technology*, 153-54.

20. For a recent accessible account, see Arnold Pacey, *Technology in World Civilization: A Thousand Year History* (Cambridge, Massachusetts: The MIT Press, 1990).

21. Mitcham, *Thinking Through Technology*, 160.

22. Mitcham lists five philosophies of technology in terms of the will: to survive or satisfy some basic biological need (Spengler, Ferré); to achieve control or power (Mumford); to be freedom (Grant, Walker, Zschimmer); to be efficient (Skolimowski); to realize the *Gestalt* of the worker (Jünger) or almost any self-concept (Ortega). (See *Thinking Through Technology*, 247-48.)

23. The economic relations of this control are secondary—witness the agreement between both Marxist and bourgeois capitalist analyses on this point.

24. Lewis Mumford, *Technics and Civilization* (London: George Routledge and Sons, Ltd., 1946).

25. Phase I itself would probably have been preceded by an earlier stage which could be called the phase of "found technology." It would have included some degree of shaping and designing the thing found to suit the use it was put to—see Don Ihde, *Philosophy of Technology: An Introduction* (New York: Paragon House, 1993), 48, 79. It would be difficult to conceive of the beginnings of human culture in its absence. It is found also in the case of certain primates such as chimpanzees—in this context, Ihde calls it "prototechnology." But with humans, unlike other primates, they soon developed beyond found technology and its minimum designing to that of maximum designing.

Human cultures in general, not only Western European culture, would also have gone through Phase I A and B. What is distinctive about Western European culture is that it went on to Phase II from the seventeenth century onward.

26. F. W. Atkins, *The Second Law* (New York: Scientific American Books Inc., 1984), 7.

27. See Hunter Davies, *George Stephenson* (Middlesex: Hamlyn Paperbacks, 1980), 19-32.

28. Mumford, *Technics and Civilization*, 108.

29. Mumford, *Technics and Civilization*, 215-16.

30. Mumford, *Technics and Civilization*, 152.

31. Mumford, *Technics and Civilization*, 217-18.

32. Mitcham, *Thinking Through Technology*, 52-53.

33. Michael Zimmerman, *Heidegger's Confrontation with Modernity: Technology, Politics, Art* (Bloomington: Indiana University Press, 1990), 181-82.

34. Hans Jonas, *The Phenomenon of Life: Toward a Philosophical Biology* (New York: Harper and Row, 1966), 189-90.

35. In Phase II, the causal arrow from theory to technology is not displaced by recognizing that technological objects spawned by theory could in turn influence the development of theory. It has been correctly observed that the design and construction of the cyclotron (and increasingly more powerful ones) have affected and continue to affect theory development itself. The point to grasp here is simply that the cyclotron would not, as a matter of fact, have been constructed in the absence of theoretical particle physics as we know it today, and that, therefore, as a technological construct and object, it is totally conceptually dependent upon the understanding of sub-atomic matter as given to us by theories and speculation prevalent in contemporary physics.

For an alternative view, see Don Ihde, on what he calls "technoscience." He writes: "Today's Big Science is so closely tied to Big Technology that one can meaningfully speak of a single, complex phenomenon which is both a scientific technology and a technological science: technoscience" (*Instrumental Realism: The Interface between Philosophy of Science and Philosophy of Technology* (Bloomington: Indiana University Press, 1991), 138). Ihde may be said to push the Heideggerian inversion to its logical conclusion, namely, that:

1. There is no "pure" science "eventually producing some 'applied' effect." Rather there is a "technology-driven science." "At the highest altitude, such a perspective was suggested most radically by Heidegger. ... But at a lower and much more concrete level we also have noted how parts of our world are instrumentally and technologically revealed and even produced" (Ihde, *Instrumental Realism*, 137).
2. Theoretical entities are much more like technological than 'natural' ones. They are "an interesting crossing of what could be called the convergence of Nature with Culture through laboratory science. The object—produced in the laboratory and reified into an independent thing—was fabricated, but fabricated in such a way that it takes on the status of 'reality.' ... Reality is what resists, at least until more powerful instruments and laboratory complexes can overthrow them" (Ihde, *Instrumental Realism*, 134).
3. "Instrumental realism" is the correct account of theoretical entities:

> By turning to the role of instruments and taking account of what they deliver for science, the very territories previously taken as theoretical domains also

> change—they shrink in size and significance. In short, instrumental realism gives some degree of type of "reality-status" to entities often taken to be merely theoretical, leaving only small areas to remain theoretical. This means, in turn, that the role of "pure" theorizing gets reduced to an even smaller area of science's activity than had previously been assumed. Theorizing becomes a special, highly speculative exercise of scientific imagination—important, but both reduced in size and open to greater skepticism—in regions outside the current reaches of instrumental possibility. (Ihde, *Instrumental Realism*, 100)

See also John M. Staudenmaier, *Technology's Storytellers: Reweaving the Human Fabric* (Massachusetts and London: The Society for the History of Technology and the MIT Press, 1985), Chapter 3.

36. See Bunge, "Philosophical Inputs." According to Mitcham (*Thinking Through Technology*, 37-38), Bunge belongs to what he calls the "engineering philosophy of technology" school as opposed to the "humanities philosophy of technology" school, to which Mumford, Heidegger, Ellul and others belong.

37. Bunge, "Philosophical Inputs," 263.

38. Bunge, "Philosophical Inputs," 265-68.

39. Scientists, if pressed, are by and large critical realists, although a few may even be naive realists. A competing tradition in the philosophy of science rejects realism, even of the critical variety. (On another front, Thomas Kuhn's *Structure of Scientific Revolutions* (Chicago: Chicago University Press, 1962), in spite of Kuhn himself, has been interpreted as an attack on the rationality of science itself.) Far from it being the case that scientific theories are attempts to capture Reality, Reality is given and constituted by them——see Appendix 2 (**Is Nature a Mere Social Construct?**). Older philosophers of science, like Duhem and Poincaré, espouse scientific conventionalism or instrumentalism. But see also Ihde (*Instrumental Realism*) who espouses "instrumental realism" (not to be confused with the Duhem/Poincaré variety), already mentioned in an earlier note.

40. Bunge, "Philosophical Inputs," 269.

41. Karl Popper, *Conjectures and Refutations* (London: Routledge and Kegan Paul), 1969.

42. Bunge, "Toward a Philosophy of Technology," in *Philosophy and Technology: Readings in the Philosophical Problems of Technology*, Carl Mitcham and Robert Mackey, eds. (New York and London: The Free Press, 1983). For a full philosophical discussion of the points raised here, see especially 69-71.

43. See *Observer*, 10 October 1993, 13. Western scientific reaction, reported there, mentioned certain ethical problems that could arise. For instance, using a retrovirus as the vector could conceivably trigger cancers.

44 Another success story of gene replacement therapy concerns the disease known as ADA (adenosine deaminase) deficiency or commonly known as the "bubble-boy disease." The boy in question had no choice but to live his short life in a plastic bubble in order to avoid infection to which he could fatally succumb. The condition was caused by some "misplaced" nucleotides on the DNA of Chromosome 20. In 1990, two girls (from Ohio) were finally cured, when physicians (W. French Anderson, Michael Blaese and Kenneth Culver) at the US National Institute of Health mixed cells containing the right molecular sequences carried by viral vectors with extracts of their own white blood cells, and then introduced the new mixture into their blood stream. Soon the transformed cells began producing the normal amount of ADA. (See Regis, *Nano!*, 272-73.)

45. The observation to follow is this author's, not Allen's. One important and interesting area of dispute today is the theory of natural selection in the light of theoretical advances in genetics. Richard Dawkins (*The Selfish Gene* Oxford: Oxford University Press, 1976 and *The Extended Phenotype* San Francisco: Freeman, 1982) maintains that the unit of selection is the single gene. Those who disagree like Ernst Mayr (*Animal Species and Evolution* Cambridge, Massachusetts: Belknap Press of Harvard University Press, 1963), Elliott Sober and Richard Lewontin ("Artifact, Cause and Genic Selection," *Philosophy of Science* 49 1982: 157-80) hold that selection is of phenotypes, not of genes or genotypes.

46. Allen, *Life Science*, xiii-iv.

47. Furthermore, patients who undergo frequent blood transfusions tend to accumulate an excessive amount of iron which causes damage to the liver and heart as well as often interfering with normal growth and development.

48. The latest development on this front, reported in late September 1997, is the success of the laboratories of the American Red Cross in Rockville in producing factor 8 in pig's milk. The pigs have been genetically modified to do so. The scientists have injected pig embryos with an artificial version of the human gene responsible for the liver in making factor 8. But to ensure that the blood-clotting protein would be found in only the pig's milk and nowhere else, the human gene has been tied in with a pig gene which only works in its mammary glands. See Andy Coghlan, "Clotted Milk," *New Scientist* (27 September 1997): 10.

49. Another example of the same progression at work concerns the condition called Gaucher's disease. Philippe Gaucher discovered it in 1882. It is inherited from two carrier parents who themselves may be free of the symptoms. The sufferer's body is unable to break down the chemical, glucocerebroside, found in the membranes of white and red blood cells, which enables macrophages, containing fatty glucocerebroside globules, to accumulate in the liver, blood marrow and spleen. This could lead to brittle bones as well as swelling of the liver and spleen. In the 1980s, Dr Roscoe Brady of the National Institute of Health (USA) identified the enzyme, which the patient lacks, responsible for such symptoms. He managed to extract the critical enzyme from placentas and administer it as a drug to patients. This first-generation drug, called ceredase, is manufactured by a company called Genzyme. But it can only be expensively produced. Dr Brady, in the late 1980s, went on to identify the gene that makes the enzyme which breaks down glucocerebroside. This gene is then inserted into cells isolated from Chinese hamsters. The cells are grown in vats producing unlimited amounts of cerezyme, the biotechnological version of ceredase. In 1994, Genzyme was on the verge of marketing this second-generation drug with the expected approval of the US Food and Drug Administration. But already the next new-generation product on the horizon is being put in place—a device to insert the actual missing gene into the patients' bodies. (See Robert McKie, "Medicine Man at the $100m Biotech Temple," *Observer*, 9 October 1994, Business Section: 6.)

Chapter Three

Independence, Human Design, and Artefacticity

This chapter examines, in the main, two issues: (a) in the first section, a conceptual clarification of the notion of nature and its cognate, that of 'the natural' in order to highlight a crucial sense, namely, 'natural$_{fa}$' which in turn is tied up with the notion of independence—the book goes on in Chapter 5, however, to argue that independence, unlike intricacy, complexity, sentience or other such attributes, is a primary characteristic and, therefore, constitutes an ontological rather than an axiological value; (b) in the next two sections, an examination of the ambiguities surrounding the notion of 'human impact upon the environment,' by distinguishing unwanted side-effects of technological impact from the deliberate and systematic transformation of the natural to become the artefactual. The final two sections are used to illustrate further some of the key points raised in the preceding ones and the relationship between them. These clarifications are attempts to put in place a further stage in establishing the fundamental thesis of the book, namely, that nature has independent value and that the most radical and critical threat to it is yet to come. The threat amounts to its elimination, both ontologically and empirically, *via* the science and technology of our modern civilization, especially when its most recent technologies—biotechnology and computer technology—will combine with certain others promised in the near future, such as molecular nanotechnology, to produce powerful synergistic effects in a profound transformation of the natural to become the artefactual.[1]

The Natural: Different Senses of 'Nature'

In the context of discussion in this book, 'the natural' is meant to be the foil of 'the artefactual.' The exploration in the last chapter has said something about the artefactual; nevertheless, an account of the natural may be said to have also indirectly emerged. This section will summarize the theses which may be identified as constituting such an account as well as explore further the differences between the two contrasting notions.

'The natural,' as here understood, refers to whatever exists which is not the result of deliberate human intervention, design and creation in terms of its material, efficient, formal and final causes. By contrast 'the artefactual' embodies a human intentional structure—its efficient, formal and final causes are understood in terms of human agency and in some cases, as we have seen, even its material cause is to be so understood. 'The natural' comes into existence, continues to exist, and goes out of existence entirely independent of human volition and manipulation.

The distinction between the natural as here defined and the artefactual can only be made by assuming that humans are not part of nature but of culture. 'Nature' in this sense then means 'nonhuman nature' or 'nature$_{nh}$' and is opposed to 'culture.' However, in spite of this fundamental opposition, it remains true that humans are living organisms and organisms are part of nature. But this other sense of nature is not what is referred to in the context of distinguishing between 'the natural' and 'the artefactual,' and so must be distinguished from 'nature$_{nh}$.' According to it, anything that exists or happens in the universe of space and time is 'natural.' Clearly, humans as organisms, as well as their artefacts, are part of nature in this sense which may be called the 'cosmological' sense, and will be written as 'natural$_c$' for short.[2]

But it is too wide. Such an all-embracing concept is distinctly unhelpful, as it ignores the crucial distinction between culture on the one hand, and nature$_{nh}$ on the other (or between the human and the nonhuman).[3] It subsumes culture under it, thereby rendering culture an empty notion. But culture is a human social creation. Any one culture may have taken a very long time to evolve but this does not alter its essential nature as a human invention. Because certain cultural practices could badly undermine nature$_{nh}$, it is crucial to uphold the distinction between culture (human) and nature$_{nh}$ (nonhuman) as an intelligible one. To conjure away the distinction by a linguistic sleight of hand would be neither honest nor clear-headed. Later, in Chapter 5, it will be argued that the distinction should not, however, be understood dualistically but as constituting ontological dyadism; while ontological dualism, in this context, is unwarranted, ontological dyadism is intelligible, fundamental, and indispensable.

In other words, for the purpose of this discussion, seven senses of 'nature' must be made distinguished and made clear:[4]

1. 'Nature$_{nh}$' is opposed to 'culture.' Culture involves human agency and its

products. The products may be intended or unintended—for instance, a piece of legislation is intended whereas the origin of language is said to be an unintended product of human agency. (But this book is primarily concerned with artefacts which are intended products.)

2. 'Nature$_c$' is far too wide as we have seen, and obliterates the fundamental dichotomy between nature$_{nh}$ and culture. According to it, the American Revolution, Hadrian's Wall, the Great Barrier Reef, the Grand Canyon are all natural events or objects, which they undoubtedly are, as they can be identified in terms of certain spatio-temporal co-ordinates. The opposite of 'nature$_c$' is the Supernatural.[5]
3. 'Nature$_p$'—what counts as natural must be pristine, unaffected in any way by the impact of human action whether intended or not. It will be argued later in this chapter that nature$_p$ *simpliciter* would not be applicable, by and large, to Earth today,and that it would be more helpful to distinguish between different types and levels of human impact.[6] Some impact may be minimal and fleeting, like walking on fallen snow. Others qualitatively change or destroy outright the ecological character of a landscape, like strip mining.
4. For the reasons stated above, one should recognize another sense of nature—'nature$_{hi}$,' that is, nature as affected by human impact.
5. 'Nature$_{fa}$' is to be understood as the foil to the artefactual, which itself is defined in terms of what is brought into material existence deliberately because of human intentions. It is this sense which is canvased by this book—'the natural' is defined as 'what is not the material embodiment of deliberate human intention' and is, therefore, independent of humans. As such it may be said to presuppose nature as 'the Other.' (The ontological elimination of nature as 'the Other' will be examined and resisted in Chapter 5.)
6. 'Nature$_{fa}$' includes both 'nature$_{nk}$' and 'nature$_f$.' 'Nature$_{nk}$' refers to what Aristotle has called second matter, that is, natural kinds. Natural kinds may be biotic or abiotic. In the main, when this book (including its title) talks about the distinction between 'the natural' and 'the artefactual,' the senses of 'natural' referred to are 'nature$_{fa}$' and, in particular, 'nature$_{nk}$,' but also 'nature$_{nh}$.'
7. 'Nature$_{nk}$' is to be distinguished from 'nature$_f$,' the analogue of what Aristotle called first matter. Nature$_f$ refers to the naturally-occurring elements in the Periodic Table. This book, however, will not address itself to another even deeper scientific project, involving the sense of 'natural$_f$,' which exercises the minds of some theoretical physicists.[7]

The relationships between the seven senses identified above may be displayed as follows:

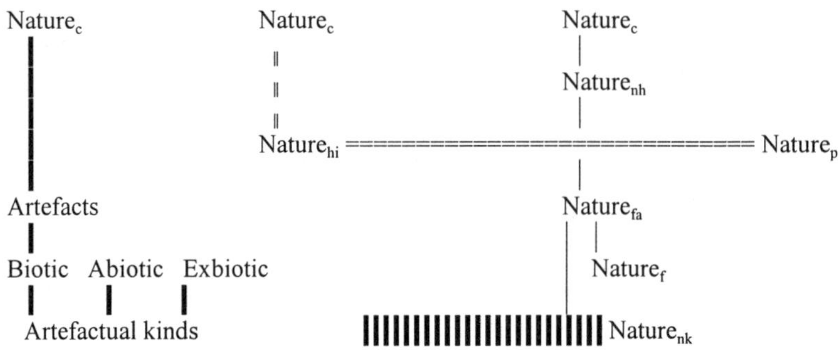

[| or || indicates the direction and degree of artefacticity toward which nature$_c$ as nature$_{nk}$ may be transformed by intended human agency; ‖ or = indicates the the way in which nature$_c$ as nature$_p$ may be transformed by the impact of human activity which is not necessarily intended; | has no directional/causal implications and is merely classificatory.]

In his essay, "Nature," John Stuart Mill distinguishes between 'nature$_c$' and 'nature$_{nh}$.' He writes:[8]

> It thus appears that we must recognize at least two principal meanings in the word Nature. In one sense, it means all the powers existing in the outer or the inner world and everything which takes place by means of those powers. In another sense, it means, not everything which happens, but only what takes place without the agency, or without the voluntary and intentional agency, of man. This distinction is far from exhausting the ambiguities of the word; but it is the key to most of those on which important consequences depend.[9]

But that paragraph also seems to hint at another sense closely related to ' nature$_{nh}$,' namely, 'nature$_{fa}$'—"what takes place ... without the voluntary and intentional agency ... of man." This sense appears to be similar to what C. S. Lewis has identified as one of the oldest, and, therefore[10]

> one of the hardiest senses of nature or natural. The nature of anything, its original, innate character, its spontaneous behaviour, can be contrasted with what it is made to be or do by some external agency. A yew-tree is natural before the topiarist has carved it; water in a fountain is forced upwards against its nature; raw vegetables are au naturel. The natural here is the Given.
>
> This distinction between the uninterfered with and the interfered with will not probably recommend itself to philosophers. It may be held to enshrine a very primitive, an almost magical or animistic, conception of causality. For of course in the real world everything is continuously "interfered with" by everything else; total mutual interference (Kant's thorough-going reciprocity) is of the essence of nature (...). What keeps the contrast alive, however, is the daily experience of men as practical, not speculative, beings. The antithesis between unreclaimed land and the cleared, drained, fenced, ploughed, sown, and weeded

> field—between the unbroken and the broken horse—between the fish as caught and the fish opened, cleaned, and fried—is forced upon us every day. That is why nature as "the given," the thing we start from, the thing we have not yet "done anything about," is such a persistent sense. We here, of course, means man. If ants had a language they would, no doubt, call their anthill an artifact and describe the brick wall in its neighbourhood as a natural object. Nature would be for them all that was not "ant-made." Just so, for us, nature is all that is not man-made; the natural state of anything is its state when not modified by man. This is one source of the antithesis (philosophically so scandalous) between nature and Man. We as agents, as interferers, inevitably stand over against all the other things; they are all raw material to be exploited or difficulties to be overcome. This is also a fruitful source of favourable and unfavourable overtones. When we deplore the human interferences, then the nature which they have altered is of course the unspoiled, the uncorrupted; when we approve them, it is the raw, the unimproved, the savage.[11]

As Mill also points out 'the artificial' as 'the artefactual' means that "we (humans) move objects, and by doing this, bring some things into contact which were separate, or separate others which were in contact: and by this simple change of place, natural forces previously dormant are called into action, and produce the desired effect."[12] Such moving about of objects, as we have seen, is accomplished by modern science (beginning in Western Europe), its method, its goal of controlling nature and eventually its technology. It has already been hinted that the deeper the science and its accompanying technology, the greater the degree of artefacticity in the artefacts produced.

What has been said so far confirms that the philosophical transformation of the natural to become the artefactual has been put in place since the seventeenth century when (a) Aristotle's thesis of intrinsic/immanent teleology was rejected to make room for extrinsic/imposed teleology involving an aggressive anthropocentrism and instrumentalism, (b) modern science with its methodology was used initially to displace the formal and final causes from its study of nature while retaining the material and efficient causes only. But ultimately its goal is to transform all four into humanly inspired, designed and executed ones. Such ultimate transformation, as promised by molecular nanotechnology, constitutes the most recent comprehensive, though not the final, expression of the aspiration to control nature. Modernity, in its boldest aspiration, tries to transform the natural$_{fa}$ into the artefactual. Even Mars would, one day, become an artefact, a human creation. In other words, nature$_{nh}$ could, in principle, be totally humanized.

But as artefacts are the deliberate products of human intelligence to serve human ends, having been engineered with features and mechanisms which are not the outcome of the processes of natural evolution, they belong to a distinctly different ontological level from that of naturally-occurring beings or things. Artefacts are (human) intentional structures, while naturally-occurring objects *ex hypothesi* are not. (The ontological elimination of nature$_{fa}$ and nature $_{nk}$will be examined in Chapters 4 and 5.)

One last observation before leaving this section—in spite of the attempt on the part of this analysis so far to assimilate those biotic artefacts which are the products of 'deep' sciences and their respective accompanying technologies to non-biotic artefacts similarly produced, there remains, nevertheless, a difference between them which cannot be explored here but which will become apparent when the Last Person Argument—see Routley, "Is There a Need?"—is applied to them in a later section in the chapter.

The End of Nature?

This section examines McKibben's claim about "the end of nature" in the light of some of the distinctions made in the last about the various senses of 'nature.'

McKibben's *The End of Nature* may be read as a powerful lament about the demise of nature.[13] That lament is predicated on the implicit assumption that one ought to respect the natural—that which is independent of us humans is a value in its own right—and in lamenting the end of nature, one is lamenting the loss of such a value in the world. This book, too, endorses nature's independence as a value, although it goes beyond the sense given implicitly to it by McKibben. It also goes further in making the assumption explicit and exploring its philosophical underpinnings in a more systematic manner.

However, in spite of the broad agreement, there are serious areas of disagreement. The last chapter has identified different senses of the term 'nature' and argued for degrees of artefacticity, issues which McKibben does not raise. This chapter will try to link these with different types of human intervention, in particular with the distinction between deliberate/direct intention on the one hand and what may be called oblique intention on the other, and to argue, in turn, that artefacticity, strictly speaking, is produced only in the context of the former and not the latter type of intervention.[14] McKibben, on the contrary, appears to assume that any human intervention, provided it produces effects which are pervasive, renders the natural as the artefactual. (The operative word here is 'assumes' as McKibben, no where, says so explicitly.) This may be too strong as it stands. It needs to be qualified.

McKibben primarily concentrates on two types of examples as evidence for his thesis: ozone depletion and greenhouse effect on the one hand and biotechnological engineering on the other. Both types of phenomena may indeed be said to have pervasive effects—the first in the sense of (i) being geographically extensive, global in scope, (ii) underlying in causal terms;[15] the second, in the sense of substituting technological conditions for the creation of life and human-made life forms for the processes of natural evolution and its evolved life forms. The first, in the terminology of this book, undermines $nature_{p}$; the second goes beyond that to undermine $nature_{fa}$. Although both forms of undermining are undoubtedly forms of human impact upon nature, nevertheless, it is very important to distinguish between them from the ontological point of view.

The use of DDT, too, has produced undesirable ecological effects worldwide.

Its presence can be detected in areas, even like Antartica, which are far from the sources where the pollutant was first introduced.[16] It may also be potent in causal terms as an underlying condition for further extensive changes. It may damage the capability of birds to reproduce successfully by thinning the shell of their eggs so that the eggs are crushed before they are ready to be hatched. It may even kill off other organisms by poisoning them outright. As it contains estrogen-mimicking chemicals, it may also cause decreasing fertility of males, whether these be human beings, cougars or sturgeons.[17] Its causal impact ultimately affects the fate of many life forms.

If so, McKibben is not quite right in pointing out that the depletion of stratospheric ozone (by the release of CFC and other gases) as well as the greenhouse effect (by the increase of atmospheric carbon dioxide through the burning of fossil fuels among other causes) are unlike the effects of DDT as a pollutant. Ozone depletion, as we know, means that harmful ultra-violet rays can penetrate Earth's atmosphere, as they are no longer filtered out by the ozone shield in the stratosphere. These rays not only can increase the rate of skin cancer among humans, but also can affect the phytoplankton on the surface of the ocean which form the beginning of a complex food chain in marine and oceanic ecosystems. The collapse of these systems in turn can cause disturbances to land-based ecosystems as well. So the threat posed by such damage to stratospheric ozone can be said to be both geographically extensive and potent in causal terms, being capable of undermining the conditions for sustaining many life forms and ecosystems as we know them today. Moreover, plankton is known to be an important factor in the carbon dioxide cycle; damage to it would then have a knock-on effect on global warming.

The greenhouse effect is equally totalizing in the way just characterized. As Earth's atmosphere warms up as predicted by the best 'state-of-the-art' understanding of the complex isssues involved, existing life forms and ecosystems will be greatly disrupted, with some of them, perhaps, destroyed for good.[18]

The new crisis that faces us, which goes beyond that posed by acid rain and such types of pollution, is well put by McKibben. To quote him:

> When the nature around us is degraded, we picture it fresh and untainted elsewhere. When elsewhere, too, it rains acid or DDT, we can still imagine that someday soon it will be better, that we will stop polluting and despoiling and instead 'restore' nature. ... In our minds, nature suffers from a terrible case of acne, or even skin cancer—but our faith in its essential strength remains, for the damage always seems local.
>
> But now the basis of that faith is lost. The idea of nature will not survive the new global pollution—the carbon dioxide and the CFCs and the like. This new rupture with nature is different not only in scope but also in kind from salmon tins in an English stream. We have changed the atmosphere, and thus we are changing the weather. By changing the weather, we make every spot on earth man-made and artificial. We have deprived nature of its independence, and that

> is fatal to its meaning. Nature's independence is its meaning; without it there is nothing but us.[19]

He is correct in pointing out the qualitative difference between relatively localized and totalizing globalized types of pollution. But he may not be right in thinking that DDT belongs to the former category. Moreover, he is keener to establish the distinction than he is to contemplate the difference between that assault on nature induced by ozone depletion and global warming on the one hand and biotechnological engineering on the other. For him, both are attempts to disrupt nature, but the latter goes further in that it may be regarded as a technological fix to solve the problems posed by the former.[20] He does, however, *en passant*, but no more, refer to one difference, namely, that the end of nature caused by the former is accidental, that caused by the latter is deliberate. He writes:

> [J]ust as the clouds of carbon dioxide threaten to heat the atmosphere and perhaps starve us—we are figuring out a new method of dominating the earth, a method more thorough, and therefore more promising, than burning coal and oil and natural gas. It's not certain that genetic engineering and macromanagement of the world's resources will provide a new cornucopia, but it certainly seems probable. ... why, then, does it sound so awful? Because, of course, it represents the second end of nature. We have pretty much, by accident, altered the atmosphere so badly that nature as we know it is over. But this won't be by accident—this will be on purpose. I don't mean that we shall end nature if something goes wrong—if, say, a strain of bacteria programmed to eat cellulose gets loose and eats every tree and weed in sight. ... It is the simple act of creating new forms of life that changes the world, that puts us forever in the deity business. We will never again be a created being; instead we will be creators.[21]

McKibben's lament about the first and second ends of nature obscures numerous points. While McKibben clearly recognizes that any action on the part of humans has impact upon nature$_{nh}$, he is not so clear about where the line should be drawn between those impacts which could be said to constitute the end of nature$_{p}$ and those which constitute the end of nature$_{fa}$.

He characterizes the first end of nature, represented by global warming or stratospheric ozone depletion, in two conflicting ways. On p. 58—cited earlier—he says: "By changing the weather, we make every spot on earth man-made and artificial." But on p. 166, he writes: "We have pretty much, by accident, altered the atmosphere so badly that nature as we know it is over." His first characterization is definitely wrong. Precisely, because the alteration has not been designed, it would be incorrect to imply that it involves the demise of nature$_{fa}$ when it involves the demise of nature$_{p}$ (as the impact is global and causally pervasive).

Although he is more correct in his second characterization, all the same, he is not clear as to what could be meant by 'accidental.' And when that clarification has been made, it is not obvious that such a demise is really 'accidental.'

Something, A, is said accidentally to have happened when the following conditions jointly obtain: (i) the agent did not deliberately intend A to happen and (ii) either A was caused by another event which was not initiated by the agent but by some other agent—the precious vase the agent was carrying was knocked out of her hands by a charging Alsatian, or A was caused by something which was entirely beyond the agent's control so that s/he could not be said to have acted, rather it was something which simply happened—all of a sudden, she suffered momentarily from a black-out or an epileptic fit, and as a result the vase fell from her grasp, shattering to pieces.

But the first end of nature as identified by McKibben cannot be said to be accidental in the senses worked out above. It may be true that the agents involved do not deliberately intend to produce the greenhouse effect or the ozone hole. The effects satisfy condition (i) above but not (ii) in either form. They are anthropogenically, though not deliberately, produced. They are the accumulation of the unintended consequences of innumerable but separable individual acts of fossil burning, growing paddy rearing cattle, etc. No one, in designing or using a car or keeping cows, deliberately and directly intends to cause global warming. Nor is each act of consuming fossil fuel, using a car or growing a few hectares of paddy on its own, sufficient in causal terms to produce the end of $nature_p$. However, that end may be said to be obliquely or indirectly intended, but it would be misleading to say that the outcome is a purely accidental one, given the knowledge we now have about its provenance and its cause. This obliquely intended outcome is then no different from that posed by DDT pollution. There, too, no one deliberately intended to cause failure in the reproduction of certain birds, to poison the water table, etc. But these effects, nevertheless, also occurred cumulatively. Given knowledge of such outcomes today, humans may collectively be held responsible for them, even though it is true that the model of individual responsibility is not applicable in these contexts.

If McKibben's first end of nature as $nature_p$ cannot be identified in terms of what counts as accidental, could he, instead, rely on the criterion of pervasiveness as clarified earlier on? But pervasiveness may turn out to be only necessary, not necessary and sufficient, to cause the end of nature. He is correct in saying that anthropogenic changes to the weather and the atmosphere can bring in their wake profound disturbances to $nature_{nh}$. But $nature_{nh}$, in its history, has endured many profound changes in its weather which has nothing to do with human agency—ice ages have come and gone, bringing with them severe changes to flora and fauna in geological history. This leads to the view that as far as organisms and their ecosystems are concerned, the severe changes they have to endure are no different in quality whether the disruption is anthropogenically caused or not—both forms may be sudden and abrupt like volcanic eruptions, asteroids crashing into Earth's surface, clear-cutting an ancient forest on the one hand, or gradual and cumulative, like evolutionary changes or the emergence of global warming today on the other. Eventually, after a period of time—sometimes quite short, but often long or very long—organisms, some old and others new, would establish new niches and new ecosystems. Anthropogenic disturbances then amount to the loss of pristine nature. The lament then is not so much about the end of nature *tout court* but the end of $nature_{p}$.[22] This is a genuine lament but it is not quite what McKibben has portrayed.

$Nature_{nh}$ in the context of global warming or ozone depletion, though no longer absolutely pristine, survives and, moreover, survives independently of us humans as long as the processes of natural evolution remain intact in spite of the anthropogenically induced changes to the weather and the atmosphere. $Nature_{nh}$, as we have seen, copes, no differently, when the changes to the weather and the atmosphere are not anthropogenically induced. The claim that $nature_{nh}$ can survive independently of us even in such a context may be supported by empirical evidence—for instance, the eruption of Mount Saint Helens in the Gifford Pinchot National Forest, Oregon, in 1980, devasted the ecosystems which lay in its path but since then, slowly but surely, new ones have begun to establish themselves. Furthermore, life on Earth in its long history well before the appearance of *Homo sapiens* had suffered five major extinctions; each time $nature_{nh}$ had recovered, although recovery times varied from 100 to 20 million years.[23] It can also be supported by the following thought experiment—imagine the removal of the human species immediately after it had induced the meteorological changes. It is likely that $nature_{nh}$ will continue to evolve in the absence of humans.

Nature's Independence

The test of independence proposed is different from that provided by McKibben. He appears to say that any change of a pervasive kind which is humanly induced destroys $nature_{nh}$'s independence. But his claim may be too strong, as it stands. It may be said to have compromised $nature_{nh}$'s independence but not to have eliminated it for good. From this point of view, it may be misleading to talk about the end of $nature_{nh}$ *tout court.*[24] Nor does such compromising of $nature_{nh}$'s independence amount to transforming the natural into the artefactual (given the way these two terms have been defined by this book). In contrast, as we shall see, such a transformation would, indeed, be at work, for instance, in the case of terraformation, the project to give Mars and other planets like Venus an Earthen atmosphere and then to introduce specially genetically engineered organisms to produce a biosphere.[25]

However, McKibben has more accurately characterized his second end of nature. He says that this end, unlike the first, is not accidental but intentional (directly and deliberately intended). Moreover, biotechnological engineering allows, at least in principle, the radical substitution of human-designed and controlled processes of reproduction for the natural processes of reproduction and evolution. As such, it could, indeed, amount to the end of nature. The nature that is undermined here is, more precisely, $nature_{fa}$. What, perhaps, McKibben's account lacks but which one is trying to make good is the thesis that $nature_{fa}$ is being transformed to become artefacts, spelling out the various degrees of artefacticity involved. As we have seen, an artefact involves conscious human design in terms of its final, formal, efficient as well as even the material causes in certain cases. As such, genetically engineered organisms possess an extremely high level of artefacticity. Processes which could not occur in nature, such as the transference of genetic material belonging to certain species to others, can now take place, the end result of which would be the substitution of human processes for the natural processes of evolution in the creation of organisms and life, as McKibben has pointed out. In contrast, the organisms and their ecosystems which emerge in adapting themselves to the changed weather conditions—anthropogenically induced—cannot, strictly speaking, be said to be artefacts. As McKibben himself has pointed out (at least in one passage, if not consistently throughout the book), these atmospheric changes are not deliberately induced, although it is true that they are the cumulative outcome of certain sorts of human actions and so, therefore, cannot be said to be accidental either. Nor do humans have further control (in the strong sense of control) over the weather and the atmosphere once they have set certain initial reactions in motion. Furthermore, they also have no or very little control over how organisms and their ecosystems will eventually adapt (or not as the case may be) to the changes in question. Neither the changed atmosphere nor the response of organisms to it conforms to the paradigmatic requirement of being an artefact *par excellence*—they are not the product of conscious deliberate human design, with humans in control at every stage of the development.

In contrast, deliberate human interventions, though on a less than global but still extensive scale, would display a far higher degree of artefacticity. For instance, the now defunct Soviet Union had plans to alter the courses of some of its north-flowing rivers, diverting them to the south to provide irrigation to the drier areas of its empire. However, it is true that the ecological consequences of such profound restructuring of large river systems remain at best partially foreseen and known.

In the light of the above comments, it would be useful here briefly to clarify the ways in which nature may be said to be independent of humans, and the corresponding notion of humans being independent of nature, may be understood:

1. In a fundamental sense, humans can never be independent of nature, in spite of what some theorists might have said. To quote one such view: "Man has probably always worried about the environment because he was once totally dependent upon it,"[26] implying that now with powerful technologies in place induced by scientific theories, humans have become freed from their dependence on nature.[27] But humans cannot create anythiny *ex nihilo*. They can only transform matter and energy from one form into another with the help of yet other forms of matter and energy. It is true that their scientific knowledge and their increasingly powerful technologies permit them radically to transform one kind of matter/energy into another kind. But nature, as first

or primary matter, is ontologically independent of humans, and until the promise of molecular nanotechnology is made good, it is not the 'stuff' upon which humans directly operate and manipulate in transforming it to suit their own purposes and goals.

2. But technology in its history has enabled humans very successfully, by and large, to transform nature as second matter to suit their ends, such that what starts off as being ontologically independent of humans becomes ultimately their artefacts. As we have seen, at the minimum, it has allowed humans to interfere with the existence and lifestyles of naturally-occurring beings without necessarily rendering them into artefacts. For instance, humans may simply delimit or enclose a space within which animals are free to roam, to breed, to nurture their young. The animals may then be subjected to seasonal hunting either for pleasure or food or both. In these ways, of course, the animals' existence is affected and determined by human actions. But in other ways, they are independent of humans. Their absolute independence may have been undermined to a certain extent, but it remains meaningful to say that they have an existence and a *telos* which are independent of humans. However, there are also plants and animals, which through careful deliberate human intervention and selection, have been bred so that they are no longer viable on their own without human sustenance and maintenance. Cultivars, like maize and other cereals, could not survive and evolve in the wild because they have been specially bred to possess a characteristic considered highly desirable from the human point of view, namely, that their seeds stay attached to the ear. In the wild, this would spell reproductive disaster, as their seeds could not disperse for germination. The camel, too, has been so successfully artificially selected that it can no longer reproduce without human help. These organisms, even without the benefit of biogenetic engineering, are human artefacts to a large extent. They have lost their independence in the most crucial aspect, namely, they cannot successfully reproduce and survive as a species without humans aiding and abetting them.
3. By altering the weather and the atmosphere in a way which amounts to an oblique, though not a direct, intention to do so, humans, no doubt, will affect the extant distribution of fauna and flora in the world. Some will die off and others would not prosper. Yet others have to migrate in order to survive the climatic disturbances. Of the latter, like the animals in the park, their absolute independence would have been undermined by the impact of human actions. But this sense of nature's independence, as we have seen, is equated with nature$_p$. Pristine nature is compromised in varying degrees by the type and scale of human impact—in the extreme case, where the impact is global and near total, nature's independence is more or less destroyed, and nature$_p$ (on planet Earth, at least), by and large, is hard to locate. This sense involved in McKibben's first death of nature is borne out by the quotation from p. 58 of his book cited earlier on. On that account, any human impact on nature,

whether intentional or not, is said to destroy nature's independence. (Other commentators also read him in that way—for instance, Borgmann writes: "To begin with, nature as McKibben and many environmentalists think of it in its healthy condition is characterized by its independence. It is unaffected by humans."[28]

4. But there is a more fundamental sense of nature's independence which is not captured by McKibben's account. This sense has to do with nature's origin, its maintenance, its continued existence, and its further evolution being **in principle** entirely independent of humans.[29] A thought experiment would make this point clear. One can unproblematically entertain, both on the conceptual and empirical levels, the possibility of Earth—that is, nature$_{nh}$—without humans, whereas one cannot entertain the possibility of humans existing in the absence of nature$_{nh}$. This shows that nature is independent and autonomous of humans in a deep sense. (However, 'autonomy' here is not used, of course, in the sense of Kantian autonomy, according to which, humans with freewill choose and legislate as norms of moral conduct whose maxims can be universalized without contradiction. Instead, it refers to nature's ability to get on with its own 'business' or trajectory, so to speak, if it were left alone by humans or in the total absence of humans.[30]) It is this sense of nature's independence which could be said to be involved in McKibben's second death of nature. Indeed, as the chapters following will show, the success of biotechnology in conjunction with microcomputer technology and molecular nanotechnology (admittedly, the latter two technologies have not been commented on by McKibben), **in principle**, would amount to the ontological elimination of nature$_{nk}$ (as well as nature$_{fa}$) and in practice may result in replacing natural kinds—biotic and abiotic—with artefactual kinds.

The Garden

The notion of the garden will next be explored in order to reinforce the analysis above about the relationship among the independence of nature, the type of human intervention and the degrees of artefacticity.

For the purpose of this discussion and to simplify matters, the notion of the garden will be narrowly confined to that of the ornamental garden rather than more broadly to include gardens and preserves that serve both aesthetic and material purposes. It is irrelevant to the present concern whether gardens (such as European medieval or indeed even neolithic ones) historically were intended to be functional rather than aesthetic or both.[31] Herbs, after all, can be nutritionally, culinarily and medicinally useful as well as attractive as plants in their own rights. Needless to add, except for royality which could afford to put aside land entirely for pleasure, most people, most of the time in human history, could not afford to do so anyway. But this would not necessarily prevent them from assembling the contents of their garden in such a way as also to satisfy their sense of aesthetics. But whether solely as a functional or an aesthetic conception, a garden is an artefact. It is a deliberate assemblage of certain biotic and abiotic elements within a certain designated, defined and delimited space to satisfy utilitarian and/or aesthetic needs.

From the perspective of this book, the (ornamental) garden may be categorized in terms of two dimensions—the space delineated and the flora (primarily) found in them to produce an ascending order of artefacticity. However, these categories are meant to be analytic, not historic, although what they refer to may bear some recognizable resemblance to actual gardens or types of gardens which have existed or still exist. The following order may be spelled out.

First, for what may be called the 'wild' garden, human intervention is minimal. For a start, the gardener—the efficient cause—may have chosen a site with natural boundaries, such as a stream at the bottom and a hillock at the top. Within that space, trees already in existence would be left alone. Additional ones may be planted both for aesthetic and ecological purposes—in the case of the latter, soil permitting, for instance, a buddleia or two may be planted to attract butterflies. Whatever plants happen to seed themselves in the garden would not necessarily be rooted out as weeds—tiny, wild pansies may wedge themselves in the cracks of a rock or an evening primose may suddenly appear. Moreover, only wild varieties of plants rather than the intensively artificially selected ones would be chosen. There will be no manicured lawns. Grasses would grow which may or may not be seasonally cut or kept down by a sheep or two. Such a garden is indeed an artefact. But it displays minimal artefacticity from the standpoint of extrinsic/imposed teleology as the organisms, in the main, are left to get on with their lives and to unfold their own *tele*. Yet, for some people, such a garden could be aesthetically very pleasing and satisfying.

Next is what may be called the 'Capability Brown' garden.[32] Here the site itself is carefully designed and created—a stream diverted, a lake built, a cascade engineered, a vista manufactured. Bowers are erected, statues put in place, pavilions, or follies even, constructed. Plants may be segregated and not allowed to mix for effect. For instance, a rose garden cannot tolerate non-roses within its space; indeed, a white rose garden cannot tolerate non-white roses. Unlike the 'wild' garden, the plants chosen would be artificially selected varieties for their display value—for instance, ornamental flowering cherries would be planted rather than other less showy wild varieties.

Third is what may be called the formal garden as exemplified in Versailles. An unnatural geometric space is structured—rectangles, squares, circles, quadrants are imposed as a grid. Mathematical regularity is the desideratum, the opposite of the 'Capability Brown' landscape which confines itself, on the whole, to exploiting natural contours or creating them where none exists. The one celebrates Cartesian, geometric regularities, the other, the more fluid, irregular shapes, boundaries and contours found in nature. For instance, it is difficult to imagine a formal garden being constructed on ground which is not more or less flat—rectangles could not readily be imposed upon undulating ground and what is not sufficiently flat would then have to be flattened. Topiary is a 'natural' aspect of such a formal garden—plants are shaped into geometric figures, birds, vases, heraldic patterns even. For its heightened effect to be achieved, everything has to be in place, kept in place, neatened and straightened with boundaries scrupulously defined and maintained. It would be a mistake to think that such a garden must be on a grand, public scale. On the contrary, its spirit permeates many a humble suburban plot (a few even boast of topiary) with everything being neat and straight as the order of the day. As nature is not neat, straight and tidy, this kind of garden requires a good deal more intervention, both at the stage of its creation and later in terms of maintenance, than even the type called the 'Capability Brown' garden.

Finally is what may be called the *bonsai* garden. Here is an attempt to create in miniature, in one's own living room, so to speak, the garden that is outdoors. It is, therefore, an artefact which is a copy of another artefact. Within the confines of space provided by a container, perhaps no bigger than that of a large serving plate, miniature pagodas, hills, bridges, rocks, streams, lakes, etc., are assembled. The plants, too, naturally, are miniatures. The art of creating such miniature plants has a long history. Its techniques have successfully resulted in distorting the *tele* of the trees. They have been deliberately dwarfed and stunted in order to satisfy the human aesthetic sense—they are the analogues of the Chihuahua in the animal kingdom, though infinitely smaller and, therefore, considered to be even more exquisite than the Chihuahua. One admires the ingenuity of this art but it remains true that the end products are the clearest expression of an extrinsic imposed teleology, displaying a very high level of artefacticity. Such plants require the maximum degree of human intervention in their creation and maintenance. In terms of formal, final and efficient causes, their own have been to a great extent eliminated and replaced by those of their human creators. In this sense, they may be said to exemplify an even greater degree of artefacticity than genetically engineered organisms. Some of these latter could conceivably survive without further human intervention and could successfully reproduce themselves without further human help, whereas *bonsai* trees cannot. Any organism which has been deprived of its reproductive powers and success may be said to be ontologically different from other naturally-occurring organisms and, therefore, lack independence and autonomy in a radical sense. We have already seen that camels fall into this category.

From the above categorization, one can see that the greater the degree of direct human intervention, the greater the degree of artefacticity, and the greater the degree in the erosion of the organism's or ecosystem's independence of humans.

Biotic Artefacts and Their Residual *Tele*

This section will raise the issue regarding the residual *tele* in biotic artefacts, even in those possessing a very high degree of artefacticity, and the implications of this residue for the notions of independence and the transformation of the biotic to become the artefactual.

We saw in the last chapter that non-biotic artefacts are straightforward human creations. In some instances, like synthetic rubber or plastic, even their material cause is a product of science and its technology. But in all instances, the material, whether existing as a natural kind or synthetically derived from other natural kinds, is constructed and executed according to human design and purpose.[33] As artefacts, they have meaning and significance alone for humans. When an archaeologist digs up an object and then identifies it as a hairpin, that object, as an artefact, is only of interest to us humans. To the organisms living in the soil which encrusts it, it is irrelevant that it is a hairpin. It follows, then, should human beings happen to become extinct in the world, all such artefacts of theirs would cease to exist as artefacts, and would simply be bits of wood which termites would break up eventually to become soil, or lumps of stone which the wind, the rain and the ice would eventually weather down, too, to become soil. All this would inevitably happen unless humans are there continuously to maintain their artefacts. And even with the utmost care bestowed on their most cherished artefacts, these, too, would eventually decline and decay. When they finally fall to pieces, their cherishers would mourn their demise.

Let us adapt the Last Person Argument—see Routley, "Is There a Need?"—and apply it to this context. Imagine that you are the last surviving person on Earth (and no equivalent type of consciousness would ever appear in the future). Just before you die, you have the technology to destroy all the non-biotic artefacts in existence. Ought you do so? Is there any moral or rational constraint upon you to desist? The answer to these questions appears to be no, for the simple reason already alluded to, that artefacts have meaning and significance alone for humans, who are their creators. Furthermore, as we have also seen, without maintenance by humans, they would just simply decay and be transformed by non-human agents into something else. Destroying them instantly in the proposed scenario makes no relevant difference to the outcome. In other words, as their identities are constructed, sustained and upheld by humans, they totally lack autonomy and independence of their creators. It is entirely up to their creators to do whatever they please with them. The last surviving person could, on a whim or out of sentimentality, spare them instant destruction; but whim apart, there is no good reason, whether moral, prudential or aesthetic, not to press the 'destroy' button, timed to detonate once s/he had expired. A world without humans is a world without (human) artefacts in any case.

Would the same hold true of biotic artefacts? Probably not. Take genetically engineered or modified organisms (GMOs). As already argued, they possess an extremely high degree of artefacticity, but, nevertheless, they appear to possess a residual *telos* which is independent of that imposed by their human creators. Of those which can survive in the absence of human maintenance and, moreover, can successfully reproduce in the wild once humans disappear from the scene, it would be less plausible to claim that there is no moral or rational constraint upon their last surviving creator(s) to desist from pushing the 'destroy' button. Their capabilities for self-maintenance, self-renewal and self-reproduction render them morally relevantly different from their non-biotic counterparts. It is true that when humans have vanished, the human purpose which they serve would also have vanished—the pig, genetically engineered to be lean so that the consumption of its meat would not cause humans to have a high cholesterol level, would no longer and could no longer fulfill that function and purpose designed for it by humans. But this does not mean that such a type of pig could not assert its residual *telos* when its creators are gone. It would eventually stop being a human artefact and strike out on a new career as pig *simpliciter*, albeit one whose historical origin and status were those of an human artefact. In the wild, the natural processes of selection and evolution would, over time, subdue, if not totally eliminate, such a history in its progeny, reverting it to the status of being naturally-occurring entities.

Notes

1. Such a role played by the combination of extant and future technologies will be examined in greater detail in the next chapter.

2. For slightly different overlapping definitions, see Frederick Ferré, *Philosophy of Technology* (Englewood Cliffs, New Jersey: Prentice Hall, 1988), 26-29. For the emergence of the notion 'nature$_c$' in ancient Greek thought, see G. E. R. Lloyd, "Greek Antiquity: The Invention of Nature," in *The Concept of Nature: The Herbert Spencer Lectures*, John Torrance, ed. (Oxford: Clarendon Press, 1992).

3. J. Baird Callicott ("The Role of Technology in the Evolving Concept of Nature," in *Ethics and Environmental Policy: Theory Meets Practice*, Frederick Ferré and Peter Hartel, eds. Athens, Georgia: The University of Georgia Press, 1994, 64-67) endorses this definition. His reasons for doing so may be reconstructed as follows: (a) Darwin has shown that humans are animals and that all animals are part of nature—in this way, Darwinism 'naturalizes' humans and 'humanizes' animals; (b) the dualism—human/nature—is part of the modern worldview which postmodernism has outgrown; (c) empirical/historical evidence shows that humans had/have penetrated all parts of the planet Earth with the possible exception of Antartica.

The standpoint adopted by this book does not challenge a. However, it does not endorse the human/nature dualism when dualism is understood as radical exclusion or hyperseparation in the way that Val Plumwood (*Feminism and the Mastery of Nature* London: Routledge, 1993, 41-68) explicitly and, perhaps, Callicott implicitly do. (On the rejection of dualism by both Plumwood and this author, see Chapter 5.) It simply argues that it is not helpful to elide the distinction between nature on the one hand and human culture on the other, when the distinction is not understood dualistically. To blur the distinction would precisely make it impossible for one to mount a critique of certain cultural practices which have the effect of destroying or undermining nature. If so, there is a need to identify and to defend another sense of 'the natural' other than 'natural$_c$'—this is the sense identified as 'natural$_{fa}$.' (See text which follows.) However, it is not the claim of this book that 'natural$_{fa}$' should never be transformed into the artefactual, only that we, humans, ought not systematically, whenever it is technologically possible, to transform the former to become the latter, and that we should be sensitive to the varying levels of artefacticity which our technologies permit us to achieve.

As for the third reason attributed to Callicott above, it implies yet another sense of 'nature' which the present discussion goes on to identify as 'nature$_p$,' that is to say, 'pristine nature,' nature that has endured no human presence or impact whatsoever in its history—see text which follows.

The sense of 'natural$_c$' seems also to be endorsed by James Lovelock as well as by Dorian Sagan and Lynn Margulis. Lovelock says: "If by pollution we mean the dumping of waste matter there is indeed ample evidence that pollution is as natural to Gaia as is breathing to ourselves and most other animals" (James Lovelock, *Gaia: A New Look at Life on Earth* Oxford and New York: Oxford University Press, 1982, 95). Sagan and Margulis write that

> there is no clear division between the technological and the biological. In the end all technological toxins are natural, biological by-products which, though via human beings, are elements in the Gaian system. Similarly, legislation and lobbying attempts, such as the recent furor in the United States over the mismanagement of the Environmental Protection Agency, are nothing more or less than part of Gaian feed-back cycles." (Dorian Sagan and Lynn Margulis, "Gaia and Philosophy," in *On Nature*, Leroy S. Rouner, ed. Notre Dame, Indiana: University of Notre Dame Press, 1984, 72)

[For a further critique of this view, see Chapter 5 section entitled **Anthropogenic and Nonanthropogenic**.]

4. For some recent attempts to distinguish between different senses of 'Nature,' see Charles Frankel, "The Rights of Nature," in *When Values Conflict: Essays on Environmental Analysis, Discourse, and Decision*, Laurence H. Tribe, Corinne S. Schelling and John Voss, eds. (Cambridge, Massachusetts: Ballinger Publishing Company, 1976), 93-113; Evandro Agazzi, "Nature and the Natural: Some Philosophical Reflections," in *The Concept of Nature in Science and Theology (Part 1)*, Niels H. Gregersen *et al.*, eds. (Geneva: *Labor Et Fides*, 1995), 3-19. See also Kate Soper's, *What Is Nature? Culture, Politics and the Non-Human* (Oxford, UK, and Cambridge, USA: Blackwell, 1995). Her aims are very different from those pursued by this book. In the Introduction, she indicates that the scope of her book is narrower than the title might suggest, as she is primarily concerned with "the politics of nature." This book would fall firmly into what she calls the "nature-endorsing" rather than the "nature-skeptical" camp.

She identifies three senses of the term 'nature,' the "metaphysical," the "realist" and the "lay"—see *What Is Nature?*, 155-56. The first seems to correspond roughly to the human/nonhuman, or the culture/nature divide which as argued in Chapter 5 ought to be understood not as ontological dualism but as ontological dyadism. 'Nature' as a realist concept turns out to underpin the metaphysics of Scientific Naturalism. But the structures, processes and causal powers she refers to under this realist concept of nature also come under the notion of nature as 'the ontological other'; this book assumes that these structures, processes and causal powers have operated and will continue to operate independently of human intervention or design—as part of nature$_{fa}$. The "lay" or "surface" sense is where Soper primarily locates "the politics of nature."

Appendix 2 (**Is Nature a Mere Social Construct?**) examines and attempts in outline to meet some of the arguments put forward by what Soper calls the "nature-skeptical" camp, primarily those theorists who maintain that nature is a socially constructed concept. Soper herself is on the whole critical of, though not entirely unsympathetic to, the thesis, but she wishes to defend 'nature' as a realist concept.

5. Soper (*What Is Nature?*, 21-25) also talks about cosmological nature; however, her sense of "cosmological" is not the same as the one used here.

6. In spite of admitting this, one should not be blind to the fact that Earth is not merely its surface. In this sense, large parts of Earth are still pristine, such as the planet in its entirety just a few meters beneath its crust.

7. The project concerns attempts to create elements in the Periodic Table beyond those naturally-occurring on Earth which have all been found, from the lightest, hydrogen, to the heaviest, uranium. In the Periodic Table, elements are classified according to their atomic number, that is, the number of protons as well as neutrons which stabilize the nucleus. For instance, hydrogen has one proton while uranium has 92 protons. Plutonium with 94 protons, as we know, was first manufactured in 1941. But physicists, up to 1970, had thought that the "superheavy" elements, between 96 and 105 onward, would not be stable, and would probably last only a fraction of a second. Element 108 would, on their calculations, last only a billionth of a second. But the situation has changed since. Theorists have now calculated that elements, around the atomic number 114, could be much more stable (indeed lasting for billion of years), and in the 1980s have made elements 107, 108 and 109 one atom at a time. At the end of 1994, German physicists at the GSI heavy ion research laboratory in Darmstadt succeeded in manufacturing elements 110 and 111. The leader of the team says that they are now hoping to make element 112. The present calculations of physicists show that it is possible to carry on until they reach element 116, which they regard as the true end of the Periodic Table—around element 118, the decay rates become really too fast to be stable at all. Yet until 1970, they thought that the Periodic Table would peter off at about element 108.

The artefactual elements are constructed by combining nuclei of lighter elements, with physicists carefully calculating the right number of protons and neutrons to use to build the new atom. A beam of atoms of one of the elements used is aimed at a thin foil made of the other element in a vacuum. In this process, when a nucleus belonging to the beam collides with a nucleus in the target material, they fuse and a new compound nucleus is formed. Element 110 was created by firing a beam of nickel atoms at a lead foil. The German physicists at Darmstadt are hoping to make element 112 using a zinc beam—see Nina Hall, "Heavy Metal Mystery," *Observer*, 12 February 1995, Life Section: 67. For the latest update on this research front, see Peter Armbruster and Fritz Peter Hessberger, "Making New Elements," *Scientific American* (September 1998): 50-55.

The above shows that strictly speaking, in the history of science and technology, the ability to create non naturally-occurring atoms has taken place earlier in time than a technology, such as molecular nanotechnology, which claims to be able (so far, by and large, in principle only) to construct instances of second matter, that is to say, artefactual kinds to replace natural kinds by arranging atoms in certain ways. In that sense, the creation and manufacture of plutonium in 1941 and since then, other 'superheavy' elements, have already gone beyond the programmatic pronouncement of modern science from its inception in the seventeenth century, to manipulate and control matter (that is second matter or nature$_{nk}$) atom by atom. Nuclear physics is even deeper and more radical than that, as it sees fit to create and manufacture elements in the Periodic Table which are not found naturally-occurring (at least) on Earth.

However, nuclear physicists apart, there are other laborers in the vineyard, so to speak, whose work has now shown that, by comparison, molecular nanotechnology of the near future could even be said to be 'conservative.' Raymond Ashoori, a physicist, and his co-researchers ("Single-Electron Capacitance Spectroscopy of Discrete Quantum Levels," Physical Review Letters 68 18 May 1992: 3088-91; "Single-Electron Capacitance Spectroscopy of a New Electron Box," Physica B 189 1993: 117-24 and "Energy Levels of an Artificial Atom Probed with Single-Electron Capacitance Spectroscopy," *Surface Science* 305 1994: 558-65) could be said to have created an "artificial atom" which establishes in principle that the electron count of such atoms, from 0 to 60, is perfectly manipulable by their human makers. According to one account:

> The "atom" in question was actually an empty space within a gallium arsenide crystal to which electrons could be moved one at a time by the application of a light magnetic pulse. In the case of an ordinary, garden-variety atom, electrons are held in space by the nucleus, whose positive charge attracts and binds the negatively charged electrons. In the "artificial atom," by contrast, electrons were held, instead, by an externally imposed magnetic field. But the final effect was much the same: a bunch of electrons whizzing around in a small space. (Regis, *Nano!*, 276-77)

In this way, atoms may be customized to be of any size and shape—long, thin ones or big, round ones. The next stage would be to try to stick these created atoms together to form artefactual molecules, and these, in turn, to form artefactual solids, liquids or gases. (See also Michael Brooks, "Take One Quantum Dot," *New Scientist* 29 August 1998: 22-25.)

8. Callicott is correct in pointing out that Mill's essay (originally published in 1874) "provides a breathtaking example by a Victorian philosopher of lasting repute of an express injunction to thoroughly conquer Nature" ("The Role of Technology," 60-61). Mill could be said to be guilty of understanding the human/nature$_{nh}$ distinction as dualism or hyperseparation. But this book is merely using the distinction (not to be understood as dualism but as dyadism—see Chapter 5 for further clarification) between 'the natural$_{fa}$' and 'the artefactual' to make quite a different point, namely, that the artefactual should not be allowed systematically to imperialize at the expense of nature$_{nh}$. It is harmless, therefore, to point out that the different senses of the term 'nature' identified here are roughly the same as those of Mill, provided it is clear that the rest of his philosophical and ideological baggage is not endorsed by the main theme pursued in this book.

9. John Stuart Mill, "Nature," in *John Stuart Mill: Autobiography and Other Writings*, Jack Stillinger, ed. (Boston: Houghton Mifflin Company, 1969), 316.

10. However, what has not been brought out by Lewis is the ontological contrast between the natural and the artefactual which this books seeks to focus on—in particular, Chapter 5 explores this ontological dyadism in greater detail.

11. C. S. Lewis, *Studies in Words* (Cambridge: Cambridge University Press, 1967), 45-46.

12. Mill, "Nature," 316.

13. Bill McKibben, *The End of Nature* (New York: Random House, 1989). Callicott maintains that McKibben is mourning that "the idea of nature has ended, not nature per se" ("The Role of Technology," 59-60). But Callicott's attribution to him of such a view would introduce a degree of confusion into his position which is not already there. As the argumentation of this section will make clear, McKibben may be somewhat confused by

not explicitly distinguishing the sense of nature as pristine nature or 'nature$_p$' from another sense as 'nature$_{fa}$' as identified in the last chapter, and by identifying nature's independence as a value with nature$_p$. But this confusion apart, McKibben is obviously bemoaning the passing of both nature$_p$ and nature$_{fa}$. Contrary to Callicott, he cannot be construed as simply mourning the death of the idea of nature and not also of nature as nature$_{nh}$. Callicott says that McKibben's idea of nature is really nature as 'the Other.' However, nature$_p$ and nature$_{fa}$ are aspects of nature$_{nh}$ and all three notions involve a conception of nature as 'the Other.' So irrespective of what he himself might have said or what Callicott attributes to him, McKibben must be lamenting the loss of nature under all these senses and, therefore, of the idea of nature as 'the Other.' Without the latter, he could not begin, conceptually speaking, to lament the passing away of nature$_{nh}$, nature$_{fa}$, or nature$_p$. But in lamenting the demise of nature in these senses, he is also at the same time lamenting the demise of the idea of nature as 'the Other.'

Callicott himself—see Callicott, "The Wilderness Idea Revisited: The Sustainable Development Alternative," *Environmental Professional* 13 (1991): 235-47; "La Nature Est Morte, Vive La Nature!" *Hastings Center Report* 22 (1992): 16-23; "The Role of Technology"—appears now no longer keen to uphold the idea of nature as 'the Other' from his own postmodernist standpoint, although he is keen to condemn changes which may be said not to be 'ecologically salubrious' ("The Role of Technology," 68). He is against Euro-Brazilian but in favor of Kayapo slash-and-burn agriculture. He is against Chicago but in favor of the Great Barrier Reef. All these phenomena are equally 'natural$_c$' (in the terminology of this book) but those he is against are simply those which are ecologically insalubrious. However, the examples cited in "The Role of Technology"—an admittedly short article—do not include a single instance of nonanthropogenic ecological in-salubriousness. This may lead one to suspect that, for Callicott, only the anthropogenic variety may involve such insalubriousness. But if anthropogenic changes alone could (on occasion) produce ecological unwholesomeness, and either nonanthropogenic ones never do or if they do, this is not a justification for intervention, then it may look as if Callicott, in spite of his own avowal, has not really dispensed with the idea of nature as 'the Other' either.

Bryan Norton ("Where Do We Go From Here?" in *Ethics and Environmental Policy: Theory Meets Practice*, Frederick Ferré and Peter Hartel, eds. Athens, Georgia: The University of Georgia Press, 1994, 235-41), in the same volume in which Callicott has written, also seems to imply a position similar to that of Callicott. Norton cited the recent Washington State Olympic National Park mountain goat controversy (as reported by V. B. Scheffer in "The Olympic Goat Controversy: A Perspective," *Conservation Biology* 7, 1993: 196-99). The National Park Service wanted to remove—through live-capture and shooting, if necessary—goats which were first introduced into the park in the 1920s and are now causing ecological havoc. The Fund for Animals (an animal liberationist group) protested, arguing that the goats are an indigenous species, not an exotic, and should be allowed to remain. Norton makes the point that both sides, contrary to what certain environmental philosophers maintain, are not standing at opposite ends with unbridgeable abstract, high-level moral principles between them, such as biocentrism or ecocentrism on the one hand and mammalcentrism on the other. Instead, both parties agree that the controversy turns upon an ecological fact, that is, the historical existence of mountain goats on the Olympic peninsula. From this observation, Norton appears to draw the same conclusion that Callicott does, namely, nonanthropogenic changes either do not cause ecological insalubriousness or if they do, one should tolerate them. (Norton may, no doubt, protest that it is precisely such unhelpful comments which render environmental philosophers irritating to policy and decision makers.) Behind either ecological fact—that the goats are indigenous or that they are exotic—stand respectively a value and a disvalue. In the former, the opposing parties are agreed that nonanthropogenic-induced ecological degradation is acceptable, and in the latter that anthropogenic-induced ecological degradation is not. This then would raise the key question: Why tolerate nonanthropogenic insalubriousness if such exists? Surely, the answer would involve a reference ultimately to nature as 'nature$_p$' or 'nature$_{fa}$,' either of which presupposes the idea of nature as 'the Other' (though not as the 'dualized other').

(For a defense of the culture/nature dichotomy, see Holmes Rolston, "Wilderness Idea Reaffirmed," *Environmental Professional* 13 1991: 370-77, and Michael Soulé, "The Social Siege of Nature," in *Reinventing Nature? Response to Postmodern Deconstruction*, Michael E. Soulé and Gary Lease, eds. Washington D. C. and California: Island Press, 1995. See also Chapter 5 section entitled **Anthropogenic and Nonanthropogenic** for a discussion of the philosophical significance of the distinction.)

14. The view taken here agrees with that of Hilpinen, whose definition of the term 'artefact' has been cited in Chapter 2.

15. McKibben is rightly not so worried about local littering and superficial (in relative terms) disfiguring of landscapes. (He refers to tin cans in clear streams—see text further on.) These clearly render nature less than pristine but their impact is less alarming than other types of impact which he focuses on instead.

16. No place on Earth could be said to be totally free from human impact either in the present or in the past. Nature$_p$ *simpliciter* could not be said to be all that relevant. What is relevant, however, is the degree and type of human impact on nature$_p$. Pre-Columbian America, for instance, appears to the Eurocentric mind to qualify as nature$_p$ in an absolute sense but in reality does not. The earlier Siberian immigrants eradicated quite a few species of megafauna including two species of elephant and, through their use of fire, also transformed the flora of the New World—see H. E. Wright and D. G. Frey, eds., *The Quaternary of the United States* (Princeton: Princeton University Press, 1965). Yet after the extinction of the Pleistocene megafauna, their descendants in Pre-Columbian days had long evolved more ecologically sensitive and sustainable lifestyles. They did not dominate the environment in the way that so much of the world today may be said to be human-dominated. As a result, the nature which confronted the Europeans upon their arrival in the New World readily appeared to them as nature$_p$ *simpliciter*. As far as Earth is concerned, it is true there may not be much of such nature$_p$ left. Of late, even the oceans which cover roughly four-fifths of Earth's surface have been polluted. (Earth's core beneath a few meters of its crust, however, is most certainly pristine.) It is helpful, therefore, to distinguish between the types and levels of human intervention and impact on the landscape.

The survey of human impact on natural ecosystems by Hannah *et al.* does not include the historical dimension. It uses the following three-fold category to indicate different levels of anthropogenic disturbance which one can identify today:

1. Undisturbed
Record of primary vegetation; Very low human population (under 10 persons per km^2 or under 1 person per km^2 in arid/semiarid and tundra communities).
2. Partially Disturbed
Record of shifting or extensive agriculture;
Record of secondary but naturally regenerating vegetation; Livestock density over

carrying capacity; Other evidence of human disturbance (e.g., logging concessions); Insufficient evidence to place in category 1 or 3.
3. Human Dominated
Record of permanent agriculture or urban settlement;
Primary vegetation removed; Current vegetation differs from potential vegetation; Record of desertification or other permanent degradation. (Lee Hannah *et al.*, "A Preliminary Inventory of Human Disturbance of World Ecosystems," *Ambio* 23 1993: 247)

In terms of these and other criteria, the survey finds 51.9% of Earth's terrestrial area—90 million km^2—to fall under Category 1; 24.2%—41 million km^2—under Category 2; and 23.9%—just over 40 million km^2—under Category 3. When these figures have been adjusted to remove rock, ice and barren land which support no human life and not many other life forms, then Category 1 covers 27.0%, Category 2, 36.7% and Category 3, 36.3% (p. 248).

So much for Earth, at the moment. But what about other planets and moons in our solar system? Our moon certainly now bears human footprints. There are also human artefacts—debris from space exploration—floating about in space. However, one could say that astronomical bodies up to now are pristine—$nature_p$ does obtain out there. But for how long though? Furthermore, should humans, given their technological capabilities, transform that nature which is both $nature_p$ and $nature_{fa}$ into the artefactual?

17. Ethynylestradiol in contraceptive pills has a similar effect. Excreted from the body in the urine which then is flushed away, it eventually finds its way to rivers and coastal waters, to be absorbed by marine organisms—see Soulé, "The Social Siege of Nature," 144, 164.

18. See *Report to the Intergovernmental Panel on Climatic Change* (Bracknell, UK: Meteorological Office, 1990); Jeremy K. Leggett, ed. *Global Warming: The Greenpeace Report* (Oxford: Oxford University Press, 1990).

19. Bill McKibben, *The End of Nature*, 58.

20. For instance, it is possible to genetically engineer plants so that they can tolerate better the new climatic changes.

21. McKibben, *The End of Nature*, 166.

22. However, the loss of $nature_p$ should not be interpreted simplistically to entail the conclusion that $nature_{nh}$ no longer exists. Such an unjustified conclusion could lead to other equally unjustified conclusions, namely, either we regard $nature_{nh}$ as having already been transformed into artefacts (such that $nature_{fa}$ also no longer exists), or we may systematically transform $nature_{nh}$ into artefacts, or finally, there is no $nature_{nh}$ or $nature_{fa}$ to be saved. (See Soulé, "The Social Siege of Nature," who makes similar points though not exactly the same distinctions between the various senses of 'nature.')

23. For a brief account, see E. O. Wilson, *The Diversity of Life* (London: Penguin Books, 1992), 22-30. Also see Chapter 5 section entitled **Anthropogenic and Non-anthropogenic** for further discussion.

24. Stephen Jay Gould says that as far as biotic nature is concerned

> the most salient feature of life has been the stability of its bacterial mode from the beginning of the fossil record until today and, with little doubt, into all future time so long as the earth endures. This is truly the 'age of bacteria—as it was in the beginning, is now and ever shall be.' These prokaryotic organisms have

> survived innumerable major natural upheavals in the last 3.5 billion years. ("The Evolution of Life on the Earth," *Scientific American* 271 1994: 65)

He continues:

> They are adaptable, indestructible and astoundingly diverse. We cannot even imagine how anthropogenic intervention might threaten their extinction, although we worry about our impact on nearly every other form of life. The number of *Escherichia coli* cells in the gut of each human being exceeds the number of humans that has ever lived on this planet. ("The Evolution of Life on the Earth," 65)

[For a recent account of the existence of micro-organisms and their tenacity to survive under the most seemingly hostile and, therefore, unlikely conditions, see Stephanie Pain, "The Intraterrestrials," *New Scientist* (7 March 1998): 28-32. Such survival and existence in turn prompt the speculation that there might have been or there might even be life after all in other planets, such as Mars buried in the Martian crust.]

However, biologists, ecologists and others are more specifically engaged with trying to save or protect whatever biodiversity there exists today, and not merely with living nature in the bacterial mode. From the perspective of this book, the loss of biodiversity is the loss of $nature_{nh}$ as $nature_{nk}$ (at least of biotic natural kinds), as well as the undermining of evolutionary processes themselves to such an extent as to destroy $nature_{nh}$'s capability to evolve new biotic natural kinds.

25. See Paul Birch, "Terraforming Mars Quickly," *Journal of the British Interplanetary Society* 45 (1992): 331-40, "Terraforming Venus Quickly," *Journal of the British Interplanetary Society* 44 (1991): 157-67, and "The Customised Cosmos," *Guardian*, 20 May 1993, 12; *Journal of the British Interplanetary Society*, 1990, 1991 and 1992; James Lovelock, *The Ages of Gaia* (Oxford: Oxford University Press, 1991); Christopher P. McKay *et al.*, "Making Mars Habitable," *Nature* 352 (8 August 1991): 489-96; Robert McKie, "Russia and America Plan Joint 2001 Mission to Drill for Life Forms on Mars: Roving Robot May Solve Mystery of Thirsty Planet," *Observer*, 24 February 1994, 3 ; *Mars Alive* published by the BBC; Tim Radford, "A Clean Pair of Wheels," *Guardian*, 28 November 1996, Online Section: 5; Robert Zubrin, *The Case for Mars: The Plan to Settle the Red Planet and Why We Must* (New York: Simon and Schuster, 1996). On the science fiction front see Michael Allaby and James Lovelock, *The Greening of Mars* (London: Andre Deutsch, 1984); Kim Stanley Robinson's trilogy—*Red Mars* (New York: Bantam Books, 1993), *Green Mars* (New York: Bantam Books, 1994) and *Blue Mars* (New York: Bantam Books, 1996).

The terraformation of the moon (Earth's moon) has recently—December 1996—cropped up all of a sudden. The scientists working in the Pentagon's Star Wars program announced that the data which their Clementine spacecraft mission to the moon had brought back two years ago now reveal a lake of frozen water in a deep crater near the lunar south pole. Anthony Cook, an astronomer at the Griffith Observatory in Los Angeles, was quoted as saying: "With water there you could have enclosed areas to grow plants, grow your own food, make your own fuel, make your own air." See Tim Radford, "Star Wars Scientists Find Water in Crater on the Moon," *Guardian*, 4 December 1996, 1. But Donald Campbell and colleagues at Cornell University—writing in *Science* 276 (1997): 1527—have challenged this possibility.

26. Anthony C. Fisher and Frederick M. Peterson, "The Environment in Economics: a Survey," *Journal of Economic Literature* 14 (March 1976): 1.

27. This criticism assumes that Fisher and Peterson are referring to the natural environment.

28. Albert Borgmann, "The Nature of Reality and the Reality of Nature," in *Reinventing Nature? Responses to Postmodern Deconstruction*, Soulé and Lease, eds., 35.

29. This sense of independence will be further explored in Chapter 5.

30. The notion of trajectory will be further explored in Chapter 5.

31. See on the former, Alicia Amherst, *A History of Gardening in England* (London: B. Quaritch, 1896), Oliver Rackham, *Trees and Woodlands in the British Landscape: The Complete History of Britain's Trees, Woods and Hedgerows* (London: Weidenfeld and Nicolson, 1995), and on the latter, Mumford, *The Myth of the Machine*.

Mumford writes:

> Within the neolithic garden, if one follows (Edgar) Anderson's shrewd reconstruction, foods, condiments, aromatics, medicines, useful fibres, flowering plants admired for their color, perfumes, or beauty of form or their place in religious ritual, all grew side by side, sometimes, like the nasturtium, equally valued for salad greens as for decoration. Observe the variety and the lack of specialization, along with the concern for quality rather than quantity: and it is not perhaps an accident that some of the most useful plants to be cultivated were those that may have been prized at first for their brightly colored flowers, like the mustard, the squash, the pepper tree, the broad bean, the wild pea or even for their perfume, like most of the aromatics. (*The Myth of the Machine*, 144)

32. In choosing such a term, there is no intention here of taking over all the usual connotations and implications of that genre of landscaping initiated and associated with the eighteenth century landscape architect, commonly referred to as Capability Brown.

33. For an interesting account of the different types of materials used in current design and manufacture as well as the processes which go into the shaping of the materials themselves, see Ezio Manzini, *The Material of Invention* (Cambridge, Massachusetts: The MIT Press, 1989). For materials in the coming century, see Philip Ball, *Made to Measure: New Materials for the 21st Century* (Princeton: Princeton University Press, 1998).

Chapter Four

Technology: Threats to the Natural

This chapter will consider (i) whether technology, both extant and of the near future, poses a threat to the natural, and if it does, (ii) what the threat really amounts to. It will argue that the most serious threat comes from the ability of such technology to transform the natural to become the artefactual, thereby rendering the natural ontologically as well as empirically redundant. The radical character of the threat becomes obvious when one looks at the implications, particularly, of technologies of the near future, like terraformation and molecular nanotechnology, for environmental philosophy. But before establishing such a claim, the chapter will first look at the less radical threat posed to the natural from extant technology.

Extant Technology and the Less Radical Threat to the Natural

Extant technology has been perceived to have had an adverse impact on the environment.[1] In brief, it may be said to have created a pollution problem if the term 'pollution problem' is understood to refer to the undermining of the integrity and functioning of the biosphere through the release of certain substances during the processes of industrial/technological production. The biosphere may be subverted in a variety of ways such as producing synthetic substances for which nature has evolved no known solvents, releasing gases like CFC which can destroy stratospheric ozone, allowing an inordinate amount of ultra-violet rays to get through into the atmosphere, damaging, among other things, phytoplankton and other plants which are the basis of complicated food chains for humans and nonhumans alike, and raising the temperature through greenhouse gases like carbon dioxide which at least in the short term—measured in geological time—upsets the equilibrium of the biosphere. Such a technology may be said to be ecologically insensitive given that the capacity of the biosphere to act as a sink in absorbing waste is a limited one. However, *contra* Barry Commoner, it is fair to argue that although extant technology plays a very important role in undermining biospheric integrity, it is not the only villain. It is one factor interlocked with others, like the pursuit of consumerism and economic growth on the one hand and population increase on the other, which together account for the observed adverse environmental impact.[2]

Three major positions may be identified in the current debate about technology and its impact upon the environment. First, technological optimism argues that technology is self-correcting. That is to say, it has the capability to put right the damage it might have inflicted upon the environment. From a historical perspective, the present predicament is a mere hiccup in the long march forward to progress. The doomsday scenario painted by the various Jeremiahs may be safely ignored.

At the other extreme is technological pessimism which maintains that the so-called capability for self-correction is no more than a pathetic faith in technological fixes. But the history of such fixes has shown that they are themselves iatrogenic—they may cure a particular environmental ill but they in turn bring forth another. For instance, sulfur-laden smoke may be removed from a particular location, such as what used to be the industrial heartland of Great Britain, by installing taller chimneys in factories. But such success in reality simply means that the nuisance is exported elsewhere, even across national boundaries, to fall as acid rain in Scandinavian countries, polluting their lakes and distressing their trees. Or the sulfur could be scrubbed out from the chimneys before the smoke is emitted so that it would not be sent down-wind to other locations. But the disposal of the pollutant as solid waste, nevertheless, remains. To solve this, some creative minds have not been slow to suggest that toxic waste in general could be exported to developing economies which would or should welcome it as a way of generating much needed foreign currency. In other words, extant technology does not genuinely solve pollution problems. It merely permits the displacement of the pollutant in question.[3] The pollutant may first exist as aerosol menacing the air, then be transformed into a solid, and if buried, would pollute not only the soil, but also eventually ground water as it dissolves or leaks, or if flushed as liquid directly into rivers and seas, would threaten the waters. The displacement may also occur across the dimension of time, not merely space—the temporary storage of nuclear waste displaces the problem on to future generations, just as the release of carbon dioxide in the burning of fossil fuel, which has sustained our extant industrial civilization since the nineteenth century, has now emerged as our problem after more than a hundred and fifty years of its first emission.

Third, technological pragmatism or realism occupies the middle ground, incorporating the valid insights of both extremes. According to this perspective, technology is a double-edged sword—it is both environment-damaging or consuming and environment-saving. While acknowledging the adverse environmental impact caused by ecologically insensitive technology, it wishes to argue, nevertheless, that there are signs that a more ecologically sensitive variety can and does arise as a response to the challenge posed by the damage which the former has caused. As the extent of the problem sinks into both political and social consciousness, more votes and more money would be directed to research to produce more environmentally friendly forms of technology.[4] To date, the advanced industrialized economies have produced a range of pollution abatement techniques which are not to be sneered at—garbage recycling, sewage treatment, soil erosion control, renewable energy projects, energy-saving devices, less reckless use of pesticides, herbicides and inorganic fertilizers in agriculture, developing a range of new materials, environmental management of parks, lakes, rivers, estuaries, coastlines, wildlife management, etc.

On this view, whether we like it or not, technology is necessarily both the cause of environmental distress as well as the solution to it.[5] Industrialization through modern technology is the only known method to increase the material standard of living. Unfortunately, it is true, the affluence is purchased at the expense of exploiting nature. But affluence at the same time makes it possible for industrialized societies to use technology to provide solutions which are capable of combating the further decline of the natural environment, not only within their own national boundaries, but also in the developing world, as the latter, too, embarks on the journey toward affluence. The double-edged nature of technology may well be the price we have to pay for indefinite material progress which, after all, has become a universal aspiration.

However, so far, one has been concentrating merely on the dichotomous nature of technology, as environment-damaging on the one hand and environment-saving on the other. But this may be an oversimplification of technological possibilities today. For it is increasingly clear that another solution lies to hand, namely, what may be called nature-replacing or substitution technology. Doeleman draws attention to this and makes the following distinction:

> [E]nvironment- or nature-saving technology counters damage to a natural environmental system by means of prevention at the source or by neutralization of potentially harmful spillovers. An environment-saving solution allows the natural system to maintain, or return to, its original state or mode of interaction. Nature-replacing or substitution technology, on the other hand, supplements or replaces a damaged function of the natural system with a man-made system. Nature-replacement technology competes with nature-saving technology. Both forms of technology are prompted in response to environment-consuming pressures resulting from the growth momentum at the macro level.[6]

As an illustration of this tendency he cites the possibility of creating an artificial, engineered environment as a solution to the air pollution problem based on the principle of air conditioning. Air conditioning has originally been designed for climate control within a relatively small confined space. But with no further technological ingenuity, it can also be used for air purification. From an air-conditioned/-purified house, one steps into an air-conditioned/-purified car to work in an air-conditioned/-purified office. To overcome the residual limitations of decentralized air-conditioning units, one could take the next logical step of enclosing whole cities within a centralized system of air quality control. This would, no doubt, have the added advantage of not only being able to deal with local air pollution problems but also with global ones as well—the destruction of stratospheric ozone might no longer appear quite so alarming for us humans, as the excessive ultra-violet rays could be filtered out so that human health would not be undermined. And as for the possible harm to phytoplankton and plants, perhaps specially genetically engineered varieties to withstand the extra dosage of radiation could be devised as substitutes. Indeed, in the long run, even human beings could be genetically engineered to withstand a far higher dose of radiation, in particular, and various toxins, in general[7]—this has the advantage of incurring cheaper costs, while at the same time retaining a greater freedom of movement than the centralized air-conditioning/-purification principle. As we have seen in the earlier chapters, humans, like other forms of living nature, are in danger of being turned into near total biotic artefacts in the presence of biotechnology as well as the potential combined presence of biotechnology, microcomputer technology and molecular nanotechnology.[8]

Doeleman believes that in terms of political and economic realities, the environment- or nature-saving technology could lose out to the environment- or nature-replacement technology for the following reasons (this account, however, is more a reconstruction than a direct summary of his position):

1. Nature-saving technology would not really be effective unless there was a dramatic and universal shift away from consumerism and exponential economic growth. But the economies in the world today, both developed and developing, on the whole, show little or no signs of repudiating the ideology of material progress and affluence.
2. Nature-saving, compared with nature-replacement technology, may be uneconomic in terms of implementation and/or opportunity costs. Take air pollution. Its sources are numerous—from the noxious substances pumped out by motor cars, to those emitted by other industrial processes of production, to a greenhouse gas, like methane, contributed by the seemingly innocuous and necessary activity of growing paddy rice particularly in the developing economies, as well as the rearing of cattle, a heavily subsidized activity in a developed economy like the European Union. To prevent air pollution demands a concerted effort on all such fronts, the costing of which becomes

a well-nigh impossibility. By contrast, the costing of implementing the nature-replacement technology for any one location or region of the world would be much more manageable, and the actual cost would also, probably, be less. Moreover, it could even be argued that the quality of the artificially maintained air would be higher than that achieved by nature-saving techniques, as the marginal cost of eliminating any residual level of pollution could be very high indeed. Furthermore, as pollution occurs in the context of legitimate activities like running businesses or pursuing leisure, measures to curb or restrain such activities would incur a high opportunity cost. And even more daunting from the political, economic and moral points of view, is the prospect of curbing such a basic subsistence activity as growing paddy rice in order to reduce, if not eliminate, the build-up of methane in the atmosphere.

3. One could argue that the above favorable comparison of nature-replacing over nature-saving technology has missed out a dimension possessed by the latter which, if taken into account, could tilt the balance in its favor, against its rival. Is it not obvious that nature-saving technology not merely preserves the bounty of nature for us humans but also nature itself? If nature's value for itself can be made to enter the benefit/cost analysis, then nature-saving technology could score over nature-replacing technology. But even if a convincing philosophical foundation can be mounted for the notion of intrinsic value in nature, in the domain of economic calculations and its presuppositions, such a value is not readily accommodated. Shadow pricing is the nearest unsatisfactory device to recognizing its validity. Economic thought has difficulty in attempting to cope with the value of nature even in anthropocentric terms, never mind with the value of nature understood non-anthropocentrically—for instance, it fails to do justice to the interests of future peoples through discounting time.
4. Nature replacement may also be self-reinforcing. When the damage to the environment is relatively limited and its recovery perceived to be technically and economically viable, nature-saving measures may have the edge in the absence of its rival. But if resources were pumped into nature-replacement technology given its attractive potentials in the light of points 1 and 2 above, more damage to the environment would occur which might make recovery in turn more expensive even if feasible, and substitution more attractive in economic/political terms.

In other words, the ultimate outcome of following this line of technological development would lead eventually to substituting the technosphere for the biosphere/nature. To the technological optimists and the technological pragmatists alike, there is nothing inherently alarming about this prospect in store for human civilization or for nature. The former, presumably, would positively welcome it in the spirit of technological triumphalism; the latter, *faute de mieux*, would buy it as the price one must pay for material progress. But the technological pessimists could only continue to criticize provided they articulate new grounds for their protest, going beyond the mere cry that technological fixes are in principle unsatisfactory. They would have to clarify for themselves what 'unsatisfactory' means. Is it to be understood in the sense of (a) being iatrogenic, that is, running the risk of causing further damage elsewhere, (b) causing further damage when that damage is assessed from the anthropocentric point of view, namely, as loss of resources and amenities to humans, or (c) causing damage when that damage is assessed from the nonanthropocentric point of view, as damage to nature itself?

Nature-replacement technology might not prove to be unsatisfactory in principle in the first two senses. Some technological fixes, it is true, are capable of producing iatrogenic damage, but surely not all. Each must be assessed on its own merit. Maybe some nature-replacement measures escape such a stricture. Maybe too, some of them could even be said to be more advantageous from the human point of view—the quality of air in an artificially controlled situation, we have seen, could turn out to be superior to that achieved by nature-saving techniques in the case of air pollution. However, it is true that no matter how well some nature-replacement devices could be said to pass the test of being satisfactory in senses a and b, they will have difficulty in passing the test in sense c. *Ex hypothesi*, nature-replacement technology ignores the intrinsic value of nature conceptually, and downgrades nature for its own sake in practice. In practice, at best it would lead to an unsatisfactory compromise of setting aside some areas to be designated as nature reserves while the rest would be made over to the technosphere. This, as Doeleman points out, is the "zoo principle," an acknowledgment by society that only residual bits of nature are worth salvaging and/or can be salvaged, to be artificially maintained and protected. If this undesirable scenario is to be avoided, then stronger efforts must be made, in the first instance, to articulate the philosophical grounds for opposing it—the fundamental task as conceived by this book.

Biotechnology and Its Radical Threat to Biotic Nature

This brief section clarifies one very important matter. It concerns the implications of biotechnology for environmental philosophy. Unlike molecular nanotechnology of the near future, biotechnology has already been in place for over two decades. We have observed that it has succeeded in pushing biotic artefacts to a deeper and more radical level of artefacticity compared to the type of biogenetic whole-organism technology based on the classical gene/chromosome theory. The organisms produced by biotechnology are near-total biotic artefacts. Transgenic organisms are the perfect embodiment of such a status.

The issue raised here is not really about the assessment and containment of risk involved in this revolutionary technology. As far as that is concerned, in the early days of biotechnology, the scientific community, to its credit, went out of its way to be extra cautious about the risks that could be incurred. However, its monitoring has led the community to relax some of the earlier self-imposed severe restrictions.[9] But while scientists in research laboratories may be conscientious in adhering to strict procedures for containing risks, can industries in exploiting the technology be similarly trusted? It is true that to date no biotechnological Harrisburg (or Chernobyl) has occurred, but then the biotechnological industry has not been around for as long as the nuclear power industry. For the skeptical, a question mark continues to hang over it.[10]

The philosophical worry expressed in this book would not be overcome even if biotechnology turns out, in the long run, to incur no greater risk of pollution and contamination than less radical techniques of genetically transforming organisms. It is not about the risk of environmental pollution as standardly understood, but the risk about the ontological supersession of the natural by the artefactual, at least as far as biotic nature is concerned.

Sagoff has also drawn attention to this tendency, that biotechnology is a nature-replacing rather than a nature-saving technology.[11] He writes:

> Maryland's former director of tidal fisheries, recognizing the benefits of genetic engineering, argued that the Chesapeake Bay "should be run more like a farm than a wilderness." He believed that the state should subsidize efforts to fabricate fish the way Frank Perdue manufactures chickens. Many experts agree that industrial mariculture, by pushing fish populations far beyond the carrying capacity of ecosystems, will render capture fisheries obsolete.[12]

He continues:

> [Biotechnologists] are engineering [fish] to withstand pollutants to which they now succumb. They have perfected a nonmigrating rockfish that need not transit the anoxic stem of the Bay. (They have also perfected an acid-tolerant trout that does well in acidified lakes.) It may not be efficient to regulate pollution to accommodate species. It may be cheaper to regulate species to accommodate pollution.[13]

Sagoff also cites the biologist Dan Janzen about the fate of tropical forests in the light of advances by biotechnology. Janzen has written:

> Tropical wildlands and most of the earth's contemporary species still exist because humanity has not had organisms capable of converting all tropical land surfaces to profitable agriculture and animal husbandry. Within one to three decades, organisms modified through genetic engineering will be capable of making agriculture or animal husbandry, or both, profitable on virtually any land surface. Agricultural inviability, the single greatest tropical conservation force, will be gone.[14]

But as the details of the more general argument will be spelled out in the following sections, there is only one point which needs emphasis here. The risk of the natural as an ontological category being made redundant is much greater than it might at first sight appear. Molecular nanotechnology of the near future which is capable of transforming abiotic nature into the artefactual will join hands with biotechnology, the end result of which is to turn biotic and abiotic entities into artefacts. The synergistic results of such an alliance will constitute a very much greater and more distinct radical threat to the ontological category of the natural than if only the one were in place but not the other.

Future Technology and the Radical Threat to the Natural

The main threat to nature from most of extant technology, as the first section shows, comes from its polluting effects which may set in train other technological fixes to correct them; but these may, in turn, as the technological pessimists have pointed out, bring further iatrogenic damage in alternative forms to other parts of the biosphere. The real flaw is simply that extant technology is by and large ecologically insensitive in this way, as we have seen.

But what if the technology of the near future were to overcome this central flaw, that it would not produce a pollution problem on a scale which could subvert the functioning and integrity of the biosphere? An environmental ethic which regards pollution as the key disvalue and, correspondingly, its prevention as the crucial value would have to welcome such a technology as a good thing—indeed, perhaps even an unqualified good thing—as it has no resources within its framework to criticize and resist it. At first sight such a technology may sound too far fetched, and so can be ignored. But before dismissing the notion out of hand, two matters should be pointed out:

1. Even if such a technology from today's vantage point may seem fanciful, this does not follow that environmental philosophy should not ponder the matter as a thought experiment in order to clarify such issues as: (i) what values are central, (ii) whether ontology has a part to play as well as axiology, and (iii) whether the ontological question precedes the axiological one.
2. In any case, the signs are that such a future technology is not altogether purely speculative and fanciful. As illustrations of such future technology, this book mentions two examples, terraformation and molecular nanotechnology. At the moment, insufficient funding may be holding back the former. With regard to the latter, some governments and research institutes the world over are already persuaded that it is one of the ways, if not the way, forward into the next century; however, it is true that, to date, it does not have much to show for beyond some relatively limited successes. But very significantly, the consensus among the scientific community is that both projects violate no known fundamental laws and principles of science. Admittedly, this is only a necessary, and not a necessary and sufficient, condition for the successful emergence of the new technologies. But with time, the readiness to invest in their research and given the ripeness of political and economic circumstances,

it is conceivable that nanotechnology, if not terraformation, will come to pass at some stage of the next century.

However, for the purpose of philosophical exploration, their actual timing is immaterial, as they pose seriously the radical possibility of technology transforming the natural to become (more or less totally) the artefactual and, in turn, of the former being superseded by the latter. This section will carry out the philosophical exploration of the issues mentioned earlier, concentrating on molecular nanotechnology as the paradigm which embodies such a technological possibility.

The proponents of nanotechnology sensibly recognize its potential for abuse as an instrument of war, and the possibility of incurring ecological disaster should risk containment fail. Moreover, any new technology incurs the further risk that unanticipated and unforeseen consequences could arise in spite of the most conscientious afforts to monitor outcomes. However, none of these threats is new. Present technology, whether as biotechnology, nuclear technology, chemical and biological techniques of killing in war, poses them too; society, so far, appears to have had some success putting in place mechanisms to cope with them. Balanced against such potential minuses are an impressive list of potential pluses which nanotechnology may possess: its anticipated capability for dealing with physical diseases, for postponing old age indefinitely, for generating resources to meet the demands of exponential growth both demographically and economically, for overcoming (relative) Ricardian scarcity of raw resources but without violating the entropic limits laid down by (absolute) ecological scarcity, for relieving inequities between the rich and the poor economies as it is envisaged to be a less capital-intensive technology to operate once pioneered and put in place.[15]

In other words, the proponents of nanotechnology appear to have addressed themselves, by and large, to the criteria laid down by what Krimsky calls "the critical school of technology assessment."[16] This approach departs from the concepts of efficiency and externalities that constitute the core of the neoclassical model of technology assessment and, indeed, challenges the supremacy of the market by rejecting the market's conflation of needs with demands. It has developed a wider framework to include factors like ecological value (which goes beyond crude resourcism), job displacement and quality of work (social values), the scale of technology and whether it can be democratically managed (political/social values), whether it relies on renewable or nonrenewable resources, and whether it produces 'waste' of such quality and quantity as to strain the finite limits of the biosphere to act as a sink (sustainability as a goal).

Krimsky has identified and formulated seven relevant factors which enter the critical school of technology assessment and developed a technology assessment index:[17]

1. Ecological Impacts (EI): direct and indirect effects on natural systems,

sustainable utilization of resources, species and habitat diversity.
2. Health effects on humans (HE): physical and psychological effects and their distribution in the population.
3. Ethical soundness (ES): impact on ethical norms, moral beliefs, or general quality of life principles.
4. Economic Productivity (EP): contributions to efficiency of production, new wealth, international markets, or new jobs.
5. Distributive Justice (DJ): contributions to the distribution of existing resources or the concentration of new or old wealth.
6. Social Needs (SN): contributions to essential needs related to health, well being, education, social order, or environmental quality.
7. Market demand (MD): fulfillment of an existing or potential market demand sufficient to justify commercial development whether or not it meets a social need.

Index Variable	Best Case	Favored	Mixed	Worst Case
EI	+	0	-	-
HE	+	+	0	-
ES	+	+	0	-
EP	+	0	+	-
DJ	+	+	-	-
SN	+	+	-	-
MD	+	+	+	-

If the proponents were to apply such a technology assessment index to molecular nanotechnology, barring unanticipated consequences, they could claim that the technology in question would yield a best case result overall. As it promises such superabundance for all and is yet virtually non-polluting, or far less polluting than extant technology, it surely may be said to herald a new dawn. Is it not then the ultimate green technology?[18] Stewart Brand, who is regarded as one of the official chroniclers of the computer revolution but also, more importantly, became, in 1968, the editor of the *Whole Earth Catalogue*, considered to be "a definitive resource book for the ecologically minded,"[19] wrote the foreward to a book co-authored by Drexler. In it, he says:

> In that enclave of environmentalists and world-savers one of our dirty words was technofix. A technofix was deemed always bad because it was a shortcut—an overly focused directing of high tech at a problem with no concern for new and possibly worse problems that the solution might create. But some technofixes, we began to notice, had the property of changing human perspective in a healthy way. ... I think nanotechnology also is a perspective shifter.[20]

In the light of this kind of response, "technofix" may no longer necessarily be a dirty word—some technofixes may be environmentally benign after all, and nanotechnology is one of them. It may seem churlish not to be grateful should nanotechnology as a benign technofix come to pass, and to decline entering graciously the new Garden of Eden. Before one is seduced to enter, it might be wise to recall that in the old Garden of Eden was a serpent. But in the new Paradise, what is the serpent and where is it lurking?

In a nutshell, the answer is this—the new serpent is the new technology itself as it represents the ontological redundancy or supersession of nature$_{nk}$. However, ontological redundancy of nature should not be interpreted to mean something silly, namely, that nanotechnology would render the processes of production totally independent of nature. Of course, no technology could perform that kind of miracle. As it has already been pointed out, nanotechnology is said to violate no known scientific laws and principles including the laws of thermodynamics—it is not magic but science applied. Moreover, it is dependent on nature$_{f}$—on the existence of naturally-occurring entities in the form of atoms of matter/energy, which happen to be in abundance, unlike other naturally-occurring entities like trees, or golden eagles which are not. The following quotation from two influential economists bear out the point, although at the time they were writing, their pronouncement amounted to nothing more than a programmatic faith in modern science:

> Advances in fundamental science have made it possible to take advantage of the uniformity of energy matter—a uniformity that makes it feasible, without preassignable limit, to escape the quantitative constraints imposed by the character of the earth's crust. A limit may exist, but it can be neither defined nor specified in economic terms. Flexibility, not rigidity, characterizes the relationship of modern man to the physical universe in which he lives. Nature imposes particular scarcities not an inescapable general scarcity.[21] Man is therefore able, and free, to choose among an indefinitely large number of alternatives.[22]

In other words, nanotechnology may be seen as an instance of the long awaited fulfillment of the ultimate promise given by modern science at its inception in the seventeenth century, but, which it has taken four centuries to make good. As we have seen, according to the metaphysics of Scientific Naturalism, matter is uniformly dead or inert, consisting of mere extension, and is itself devoid of form or *telos*. Such metaphysics is in keeping with the view that there is a general process of production which consists ultimately of the rearrangement of the elements of such matter to serve solely human ends. Hence modern science and its technology become the study of the manipulation of nature.

Nanotechnology cannot, and does not, dispense with elementary matter as atoms of the various elements which exist in nature, the analogue of what Aristotle called first or prime matter. Instead, its implied claim amounts to being able only to dispense with second matter, that is to say, natural kinds, be these biotic like species of plants and animals, or abiotic like diamond or granite. These are forms of low entropic structures which are scarce because humans may render extinct or use biotic kinds far faster than they can replace themselves. In the case of certain abiotic kinds, they are simply nonrenewable, at least in the time-span which could be relevant to the sustainability of our industrial civilization. But in a nanotechnological world, such scarcity would not be worrying. Nanotechnology appears to be able to bypass most, if not all, abiotic natural kinds, by rendering them irrelevant to the process of production. In their place, it will be able to construct new forms of second matter, new synthetic kinds. By this maneuver, not only is the scarcity of natural kinds rendered irrelevant to the industrial processes of production but the artefactual kinds may be said to supersede them. Such supersession, in turn, as we shall see, would lead to both the ontological and the physical elimination of natural kinds.

Natural kinds are entities which come into existence and continue to exist independent of human volition and agency; artefactual kinds, in contrast, are entities whose existence and maintenance are the intended outcome of human volition and agency. They come into, or go out of, existence entirely at human bidding. Technological products are artefacts, and artefacts are the material embodiment of human intentional structures. Nanotechnology, by allowing humans to assemble objects (or to disassemble them), atom by atom, with absolute precision, embodies the perfect technique for the manipulation of nature. Such manipulation amounts to near perfect, if not perfect, control and, therefore, near perfect or perfect mastery of nature. Whether such control and mastery are considered as domination is immaterial. If the notion of domination conjures up physical conquest, such as disemboweling the earth as in current mining, tearing out part of the earth as in quarrying, disfiguring the earth's landscape as in surface waste disposal, cutting down trees and destroying habitats and whole ecosystems as in massive deforestation, then such images of laying waste the land through the equivalent of scorch-earth policies are clearly irrelevant in the context of nanotechnology. But if domination is to be understood in terms of a relationship between two parties where one party (the dominator) totally and successfully imposes its will on the second party (the dominated), then the notion could be said to be appropriate. Humans in possession of nanotechnology are in a position systematically to replace natural abiotic by artefactual kinds if and when it suits their purposes to do so—humans are in total charge, the master of their own destinies, whereas natural kinds are, powerless, at their mercies. Such a situation justifies the political image of domination with which modern science has been associated.

This image is reinforced by another matter, that of the ultimate humanization of nature.[23] Under extant technologies, the process of humanization is, relatively speaking, not as profound as it could be when compared with nanotechnology. Up to now, natural kinds may have been transformed by extant technologies, to some extent, into artefacts but their degree of artefacticity is, relatively speaking, still not very deep, although biotechnology, in respect of biotic nature, is capable of increasing such depth by crossing the species boundaries. Nanotechnology claims to be able to construct *de novo* synthetic, abiotic kinds, from the design board, using the right arrangement of atoms. In conjunction with biotechnology, it could also redesign existing biotic kinds, turning them into near total artefacts.

The serpent which haunts the new Garden of Eden is not so much the serpent of pollution. On the contrary, the more perfect the control and mastery over nature, the less likely is the technology to produce polluting effects. After all, pollution has been referred to as the "naturally mediated unintended and unforeseen consequences of specific practices of activity upon nature."[24] On this criterion of perfect mastery, the more perfect the technology, the less polluting it is—perfect precision and control mean that only whatever is intended comes to be and all that is unintended, as far as possible, is eliminated.[25]

If the most fundamental environmental value is not to undermine the functioning and integrity of the biosphere *via* polluting processes and pace of production, then nanotechnology must be considered to be environmentally benign and, therefore, the ultimate green technology. It is possible, as we have just seen, for such a technology in combination with another like biotechnology, to ensure that the biosphere carries out its public service functions, namely, to act as a sink to absorb waste, to continue the great carbon, nitrogen, hydrogen cycles. But if the most fundamental environmental value is not merely that, but the preservation of natural kinds together with the processes at work in nature which ensure that natural kinds continue to exist, to change and to evolve, to maintain themselves autonomously, then nanotechnology (in conjunction with biotechnology) seems to pose a severe threat to the preservation of the natural, as it possesses the potential to humanize the whole of nature. It is to be resisted then on grounds that the natural (meaning natural kinds and the processes which generate and sustain them) could be made redundant and replaced entirely by the artefactual (synthetic kinds, whether biotic or abiotic and the processes manufacturing them).

As we have seen, the natural and the artefactual belong to two very different ontological categories. The natural constitutes 'the Otherness' for what is human. By rendering the natural redundant in principle, nanotechnology is in danger of destroying 'the Other.' To put it minimally, it is compatible with ontological impoverishment even if it does not entail either a permission or a duty to eliminate the natural, both empirically and as an ontological category.

Ontological impoverishment is to be deplored not merely because in the end it amounts to human impoverishment. It is that of course, but more importantly, it is to be deplored as yet another expression of strong anthropocentrism and of a purely instrumental attitude to nature on the part of humans. It amounts to the denial, in yet another context, of the claim that nature can be a locus, if not also a source, of intrinsic value.[26]

It is morally wrong of us humans to eliminate nature (by rendering it redundant, making it over to our image to serve our purposes), not simply because it diminishes ourselves as moral beings, but because the diminishment lies precisely in our moral blindnesss to something other than ourselves which deserve moral consideration, or could be said to be the bearer of intrinsic value. In other words, although moral blindness is clearly a human failing, it is not merely to be deplored because it constitutes a human failing, but because ontological elimination, loss or supersession is constitutive of that failing.

Some might think this indictment too hasty a conclusion to draw. After all, another possible but more comforting scenario than the supersession of natural kinds (and the processes which sustain them) could well emerge—humans who no longer need to cut down trees for constructing buildings or for paper-making may well leave trees alone and, indeed, may even plant more or permit more to grow for their own aesthetic and spiritual edification. But several things may be said in response to this possibility.

Aesthetic appreciation of nature may be based simply on regarding nature as a trigger for an aesthetic experience. If so, this is to strike an instrumentalist attitude to nature as we have argued in that chapter, even though it is not the instrumentalism of crude economic resourcism. Moreover, a suitable substitute may be found to act as such a trigger. For some people, seeing a so-called wild-life film could produce such an effect. If the aesthetic experience alone is what counts, the trigger itself becomes irrelevant—it does not matter whether the viewers actually confront wild-life in the raw or they confront it mediated by images on a screen. A conventional film, it is true, may not be a very efficient trigger, as many dimensions of the real objects are missing. However, the technology of virtual reality (itself born of modern science) promises to make good such deficiencies—by donning appropriate helmet, suit and glove wired up in certain ways, one could access to cyberspace.[27] Maybe, there is no need to walk through a forest to have an aesthetic experience of it. A simulated one could do the job just as effectively, if not better.

In other words, today, the knowledge that the camera in a conventional film is an attempt to capture real phenomena out there in nature, or even the knowledge that virtual reality is an attempt to simulate something which exists outside the world of simulation, may be a relevant ingredient in one's aesthetic experience of the natural. But in the world of tomorrow where virtual reality dominates people's lives, then such knowledge may become less relevant, if not irrelevant—the natural drops out of their frame of reference, and what is simulated becomes for them the only reality that matters. An example of an analogous process at work in another area of life may be cited to make the point clearer. In the UK, food technology has been so successful in its manufacture of mass 'ice-cream' that such a product contains no cream—indeed its main ingredients are air, water, vegetable fat of some kind, artificial flavoring and coloring (in that descending order of magnitude). Many children brought up on it reject real or proper ice-cream on the grounds that it is not 'ice-cream' and rate it as inferior to what to them is the 'real' thing. The substitute has ousted the original from its claim to reality. In the same way, the world of simulation may come to oust the original world from its claim to reality, and reassign to itself the patent right to such a claim.

However, subjective aesthetics may be understood in a different way, not as an object triggering the aesthetic experience but as a version of the Hegelian aesthetic. To find aesthetic satisfaction in nature is in many ways a historically-situated and contextualized experience. Note the profound differences between the pre- and post-Romantic conceptions of the English landscape. According to the former, the scenery of the Lake District and similar parts elsewhere was regarded as frightening and threatening; according to the latter, as beautiful. But the English romantic poets did not discover the beauty of the English Lake District. Rather they created it. By celebrating certain natural objects in their poetry, they bestow aesthetic value on them. In this sense, the aesthetic value of nature is derivative upon artistic activities.[28]

As a historical account of the development of the English sensibility toward landscape, it is correct. It just was the case that English Romanticism brought about an aesthetic revolution in the way people perceived the landscape.[29] But is the scenery of the Lake District frightening or beautiful? It is an open question which perception is appropriate. Perhaps, it depends on the conditions under which the experience takes place—on a fine English summer day, the landscape is beautiful but when the wind starts up, the mists and fog descend, the rain/sleet/hail lash against one's face, it is threatening or even awesome, but not beautiful, at least not in the normal understanding of beautiful, as pretty.

But, perhaps, subjective aesthetics—understood either to mean that the object merely serves as a trigger for aesthetic experience or as Hegelian aesthetics—is altogether too superficial a view about the aesthetic experience of nature. Beauty in nature goes beyond the mere aesthetic and/or creative response of human perceivers. It involves something more, namely, grasping that natural objects possess certain properties such as their intricacy, their complexity both of structure and of functioning, their delicately and efficiently interrelated parts in virtue of which one is justified in finding them beautiful. This alternative account may then be called 'objective aesthetics.' But nanotechnology provides no comfort to those who wish to ground their aesthetic experience in this way. The artefactual products of nanotechnology and biotechnology could be said to possess such characteristics too. Maybe human beings in the future might come to find less aesthetic satisfaction from natural kinds than from synthetic kinds. They may, in the manner of Hegelian aesthetics, create and bestow aesthetic value upon synthetic kinds. Novel artefacts, both biotic and abiotic, even more exotic and exciting than what nature has to offer, could be designed and, indeed, even customized to the taste and fancy of individuals in much the same way as gardens are landscaped and designed to order. Of course, up to now, the topography of gardens have been modeled by removing hills or building artefactual ones, re-directing streams, creating lakes and so on. As for their biotic contents, until the recent advent of biotechnology, plants have been modified by the more conventional methods of breeding, greatly improved through the understanding of genes and chromosomes given by the science of classical genetics. But the garden of the future, aided by nanotechnology and biotechnology, would possess even a greater degree of artefacticity than that—the rocks in the rockery may, for instance, no longer be of granite or limestone but of a synthetic kind of material, and the plants and animals could have been fashioned from genetic material that have crossed the species boundaries in a radical way.

In other words, the whole of nature could, in principle, become a garden, that is to say, an artefact, an human intentional structure. There is nothing incoherent about such a technological conception, displacing the natural as an ontological category, in spite of some attempts to argue to the contrary. For instance, take Benton's arguments.[30] He cites the incidence of radiant energy from the sun on Earth and Earth's own weather system as conditions which are absolutely beyond human manipulability. But in a sense this is precisely the kind of project which terraformation of planets involves; the consensus of the scientific community is that it is an enterprise which does not violate known scientific laws and principles. Indeed, the engineering for doing so is within the capabilities of extant technologies, and many scientists are rearing to go but are held back by mere shortage of funds. There are several engineering solutions proposed. A quick route to the terraformation of Venus, for instance, among other things, involves (a) cooling Venus to about 290 K by erecting a sunshade in the form of a large mirror, positioned between Venus and the sun, which will radiate the heat to deep space, cooling down the planet, (b) providing sufficient water for the planet by capturing one of Saturn's ice-moons, diverting its course to Venus, through a combination of gravity and solar power, and then crushing the ice-moon over the planet, (c) shortening the present solar Venusian day of 117 Earth-days to an Earth-day of 24 hours by mounting a soletta, orbiting in a 24-hour polar orbit, to ensure a day exactly 24 hours long.

If synthetic kinds, whether biotic or abiotic, are just as complex, just as intricately and delicately structured and organized as natural kinds, and, therefore, just as worthy of being aesthetic objects, then there is nothing to choose between them in terms of an objective aesthetic. Nor are they deficient in any way from the perspective of a subjective aesthetic. In other words, an attempt to resist the supersession of natural kinds cannot be grounded in aesthetics, whether subjective or objective. The work has to be done, as this book argues, through ontological resistance.

In epistemology, ontological deflationism is welcome as a methodological principle—Occam's razor recommends that a hypothesis postulating fewer entities is to be preferred over its rival when both are more or less adequate from the explanatory point of view. But in the metaphysics of the environment, Occam's recommendation may not be appropriate; the loss of natural kinds as an ontological category is not merely a matter of methodological simplicity but a simplification of the world we inhabit. Perhaps what is needed is not merely an Endangered Species Act, but legislation to prohibit the supersession of natural kinds and the processes which engender them.

The anthropocentrism embodied in the possible supersession of natural kinds is truly radical in the following senses: it involves all forms of instrumentalism (natural kinds as raw resource, as source of psychological/aesthetic values, as performing public functions like acting as a sink for waste); it adheres to Humean projectivism (natural kinds are neither a source nor locus of intrinsic value); it constitutes the ultimate logic of nature being humanized through the concept of *homo faber*. In such a world, nature is permitted to operate as an ontological category only in the form of atoms—in other words, precisely only within the metaphysical space delineated by Scientific Naturalism. The delineation of such severely limited metaphysical space is itself a reflection of the fundamental goal of modern science maximally to control, manipulate and dominate nature. And technologies like biotechnology and nanotechnology are the ultimate fulfillment of that goal.

Homo Faber, the Humanization of Nature, and the Naturalization of Humanity

Modernity is often associated with the celebration of abstract reason. Its elevation of 'the rational' as the epistemological authority toppled traditional authority in the form of the theological and the ecclesiastical. Although this is undoubtedly correct historically speaking, it should also be pointed out that the celebration of abstract reason is not a new thing in the history of Western philosophy—Plato stands out as an uncompromising champion in classical Greek philosophy. Modern European thought revived and reconstrued that ancient Greek notion, developing it in radically new directions to serve different ideological goals from those of the classical age.

As we saw in Chapter 1, the ancient Greeks regarded *episteme*, achieved through the exercise of reason, to be both intellectually and morally superior to *techne*. For Plato, the material world upon which *techne* operates belongs to Appearance, not Reality. The pursuit of wisdom and virtue through grasping the forms of the Good, Justice and Beauty is the ultimate philosophical goal. Anything else is a distraction. The highest intellectual/moral life is that of the contemplation and the love of such forms. The desire to dominate the material/natural world through *techne* is no part of this ideal. The role played by *techne* in such a conception of the good life and the good society is a very limited one. The highest human fulfillment and flourishing are not to be found in the mastery of the sensory world nor in the comforts that such control may bring.

The ancient Greeks held material affluence to be a form of corruption, and according to Socrates, a society addicted to it was a "feverish state." In Plato's *Republic*, Socrates, in reply to Glaucon's criticism about his characterization of the good state, says:

> The true state I believe to be the one we have described—the healthy state, as it were. But if it is your pleasure that we contemplate also a fevered state, there is nothing to hinder. For there are some, it appears, who will not be contented with this sort of fare or with this way of life; but couches will have to be added thereto

> and tables and other furniture, yes, and relishes and myrrh and incense and girls and cakes—all sorts of all of them. And the requirements we first mentioned, houses and garments and shoes, will no longer be confined to necessities, but we must set painting to work and embroidery and procure gold and ivory and similar adornments, must we not? "Yes," he said. "Then shall we not have to enlarge the city again? For that healthy state is no longer sufficient, but we must proceed to swell out its bulk and fill it up with a multitude of things that exceed the requirements of necessity in states.[31]

For Socrates, human flourishing is not predicated upon "the realm of necessity" being taken care of by *techne* (technology). On the contrary, the love of the soul for the Good will be diverted, beyond a certain point, by technological deliverance from necessity. For moderns like Marx, intellectual and artistic flourishing is only meaningful and possible if structured upon a material base of abundance, if not superabundance.[32] The Socratic life, however, is predicated upon sufficiency, not abundance or superabundance, and must necessarily exclude material affluence, as the latter nourishes the weaker, less perfect side of human nature, undermining any genuine pursuit of the soul's desire for perfection, for the Good.[33]

Aristotle, too, shares Plato's attitude to *techne*. In *Metaphysics*, he distinguished between the theoretical, the practical and the productive sciences. The first type, like physics, mathematics and metaphysics, is concerned with principles of being which cannot be other than what they are, and whose investigation primarily satisfies the intellectual faculty for understanding and making sense of the cosmos. The second includes ethics and politics. The goal of ethics is to attain *eudaimonia* (happiness is not understood in the utilitarian sense of pleasure as a sensation, but as well-being, issuing from an integrated life of fulfilling pursuits). The goal of politics is the realization of the common good, which includes the ethical goal of the pursuit of *eudaimonia*. We try **to do** things well in the practical sciences. But in the productive sciences, we **make** things instead. For Aristotle as for Plato, the *bios theoretikos* (the life of contemplation) constitutes the highest plane of activity, as it addresses the divine element in human beings.[34] *Bios praktikos* occupies the level below. The life of fabrication, of making things, is the lowest form of activity. It has no intrinsic value. It is not worth doing for its own sake. It has only instrumental value as it supplies the basic necessities without which life, and *ex hypothesi*, the good life, is not possible. In other words, for Aristotle in particular and the ancient Greeks in general, the life of fabrication may be a necessary subsidiary preoccupation but cannot itself constitute a focal point in human life, metaphysically, spiritually and morally.

Modernity may also celebrate abstract reason but does so in an entirely different ideological context. Here, abstract reason, as manifested in scientific methodology in particular, is harnessed to the goal of controlling nature. While there is some emphasis on the pursuit of scientific knowledge out of pure intellectual curiosity, the predominant motivation, as we saw in Chapter 2, is to translate that knowledge into technological terms to serve the overwhelming goal of improving the well-being of humans. The life of fabrication, far from being a subordinate, subsidiary, though necessary human activity, has become a central preoccupation, so central that fabrication becomes the very essence and constitution of human nature. In a word, the notion, *homo faber* captures the spirit of modernity itself.[35]

It would not be unfair to use Marx as an exemplar of modern European thought in spite of the present eclipse that he is suffering. The vicissitudes of his reputation and his influence do not touch the fact that he is an intellectual giant whose ideas cannot be dismissed. One might disagree with him but noone could seriously ignore what he has said. Central and pervasive in Marx's works is the notion of *homo faber*, in spite of the uneven emphasis he gave to it at different periods of his writing.[36] It figures predominantly in the Marx of *The Economic and Philosophical Manuscripts of 1844*, although in the mature Marx of *Capital*, it is somewhat muted and qualified.

In his later writings, Marx famously distinguished between the forces of production (tools, technological means including scientific knowledge) and the relations of production, consisting of the legal arrangements (underpinning private property and class) which govern the ownership and control of the forces of production, as well as the distribution of the results of the production process. In Marx's analysis, it is the capitalist relations of production which is the fundamental cause of alienation—the laborers having sold their labor power as a commodity to the capitalists are, thereby, treated as no more than a production input like other raw resources, and are severed from the products of their labor. Marx's main prescription for overcoming alienation lies in a change of these relations, from private to common/public ownership of the means of production. In other words, in his view, the forces of production, once ensconced in a new set of legal relations of production, will be on course to benefit all humankind, instead of benefiting only the capitalist class at the expense of the working class. This means that the change from capitalism to socialism/communism does not require any radical change in the forces of production themselves. Indeed, science and technology, as the major forces of production, by satisfying the demands of the realm of necessity, enable human beings to reach the realm of freedom, of truly free play and creative activity. Or to put the same point in another way, on the back of *homo faber* rides *homo ludens*.

Put this way, thus far, it may look as if that for Marx, the true essence of human nature is given by *homo ludens*, not *homo faber*. Could one not maintain that like the ancient Greeks, what is important to Marx is not the life of fabrication (of work) but the life of intellectual, artistic creativity—if not *bios theoretikos*, it is at least the life of freely chosen activities? Just as the ancient Greeks regarded fabrication as a mere means to sustain other truly important ends in life, may not Marx, too, be said to regard fabrication as a mere means to sustain the really important end of freedom and creativity?

But the analogy does not hold for several reasons. First, Marx, as we know, scorns utopian socialism for being silly and naive in thinking that socialism could meaningfully be built upon anything less than a set of vigorously and continuously advancing industrialized, technologized forces of production.[37] As an apostle of progress and modernity, Marx stood in awe of the might and mightiness of industrial civilization, even as he vociferously objected to the capitalism which had captured and harnessed it for its own end, thereby causing exploitation and alienation. The supreme task of scientific socialism is to unhitch industrial civilization from (bourgeois) capitalism. The cause of alienation is not industrial civilization *per se*, but the legal framework of private property and class within which it is embedded. The young Marx was confident that the realm of necessity would be superseded by the realm of freedom under communism. The older Marx was less sure and conceded that although less and less labor would be required, and more and more of external nature would come under the control of technology (and science), nevertheless, labor could not be totally dispensed with, and material nature would provide a residual resistance, leaving the dialetic between human and external nature never fully resolved.

Second, Hegel and the young Hegelians cast a deep shadow on Marx's thinking, in spite of Marx claiming that he has turned Hegel upside down and outgrown him. Hegel's influence is, of course, most obvious in the writings of the young Marx where, as we have seen, the notion of *homo faber* may be said to play a key part. In Marx's account of the relationship between humans and nonhuman nature, what he calls "external nature" merely serves as a means by which we, humans, can express and realize ourselves through our labor. To see the full force of the Hegelian influence, we need to distinguish between five components in the human/external nature relationship—the human agents as laborers, their tools (technology), their labor, external nature and the products of their labor.

Clearly, the products of human labor, according to Marx, are consciously and deliberately planned. They are artefacts and are, therefore, intentional structures. They have come into existence as fulfillment of the ends of their creators, the laborers. In their labor, in the process of fabrication, the human agents are realizing their own ends, thereby realizing themselves. Human agents who rely only on their limbs and other bodily organs as tools in fabrication would only be able to make a very limited range of artefacts. As a result, many ends that they might entertain could not be realized and correspondingly, their self-realization is also limited. But with the help of tools, like flint or needle, they could expand the range of fabrication. Modern technology is but another tool, more powerful than those which existed before, enabling them to increase the range of fabrication. Before the spaceship, humans could only dream of flying to another heavenly body. Now that science and technology have enabled us to make that spaceship, humankind can realize the goal of leaving Earth and achieve self-realization in space exploration. In other words, an artefact, the product of human labor, is a concrete embodiment of human self-realization (both collectively, and in most cases also, individually) through transforming (some part of) external nature by means of tools/technology.

This way of presenting Marx's thought is faithful to his view, borrowed from Feuerbach, that humanity realizes its "species being" or essence through its labor in transforming external nature. In other words, fabrication constitutes human essence and identity—man is *homo faber*. Feuerbach, who was an atheist and a materialist, had repudiated Hegel's idealism that the subject of history was the Absolute Idea or Spirit. Instead, Feuerbach claimed humanity itself to be the subject, and that it would only realize its essence when it had overcome alienation from itself and achieved, through expansion of its consciousness, the realization of the unity of subject and object, namely, humanity and external nature. But Marx, in turn, repudiated Feuerbach's account of the relationship between humanity and nature on the grounds that it was too passive. He conceived it, instead, as a creative activity on the part of humanity through its labor in transforming external nature.

However, in maintaining that "species being" is realized through fabrication, Marx appears to presuppose and echo another Hegelian/Idealist theme, namely, that nature itself has no **being** until humanized by us. On this view, nature as such is in a state of indeterminate negativity, without being or value. In other words, it is ontologically and morally latent but with the potential to be and to be valuable. But this potentiality can only become actual when humans work upon it and infuse it with value and spirit. According to Hegel, there is a law of progress, led by Reason, which permits humankind to advance to ever higher levels. Reason is the essence of value, of spirit. By infusing nature with value and spirit, such humanized nature progresses to higher levels of being and rationality. Untouched nature in the form of wilderness lacks determinate being and value. By turning it into, say, a park or cultivated land, we have freed nature from its ontological and moral void.

Fichte, for instance, expresses this idealist conception of the relationship between humankind and nature most forcefully and succinctly, as the following quotation bears out:

> Nature must gradually be resolved into a condition in which her regular action may be calculated and safely relied upon, and her power bear a fixed and definite relation to that which is destined to govern it — that of man. In so far as this relation already exists, and the cultivation of Nature has obtained a firm footing, the works of man, by their mere existence, and by an influence altogether beyond the original intent of their authors, shall again react upon Nature, and become to her a new vivifying principle. Cultivation shall quicken and ameliorate the sluggish and baleful atmosphere of primeval forests, deserts and marshes; more regular and varied cultivation shall diffuse through the air new impulses to life and fertility; and the sun shall pour his most animating rays into an atmosphere breathed by healthy, industrious, and civilized nations. Science, first called into existence by the pressure of necessity, shall afterwards calmly and carefully

> investigate the unchangeable laws of Nature, review its powers at large, and learn to calculate their possible manifestations; and while closely following the footsteps of Nature in the living and actual world, form for itself in thought a new ideal one. Every discovery which Reason has extorted from Nature shall be maintained throughout the ages, and become the ground of new knowledge, for the common possession of our race. Thus shall Nature ever become more and more intelligible and transparent even in her most secret depths; human power, enlightened and armed by human invention, shall rule over her without difficulty, and the conquest, once made, be peacefully maintained. This dominion of man over Nature shall gradually be extended, until, at length, no farther expenditure of mechanical labour shall be necessary than what the human body requires for its development, cultivation and health.[38]

It is clear that, for Fichte, nature has neither being nor value except for what we, humans, choose to create out of it and to endow it with by means of our labor, science and technology.

Hegel, the intellectual giant of this tradition, has written: "A person has as his substantive end the right of putting his will into any and every thing and therby making it his, because it has no such end in itself and derives its destiny and soul from his will. This is the absolute right of appropriation which man has over all 'things.'"[39]

In other words, the concept of *homo faber* in defining "species being" or human essence embodies two interrelated themes which give in turn the key notion of the humanization of nature. These are: (a) that humans realize themselves through fabrication, that is, through imposing their ends and values on nature *via* their labor and their tools/technology; (b) that nature itself is bereft of being, of value, until humans work upon such a blank canvas to endow it with being and with value. This means that the two related concepts—*homo faber* and the humanization of nature—constitute the most extreme form of anthropocentrism possible. They go beyond the more usual Humean/Cartesian theses, namely, that:

1. Human consciousness is the source of all values.
2. Human consciousness is the sole locus of intrinsic value.
3. Nonhuman nature is only of instrumental value to humans.

Another two are added:

4. Nonhuman nature is not only morally, but also ontologically, latent or void, with the potential to be, and to be valuable, only when acted upon and transformed by human labor.
5. Humans themselves achieve their true being *via* the continuous creative process of fabricating artefacts out of the morally and ontologically latent nonhuman nature.

Thus understood, the law of progress will reach its ultimate level when the whole of nonhuman or external nature has become humanized. As we saw, Marx, in his later writings, was less optimistic about such an outcome. But Marx died in the last century. Technology has advanced since then. Technology of the twenty-first century could, in principle, overcome the obstacles lying in the path of the ultimate humanization of nature project. All that is natural can, in principle, be transformed into the artefactual through fabrication using extant and/or near-future technology.

Moreover, ironically for Marx, the metaphysics of Idealism meshes in perfectly with that of Scientific Naturalism to underpin the humanization of nature project. We saw that modern science at its very beginning in the seventeenth century has granted extremely limited ontological space to nonhuman nature—as *res extensa*, nature exists and is real, only as atoms and molecules, and in terms of so-called primary qualities. Its being is that of first or prime matter, not that of second matter. As we saw, on this view, natural entities and kinds, in the forms of individuals and species, are not as real as the atoms and their arrangements to which natural entities and kinds may be reduced. The last section explored the possibility that the technology of the future—based, say, on an alliance between biotechnology and nanotechnology—would make it possible for humans to synthesize and create artefactual entities and kinds to replace naturally-occurring ones. The metaphysics of Idealism and of Scientific Naturalism is, then, not that dissimilar after all, except in one respect—namely, that while the former regards nonhuman nature to be bereft of being and reality altogether in the absence of human fabrication, the latter grants nature a limited reality and being in the form of atoms and their molecules. But from the standpoint of the humanization of nature project, this philosophical theoretical difference does not appear to be critical. Of course, as far as *praxis* is concerned, only science and technology undertaken within the metaphysical framework of Scientific Naturalism could make the humanization of nature project actually come to pass. From the Marxist perspective, provided the labor which relies on such a technology to humanize nature is no longer estranged labor, but labor emancipated from the capitalist legal relations of production, human beings would spontaneously and universally produce under conditions of freedom, engaging in activities that are self-determined rather than determined by the requirements of (private) capital. The forces of production having been unfettered by the relations of production, humans would no longer exploit fellow humans, but together, they would march inexorably on the road to progress, realizing themselves through fabrication, through humanizing nature.[40]

It may be a suitable point in this account of the centrality of the notion of *homo faber* in Western thought to introduce another strand, namely, Locke's contribution *via* his labor theory of value. Locke is no Idealist. His empiricist ontology and epistemology would incline one to put him in the materialist camp. So while Locke does not maintain that nature is totally ontologically void, he does hold that (natural) material objects are to be understood only within the reduced metaphysical space assigned to them by Scientific Naturalism. He officially incorporated Galileo's distinction between primary and secondary qualities into the philosophy of empiricism—secondary qualities, unlike primary ones, are not real and do not reside in the objects themselves. This qualification apart, Locke's thoughts, however, appear to mesh in with the other Idealist assumption mentioned above, namely, that nonhuman nature is morally void until worked upon by humans.

In Locke's hands, this is presented as the labor theory of value, which serves to reinforce the main anthropocentric theses that human consciousness is the source of all values as well as the sole locus of intrinsic value, and that nonhuman nature is only of instrumental value to humans. Locke implies that natural items, like nuts, berries and peat, in themselves, have no intrinsic but only potential instrumental value for humans. Before human involvement, they are, strictly speaking, almost morally void—but see below for further qualification. However, with human involvement, they acquire moral significance. Not only is their potential instrumental value actualized, they also acquire a kind of derivative "intrinsic value" when humans mix their labor with them. Human labor, emanating from a being with intrinsic value, and, therefore, itself embodying intrinsic value, has the power to impart to objects such a value. That is why for Locke, objects infused with such derivative intrinsic value may even be said to be a part of the identity of the individual who has mixed his labor with them. In other words, his labor theory of value plays a dual role in his system of thought—not only does it yield a principle of distributive justice and a doctrine of property rights (he who mixes his labor with a natural item is entitled to it and to the fruits of his labor) for his political philosophy, it also serves to lay the metaphysical foundation in terms of the strong anthropocentrism required for the kind of economic theory emerging in the modern world.[41]

Marx, as we know, borrowed Locke's labor theory of value. In so doing, Marx, too, would have taken over his predecessor's presupposition that nature *per se* is morally void. While Locke was working to found a bourgeois system of economic thought ultimately with the capitalist class justifying the fruit of the labor of the working class through the Lockean doctrine of individual private property rights, Marx, later, argued that the fruit of the labor of the working class belonged to the workers alone. Bourgeois economics, for Marx, has assigned ownership of the fruit of labor to the wrong class in society. But both Marx and Locke would have seen eye to eye on the fundamental point that nature itself has no value apart from realizing instrumental value for humans, and at the same time being infused with derivative intrinsic value imparted to it by human labor.

Given the crucial role the labor theory of value played in Locke's political as well as his economic philosophy, it may not be too far fetched to say that the ideas of *homo faber* and of the humanization of nature are implicit in that theory, even though he, himself, did not bring them to the fore. For Locke too, nature is nothing but "standing reserve" in Heidegger's words, and it is man's destiny to mix his labor with it to turn something which is valueless into something of value. In so doing, to a greater or lesser degree, depending on the technology available, man is able to turn the natural, which is valueless, into the artefactual, which is valuable.

But surely at this point a critic might retort that mixing one's labor with an object could hardly be said to be intrinsically involved with the idea of *homo faber* and its related idea of the humanization of nature, even if one were to grant that Locke does hold that it is humans who endow nature with value through their labor. Of course, Locke's own examples of picking nuts and fruits as instances of mixing one's labor with natural objects may provide ground for the complaint. But it would be a mistake to think that this is the only relevant example. It serves him as a clear case involving human labor alone, with no tools used other than the human hand, in such an act of appropriation. But his mention of peat gathering introduces the complexity of tools—one presumably would need a sharp cutting implement of some kind to do the job properly. *Homo faber* makes such tools which are artefacts. It is true that *homo faber* confining himself only to tool making to appropriate nature for his basic needs, leaving as much and as good for others, would not necessarily be embarking on a process of humanizing nature on a systematic scale. But that is only part of the Lockean picture. After all, the society Locke legitimates is not a society of subsistence existence, but modern capitalist society, ultimately resting on science and technology to yield the tools for appropriating nature, to advance the ends of property acquisition and capital accumulation. And it is obvious that since the seventeenth century, technology (whether directly born of science or not) has enabled industrial *homo faber* to humanize nature on a large and systematic scale throughout the world.

Like the Idealists and Marx later on, Locke held that man's true being is to endow nature with value through labor. But unlike the Idealists, he was not prepared to say outright that nature, without the investment of human labor, is absolutely worthless. Instead he said that it has some miniscule worth, but so miniscule as to be almost non-existent. The following quotations bear these points out very clearly:

> Nor is it strange, as perhaps before consideration it may appear, that the Property of labour should be able to over-ballance the Community of Land. For 'tis Labour indeed that puts the difference of value on everything; and let any one consider, what the difference is between an Acre of Land planted with Tobacco, or Sugar, sown with Wheat or Barley; and an Acre of the same Land lying in common, without any Husbandry upon it, and he will find that the improvement of labour makes the far greater part of the value. I think it will be but a very modest Computation to say, that of the Products of the Earth useful to the Life of Man, 9/10 are the effects of labour: nay, if we will rightly estimate things as they come to our use, and cast up the several Expences about them, what in them is purely owing to Nature, and what to Labour, we shall find, that in most of

them, 99/100 are wholly to be put on the account of labour.[42]

And: "Nature and the Earth furnished only the almost worthless Materials, as in themselves."[43] Furthermore, for Locke, humans are under a divine mandate to transform nature, which is as good as worthless, into valuable items—valuable because these embody their labor, the source of all value.

His views on labor and nature, undoubtedly, then laid the foundation for a nascent economic philosophy, which eventually underpins the science of economics, to be more fully developed later, by Adam Smith and others. Economics has as its object the creation of wealth, which is to be justified as a means to improve human material well-being, the ultimate over-arching value of humanism and of the Enlightenment project. But if the means to achieve this ultimate end is through fabrication, humanizing nature *via* science and technology, turning the natural into the artefactual, then it looks as if Locke has also initiated the process of conceiving man as *homo faber*—man at one and the same time not only achieves well-being, both materially and in terms of the spirit of freedom and self-realization, but also successfully endows what is ontologically reduced, and almost morally void, with significance.

Back to Marx. From this perspective, Marx seems to have appealed to the various different strands in modern Western philosophy which together constitute the extreme anthropocentrism embedded in it. These strands include the contribution of: (a) Idealism especially in his earlier writings, (b) Scientific Naturalism, which underpins our industrial, technological civilization, and which he celebrates, (c) Locke's labor theory of value, which, in turn, combines the severely reduced ontological space accorded to nature by Scientific Naturalism with the notion (later identified with Idealism) that nature is devoid of any moral significance in the absence of the human infusion of its own value into nature, through mixing human labor with it. In his eclecticism, Marx unwittingly shows how "mainstream" his own system of thought really is. In other words, in understanding Marx in this respect, one comes to grips more effectively with the philosophical foundation of modernity.

Marx not only talked about the humanization of nature but also about the naturalization of humanity in the process of fabrication. It is obvious what Marx meant by the former, as we have seen. But what could he mean by the latter? He is taken to mean that while humans transform nature through their labor, humanized nature in turn transforms humans. There is no static human nature whose needs also remain constant and permanent.[44] Instead, humans develop new and growing needs, entertain new and different projects of self-realization as technology enables them to transform and humanize nature in new and different ways. In other words, humanized nature opens up new paths for human self-realization. As we mentioned earlier, before science and technology enabled spaceships to be constructed, human beings could only dream of flying through space but not actually realize themselves and their potential as astronauts. Human needs, on this view, are ever-expanding because projects for self-realization are constantly being reformulated in the light of labor and developing technology opening up new frontiers and new vistas for exploration.[45]

Nanotechnology especially in harness with biotechnology and microcomputer technology can turn human beings themselves into near-total artefacts—already humans could now in principle have germ-line therapy. Naturally, Marx could not have foreseen the possibility of technology developing in this radical direction by the late twentieth and twenty-first centuries. But such a possible development would be in keeping with the concept of the naturalization of humanity. Humans, under this new technological development, are really no different from any other form of biotic nonhuman being—both can be turned into near-total artefacts. "Designer" mice (like the oncomouse) or "designer" babies are perfectly possible already. In the pipe-line are attempts, for instance, specially to engineer certain human genes into pigs with the aim of using their hearts eventually in human heart transplantation.[46] Technical difficulties, of course, remain to be overcome but research laboratories throughout the world are certainly not without hope. On another longer front, indefinite postponement of death (even if it is not to be called 'immortality') and what may be called "punctuated existence" are being promised.[47] As one type of near-total biotic artefact among others, humans, too, are but particular arrangements of atoms and molecules, much the same as any other plant or animal around.

However, there is a difference between human and nonhuman biotic artefacts—in the case of the latter, humans simply transform nature through fabrication, thereby humanizing it, but in the former, humanized nature and naturalized humanity coincide, as the nature that is being transformed is none other than human nature itself. Ultimately even the *homo* that plays, that dreams, is naturalized humanity itself within a world which has been humanized by humanity's continuous process of creative fabrication through its labor and its tools/technology. In other words, *homo faber* reigns supreme.

As the case is sometimes put, the cultural history of humankind is indeed the history of its technology and the different artefacts which different forms of technology bring forth. Bergson has articulated it well:

> In thousands of years, when, seen from the distance, only the broad lines of the present age will still be visible, our wars and our revolutions will count for little, even supposing they are remembered at all; but the steam-engine, and the procession of inventions of every kind that accompanied it, will perhaps be spoken of as we speak of the bronze or of the chipped stone of prehistoric times: it will serve to define an age.[48]

The thesis that the identity and essence of humanity is constituted by the activity of fabrication amounts to this: it is not the claim that humans only manufacture machines and consumer goods and never play, make love, eat, write novels outside of the activity of fabricating things. Rather, it is to say that even when people play, make love, eat, write novels, they are doing these things in the context of a fabricated, artefactual world. Take writing. In the past, it was done with reed and papyrus, equipment with a low level of artefacticity. Today many (including this author) use the computer as a word processing machine, a technologized artefact *par excellence*. Fabrication penetrates every activity we engage in, including the act of breathing, as the air we breathe may increasingly be air-conditioned, filtered and purified, as we have seen. In this sense, *homo faber* is fundamental because it underpins even *homo ludens*. As technology develops more radically and powerfully, *homo faber* structures and creates a more and more artefactual world within which all human activities necessarily take place. Even the activity of walking in the mountains or in the wilderness is not exempt, although walkers may convince themselves that it is. It is true that they are using their feet, their own power of locomotion. But the boots on their feet, the socks which encase the feet, the waterproof clothes they wear, the sleeping bag they crawl into at night, the rucksack on their back are probably made of synthetic substances of one kind or other. As walkers, workers, lovers or whatever, we are what we are capable of making through our labor, our science and our technology. The process of fabrication and its products transform nonhuman nature, and in so doing, also transform ourselves. Whereas other civilizations at other times in human history have chosen some other route to self-realization, be it political or theological, and, as a result, subordinated the process and activity of fabrication to these other ends and goals, modern civilization has committed humanity to self-realization *via* fabrication itself. The price humanity pays for being *homo faber* is, increasingly, in the terminology of Marx and others, the near-total, if not the total, humanization of nature and the naturalization of humanity itself.

It is fitting to end this section with two quotations from Bergson, who has captured so perfectly the spirit of modernity, which construes *homo faber* as constituting the essence of human nature. In the first, he says:

> If we could rid ourselves of all pride, if, to define our species, we kept strictly to what the historic and the prehistoric periods show us to be the constant characteristic of man and of intelligence, we should say perhaps not *Homo sapiens*, but *Homo faber*. In short, intelligence, considered in what seems to be its original feature, is the faculty of manufacturing artificial objects, especially tools to make tools, and of indefinitely varying the manufacture.[49]

In other words, human intelligence is to be understood solely in terms of that type of intelligence embodied in instrumentation and manufacture of artefacts, and in the scientific reasoning which informs these activities. All other forms of intelligence displayed in activities like joke-telling and story-telling are written off as marginal. It is the intelligence of instrumentation and of manufacture, belonging to human consciousness alone, which ensures that "man comes to occupy a privileged place."[50]

In the second quotation Bergson writes:

> We have said that intelligence is modelled on matter and that it aims in the first place at fabrication. But does it fabricate in order to fabricate, or does it not pursue involuntarily, and even unconsciously, something entirely different? Fabricating consists in shaping matter, in making it supple and in bending it, in converting it into an instrument in order to become master of it. It is this mastery that profits humanity, much more even than the material result of the invention itself. Though we derive an immediate advantage from the thing made, as an intelligent animal might do, and though this advantage be all the inventor sought, it is a slight matter compared with the new ideas and new feelings that the invention may give rise to in every direction, as if the essential part of the effect were to raise us above ourselves and enlarge our horizon. Between the effect and the cause the disproportion is so great that it is difficult to regard the cause as producer of its effect. It releases it, whilst settling, indeed, its direction. Everything happens as though the grip of intelligence on matter were, in its main intention, to let something pass, that matter is holding back.[51]

It is significant that nonhuman nature is only of instrumental value to humans. But for Bergson, it is even more significant that human consciousness and intelligence—as embodied in instrumentation and manufacture as well as the scientific reasoning which stands behind these—so effectively breaks down any recalcitrance and resistance of nonhuman nature against the penetration and imposition of human intentions upon it. In other words, the transformation of the natural to become the artefactual truly affirms humanity's mastery of nature. *Homo faber* reigns supreme.

Homo Faber, Artefacts, and the Language of Machines

In the pre-modern world, houses or statues are ready paradigms of an artefact. But in the modern world, a machine is the ready paradigm. A dictionary definition of a machine is as follows: "any system, usually of rigid bodies, formed and connected to alter, transmit and direct applied forces in a predetermined manner to accomplish a specific objective, such as the performance of useful work."[52]

Three elements which constitute its make-up may be distinguished:

1. The rigid components.
2. The arrangement of these components *via* parts like crank or shaft such that the ensemble can perform useful work.
3. The force or power which moves the parts to perform the work in question.

But to the above, a fourth implicit assumption may be added:

4. The three elements listed are nonbiotic. Where all three are biotic in character, on this account, the entity in question does not count as a machine. For instance, a *punkawallah* does not count as a machine, although the *punkawallah* has rigid or semi-rigid parts, linked by other rigid parts to perform useful work—the useful work in question, being to drive away flies

and mosquitoes, which would, otherwise, make the life of the *sahib* and *memsahib* insufferable, by waving a fan intermittently (usually a large dried leaf). A *punkawallah* is a human, a biotic being, **not** a machine. (This point will be discussed later.)

Machines are meant to be the substitutes or part substitutes for humans in performing useful work for humans. Animals historically provide the most obvious motive force to move the pre-arranged parts, as in a cart and horse arrangement. In a windmill, the motive force is a force of abiotic nature, namely, the wind. Sometimes, as with a bicycle, humans provide the motive force, which acting on the machine, enables it to perform a greater amount of work than the human power on its own could do. But a car relies on the energy which is released from the combustion of petrol—the human driver merely provides direction rather than motive force. The newest arrival in the family of machines is the automated machine in which humans are no longer even needed to provide direction and regulation, as it has been constructed to be self-regulating.

Machines may be divided into two types, those belonging respectively to the two phases, I and II, in the history of technology as set out in Chapter 2. A flourmill-powered-by-a-windmill is a machine complex which falls into the earlier, while an aeroplane falls into the later, phase. A windmill can only perform work if the wind is there, but not when the air is calm. But an aeroplane is powered by a refined product of the crude oil, itself the product of fossil remains mined from the ground; as such, its ability to take to the air is not dependent on the vagaries of the weather, although its take-off or landing could be curtailed by bad weather. Machines of such a kind are subject to greater human control, and hence, considered to be more desirable than those which only perform when a set of favorable natural conditions obtains.

Having sketched a skeletal account of machines in the mechanical engineering sense, the discussion will turn to the question whether, given this perspective, organisms of one kind or other can at all be regarded as machines. Drexler (usually regarded as the father of molecular nanotechnology) says that dandelions and rabbits are machines because they are capable of performing work, that is, putting atoms and molecules together in a certain way to maintain their own structures and their functioning as well as to reproduce themselves.[53] But are they? For a start, organisms, like dandelions and rabbits, outside a cultivated, domesticated context, are not artefacts—they are naturally-occurring beings. What they do when they eat (absorb nutrients), defecate (expel waste generated in the processes of metabolism), mate and reproduce, they do entirely **for** themselves, although it is true that we, humans, may find them and/or the products of their activities of use to us.[54] From this perspective, not even biotic artefacts are machines, although they may have been genetically engineered today to advance a human goal.

The history of patenting reveals that up to 1980, no one could be sure in any country with a Western-type of legal system that patents could be granted to any living organism which claimed to have been made by humans. The situation, however, altered with the decision of the US Supreme Court in the case of Diamond v. Chakrabarty. Chakrabarty, a scientist, working for General Electric submitted an application to the US Patent Office in 1972 for a new strain of the bacterium *Pseudomonas* and was eventually granted one. Following that, rDNA organisms have similarly been granted patents in the last decade. The bestowal of patent status on human-modified organisms is clear recognition that they are biotic artefacts. But it would be a mistake to infer from their status as artefacts that they are machines. The inference would amount to a redefinition of "machine," such that any entity, which has been selected or constructed to advance a human goal, would be a machine. On one view of definition, it is true that all definitions are arbitrary and purely stipulative—as Humpty Dumpty says, one can make words mean what one wants them to mean. But even within such an extreme nominalist framework, there may be good reasons for preferring one definition to another. In this instance, the redefinition of "machine" would make the term so broad as to be positively misleading, and even a specially bred sniffer or St. Bernard dog would have to be classified as a machine as it performs a chore for humans.[55]

There is an ambiguity in the expression "useful work" which Drexler and others (including Maturana and Varela, who will be discussed later in this section) appear to have overlooked. A machine, in the mechanical engineering sense, is a device with certain characteristics designed "to perform useful work." "Work" in this context is a term in physics and is "a measure of the result of transferring energy from one system to another to cause an object to move."[56] Work is done when a spring is being stretched or compressed, a weight is being raised or lowered, or a gas is being compressed in a cylinder. But "useful work" also refers to an activity leading to an outcome which we consider to be desirable. For instance, if the plates are dirty, they have to be washed should we want them clean. The washing up is the work that has to be done. Someone has to do the work and members of a group may squabble as to whose turn it is to perform it. In principle, one could calculate the amount of work (in the technical sense) that is done whenever a human washes a dirty plate, although in practice, there would not be much point to the calculation. The washing up in most households today (at least in certain segments of society in the contemporary world) is done by a machine. The dishwasher performs "useful work" both in the technical as well as in the lay senses. The link between the two senses lies in this—any activity on our part necessary for achieving a desired result requires the expenditure of energy which we (modern people) tend on the whole to regard with disfavor. So we design machines with an external energy source to perform the activity which we otherwise would have to undertake ourselves. The machines deliver us the desired result, dispensing with input of our own energy. So they necessarily perform useful work understood both in terms of physics and in terms of doing a job of work for us. In a slave society, slave owners had slaves to perform useful work (understood in the lay sense) for them. In modern society, it is said machines have become the substitutes for slaves. Slaves, no matter how wretched their predicament, were (are) not machines—they were (are) humans suffering degradation and indignity through being treated as if they were (are) machines, with no goals or projects of their own independent of those imposed upon them by their masters. Machines, *ex hypothesi*, as artefacts embody the ends and projects of their human designers.

Up to now, a true machine—that is, what conforms to the mechanical engineering sense of the term—*per se* is not ontologically threatening although it could turn out to be environmentally worrying. A car as a machine-artefact does not, as such, encroach upon the ontological status of other beings as naturally-occurring beings. It is true it threatens their existence, not directly but indirectly, but not their very ontological status. Cars require roads or motorways to run on. They use up space which might otherwise be left for other human purposes and/or to nonhuman naturally-occurring beings. The fossil fuel they use produce toxic substances of one kind or other which could harm both human and nonhuman beings. As we saw in the first section of this chapter, cars embody a type of technology which poses pollution problems to the environment and is nature-damaging. But machines as such, up to now, do not render nonhuman naturally-occurring beings necessarily redundant.

But in the technological world of the not so distant future, the danger of ontological redundancy or supersession comes from the possibility of a new type of machine-artefact, namely, nanomachines which use "micromanipulation techniques" such as the STM, the scanning tunneling microscope, or the AFM, the atomic force microscope. If and when such and similar machines become available, natural abiotic kinds could be rendered superfluous, as we have seen. We have also seen that humans, as one biotic kind, could become near-total artefacts. Furthermore, we have seen that in such a future, the forces of bio-technology, molecular nanotechnology and miniaturized computer technology would combine—the alliance of these forms of engineering, genetic and mechanical, would ensure that the project of the humanization of nature, not only in terrestrial but also in astronomic space, would be unfolded and carried out to its logical conclusion.

In the light of the analysis above, it is clear that machines are only one type of artefact. Ever since its invention to the present, the mechanical clock has been a very powerful and potent referent of the notion of artefact. However, biotechnology in the last quarter of this century has established that an equally potent alternative referent could be the chimera or the transgenic organism. In other words, from now on, the language of the artefactual does not need to coincide with the language of the machine, as the biotic itself could be transformed into near-total, if not total, artefacts. However, the very predominance of the language of the machine itself, as well as its identification with the language of the artefactual, have, as we have seen, distorted Drexler's account of potential nanomachines. As a result, he maintains that nature has already successfully designed its own nanomachines—these are individual organisms. But naturally-occurring organisms are just not artefacts. *Ex hypothesi*, they are not and cannot be machines, whether nano or not. They are **the foil** to machines. Drexler's troubled account of what constitutes nanomachines is the result of muddled, erroneous reasoning:

1. He mistakes nature's organisms for artefacts. He fails to grasp and, thereby, implicitly destroys the ontological distinction between the natural ($_{fa}$ and $_{nk}$) and the artefactual.
2. By doing so, he also seems to have overlooked that the transformation of the natural into the artefactual is effected *via* tools/technology (which since the nineteenth century is backed by fundamental scientifc advances). This is ironic, given his own vision of the nanotechnological future.
3. He fails, too, to appreciate that the biotic can be transformed into artefacts, without becoming machines, as machines are only one type of artefact, among others.

Drexler, of course, is not alone in confusing organisms with machines and in being obsessed by the language of machine. He is simply following a well-established tradition—for instance, Karl Ereky, considered to be the father of the term "biotechnology" (which he used in a series of statements he wrote in the years 1917 to 1919), distinctly regarded the pig as a machine. For him, the difference between the industrial and the peasant approaches to pig rearing was not that the former used electrical pumps and automated feeding and the latter did not. Rather it was because under the former, the pig was regarded solely as a converter of so much feed input into so much meat output. In a similar vein, even earlier, Max Delbrück began a lecture in 1884 with the statement "Yeast is a machine."[57]

The language of machine, as used by Delbrück, Ereky and Drexler, betrays a deep-seated strong anthropocentrism in their attitude to biotic nature. Naturally-occurring organisms are regarded as merely possessing potential instrumental value for us, humans. To say that yeast is a machine amounts to casting an exploitative eye on the microorganism as a being whose value lies in doing a job of work for us. Biotechnology, in the various senses of the term whether old or new, is precisely that technology which harnesses the metabolic and/or reproductive capabilities of organisms to carry out tasks we have set for them. A pig eats food, and through its metabolism, converts nutrients into a form which can be absorbed by its physiology to sustain, maintain and reproduce itself—the pig as an individual organism *qua* organism eats for itself, not for us. But we are like the virus. The virus injects its DNA into bacteria to get the latter to reproduce on its behalf. It hijacks the bacteria's reproductive capability for its own end. Under biotechnology (the new use of the term, referring to those techniques induced by molecular genetics and molecular biology), we, too, inject alien DNA into bacteria and other organisms, thereby also hijacking their reproductive and/or metabolic capabilities to perform tasks for us.[58] To call such organisms, thus rendered artefactual, "machines" because they carry out our ends amounts to yet another affirmation of at least two of the theses which make up strong anthropocentrism, namely, that humans are the sole loci of intrinsic value and nonhuman nature has mere instrumental value for us, humans.

However, defining organisms (whether naturally-occurring or modified through biotechnology) to perform useful work on our behalf as machines goes beyond merely treating them as possessing only (potential or actual) instrumental values for us. The procedure at the same time involves a misconstrual of the ontological status of organisms as naturally-occurring beings, and reflects a determination to change their status to that of artefactual beings. The two aspects—their sole instrumental status in terms of use values for humans and their ontological status as artefactual beings—are clearly, inextricably linked. In other words, the anthropocentrism, so deeply embedded in the philosophy which underpins modernity, must be grasped, not only *via* its axiology, but also its ontology.

One could say that the modern human ultimately does not feel at ease among naturally-occurring beings or entities, be these biotic or nonbiotic. We only feel at home when home is the world of humanized nature—in other words, only when the natural has become transformed into the artefactual. Artefacts embody our labor, our intentions and purposes. By creating them, we have imparted and infused our own intrinsic value into them, as already observed. As a result, they could even be said to be an extension of ourselves and part of our own identity. We are most at home when we and the artefactual world have become one. If so, this amounts to nothing less than the elimination of nature as 'the Other.' Ontological impoverishment is the price that has to be paid to enable humankind to be thoroughly at home.

So deep-seated is this modern worldview that it has infected theoretical biologists who fall prey to using the language of machine even in their explanatory account of how organisms are organized as living entities. Just one example will be cited to illustrate this tendency, namely, the work of Maturana and Varela in theoretical biology. It has been chosen because of the significant influence it has on environmental thought itself of late, especially through their popular book, *The Tree of Knowledge*.[59]

The appeal of Maturana's and Varela's writings for environmental philosophy lies in the concept of "autopoiesis." The word itself comes from the Greek *autos* for "self" and *poiein* for "to produce" or "to bring forth." It is used to characterize that property peculiar to a living entity, existing as an organizational unity and maintaining its identity through self-renewal, self-regeneration and self-generation. However, although the term "autopoiesis" may be new, the concept itself even for environmental philosophy, strictly speaking, is not so. The concept may be traced to systems analysis.[60] It is also related to the notion of dissipative structures as introduced by the physicist Prigogine and his co-author Stenger.[61] Capra relies on it:

> [A]n organism is primarily engaged in renewing itself; cells are breaking down and building up structures, tissues and organs are replacing their cells in continual cycles. ... All these processes are regulated in such a way that the overall pattern of the organism is preserved, and this remarkable ability of self-maintenance persists under a variety of circumstances, including changing environmental conditions and many kinds of interference.[62]

Warwick Fox (*Toward a Transpersonal Ecology*) and Robyn Eckersley (*Environmentalism and Political Theory*) both welcome the concept as a possible bearer of intrinsic value in nature. An entity which is concerned with its own self-renewal is an entity with interests, thus qualifying it for moral consideration. It may also be said to constitute an end in itself and therefore have intrinsic value. Furthermore, the autopoietic perspective makes it possible not only to ascribe intrinsic value to individual organisms but also to species, ecosystems, the ecosphere as Gaia; the definition of autopoiesis covers any entity or system which maintains its own organizational unity through self-maintenance. Hence, Eckersley says it provides a more satisfactory theoretical basis for the ecocentric approach in environmental philosophy.

Put thus, the approach clearly has merit, and as we shall see, overlaps to some extent with that pursued in this book *via* the notion of the natural understood as what is autonomous and independent of humans, both in terms of its origin and its continuing existence, giving rise to a notion of intrinsic value to be further explored and developed in detail in the next chapter. The purpose of this section is different. It is to bring out the fundamental distortion involved in the autopoietic approach to biotic nature owing to the predominance of the language of machines built into it by Maturana and Varela.

Maturana and Varela assert, in no uncertain terms, that all living systems are machines. By this they mean the following:

> First, we imply a non-animistic view which it should be unnecessary to discuss any further. Second, we are emphasizing that a living system is defined by its organization and, hence, that it can be explained as any organization is explained, that is, in terms of relations, not of component properties. Finally, we are pointing out from the start the dynamism in living systems and which the word 'machine' connotes.[63]

Their exclusion of vitalism is clearly uncontroversial. But one cannot be so sure with regard to their second and third theses.

They start off by giving a standard account of machines: "Machines are usually viewed as concrete hardware systems, defined by the nature of their components and by the purpose that they fulfill in their operations as man-made artefacts."[64] They then go on to reject it as a naive account, as it concentrates on the nature of their components rather than on the relations between the components which integrate the machine as a unity, as an organization. The shift from structure (which is concerned about the properties of the components) to organization enables them to give an abstract generalized account of machines which is applicable to all machines irrespective of the type of machines they are, and of the components which enter into their concrete realisztion as machines. As a result, machines, under their dispensation, need no longer be "concrete hardware systems, defined by the nature of their components." In other words, machines are no longer necessarily objects constructed out of abiotic or exbiotic nature. This leaves room presumably for extending the notion of machines to living organisms, to regarding dandelions and yeasts as machines.

Next, they go on to remove the second element which constitutes the normal, but in their opinion naive, account of machines, namely, "the purpose that they fulfill in their operations as man-made artefacts." According to Maturana and Varela, their purpose in this crucial sense is not part of the organizational unity of the machine and is relegated to what they call "the domain in which the machine operates," "the domain of observation." Furthermore, they also contend that the notion of purpose or function of a machine is simply an invitation to invent the machine one is talking about, but is irrelevant to a characterization of the machine organization. In other words, they seem to imply that to understand a car as a machine with a certain organizational unity, there is no need to talk about the function or purpose it supposedly serves in inventing and manufacturing it.

This surely should strike one as rather odd, even a "naive" reaction from their perspective. A car's organizational unity is controlled by the purpose it is designed to serve, namely, to get us from point A to Z by moving both itself and us, the driver and passengers, through a certain portion of space. The machine's organizational unity, one would have thought, would be different—the components would be differently related to one another—if the purpose it is designed to serve were different. Moreover, if the purpose were indeed different, even the components themselves and their properties might have been different. Suppose the machine were not designed as a conveyor of people and their possessions from one place to another but for some other purpose, such as to drill a hole underground. Would the machine have the same components with the same properties and be connected up with one another in the same way as the components which enter into the make-up of a car? Far from the standard account being naive, it is their account which appears to be so. An artefact's purpose or function is not detachable from its organizational unity. On the contrary, its function or purpose informs the very way in which its components are put together as such a unity. Its construction as well as its existence as an organizational unity cannot properly be grasped without a reference to its purpose or function, which enters intrinsically into any adequate account of it, both at the conceptual and explanatory levels.

The point may be pursued further *via* a thought experiment. Imagine an archaeologist in the distant future, after our industrial civilization has crumbled away, excavating an object, like a car, buried in one of today's landfills. An approximate account of its organizational unity could probably be arrived at by careful experimentation using models which are reconstructions of it. But the experimentation itself must be guided by a series of hypotheses about the object, one of which surely is that it is a machine-artefact designed to serve a certain end in mind, namely, to convey people from one point to another over land, but not across water or through the air. Moreover, there are cars and cars. Some are designed and constructed as racing, others as family cars. In the case of the former, as maximum speed is of the essence, this purpose is built into the organizational unity of such a machine. In the case of the latter, as speed is not the sole *desideratum*, others more important, such as capacity for passenger-load, comfort, safety, economic costs and so on, together with a compromise between these relevant *desiderata*, are similarly built into its organizational unity. In other words, the purpose a machine serves is not a factor that can be detached from, or considered as extraneous to, any adequate or complete attempt to explain and account for their organizational unity. That aspect of a machine cannot be relegated simply to the domain of observation, as Maturana and Varela maintain.

Furthermore, when a machine, like a car, breaks down, it is facile to say that its failure to discharge its function also belongs merely to the domain of observation. The disintegration of its organizational unity—a broken crank shaft—manifests itself as failure to discharge its function, that is to move at all. The broken crank shaft is the cause of the car's inability to move, and, therefore, of its failure to carry out its function. The cause and the effect together constitute the disintegration of the car's organizational unity. To restore the latter, the mechanic must be guided by the purpose the machine is designed to carry out, which leads her/him to identifying the broken crank shaft as the cause of its immobility.

Maturana and Varela define allopoietic machines such as cars as "machines that have as product of their functioning something different from themselves."[65] "That something different from themselves" is precisely, in the case of a car, its ability to move across space over land, carrying people in it, the very purpose it has been designed to serve. Its success or failure as an entity with a certain organizational unity can only be judged in terms of the purpose conceived by the humans who construct the artefact in accordance with it. Moreover, in engineering terms, no machine can be designed simply from the point of view of its organizational unity—whatever that might mean—but is, and must be designed with a whole host of other considerations in mind. For a start, as Chapter 2 has shown, technology, unlike pure science, works under a different epistemological target. Pragmatism and efficiency, rather than truth or approximation to the truth, guide its choice of solutions to a problem. It is also opportunistic in using knowledge and theories from any domain. Compromise with scientific purity apart, technology also works under other crucial constraints already mentioned, such as the cost of the end product, its aesthetic aspects, its "user-friendliness," its safety and so on. All these are woven into the make-up of the technological product which cannot be detached from its so-called organizational unity. Or to put the same point differently, the organizational unity exemplified in a machine is not "pure" in the way Maturana and Varela seem to think it is.

The two authors have seemingly failed to appreciate that their revised definition of the term "machine" has destroyed what is most distinctive about it. A machine on the standard view is an artefact, designed, constructed by humans to serve a distinctive human end. On their account, a machine is no longer an artefact, designed for a specific purpose. Instead, it is any system with an organizational unity to it but physically expressed. (A system with an organizational unity but conceptually expressed would, presumably, not be a machine on their view.) As such it is by no means clear from their account that their term "machine" cannot meaningfully be used to refer to a static organizational unity like a crystal.[66] However, from their remarks, this much is clear—that they are interested in dynamic "machines," not static ones, and in particular, autopoietic dynamic "machines," not allopoietic dynamic ones, like cars. We have seen how they define the latter. They define the former as follows: "a machine organized (defined as a unity) as a network of processes of production, transformation and destruction of components that produces the components which: (i) through their interactions and transformations regenerate and realize the network of processes (relations) that produced them; and (ii) constitute it as a concrete unity in the space in which they exist by specifying the topological domain of its realization as such a network."[67]

It is also clear from their remarks that the individual organism is a living system, that all living systems are physical autopoietic machines and that all physical autopoietic machines are living. In other words, "autopoiesis is a necessary and sufficient condition for a system to be a living one."[68]

So far it has been argued that in the case of machines, standardly understood, such as cars (but in Maturana's and Varela's termininology, they are allopoietic machines), no complete or adequate account of them in explanatory/organizational terms can be given without a reference to the purpose or use to which they have been designed, that such a purpose cannot simply be relegated to the domain of observation or description. At first sight, it might appear that even if the two theorists are wrong in their account of allopoietic machines, they must surely be right with regard to what they call autopoietic machines, that is to say, living organisms, when they maintain that the notion of purpose plays no part whatsoever. However, on closer examination, it might not be so easy to dispense with the notion of end, if not of purpose, altogether.

But to prevent confusion, several meanings of end or purpose need to be borne in mind. Chapters 1 and 2 have anticipated this problem by clarifying the notion of teleology itself in terms of three different theses. To recap, these are:

1. External teleology, which may be identified in two forms:
   (a) The more finely differentiated variety involves a full-blown hierarchy of natural beings, with humans at the top. Beings lower down the hierarchy serve those above them. Plants serve animals while both serve the ends and

purposes of humans. This view was prevalent in ancient Greek thought as represented by Aristotle, and dominant as Aristotelianism in the medieval cosmology as the Great Chain of Beings. However, we have seen that, paradoxically, this variety historically has been associated with a less aggressive and more passive form of anthropocentrism in ancient Greek as well as medieval thought.[69]

(b) Modern thought opted for a simplified hierarchy, namely, that all of nonhuman nature is only of instrumental value for humans which alone possess intrinsic value. The modification, however, occurs within the context of an aggressive, strident anthropocentrism.

2. Intrinsic/immanent teleology—for Aristotle, explanation is in terms of the four causes, material, efficient, formal and final which in the case of living beings are fused, and may only be separated for the purpose of intellectual analysis. Every living being possesses its own *telos* which informs its identity, and governs its attempts at self-renewal, self-maintenance, its processes of growth, development, maturity, reproduction and finally decay.
3. Extrinsic/imposed teleology—modernity has discarded final and formal but retained material and efficient causes in the explanation of natural phenomena. This metaphysical shift makes it possible for inert and living matter to be regarded in purely mechanistic terms, both equally devoid of *telos* and value. However, at the same time, in the domain of fabrication (as opposed to the domain of the explanation of natural phenomena), the four causes are retained, but separated out and assigned to four different and distinct sources. The *telos* of an artefact is endowed by humans, and as argued, it is this—*contra* Maturana and Varela—which informs its organizational unity and renders it intelligible. In the case of biotic artefacts, the organism's own intrinsic/immanent *telos* has been displaced by just such an extrinsic *telos*. Biotechnology, operating at a more fundamental level of manipulation than Mendelian whole-organism biogenetic technology, enables us to fabricate living organisms. Parts of their genetic components—their material cause—come from another organism, that is to say, an external source. As a result their form may also alter—wingless chickens could be genetically engineered. Their ability to grow and maintain themselves has also been commandeered by humans to carry out human ends. To all intents and purposes, humans are their efficient cause.

To understand and explain a living organism in terms of theoretical biology, external teleology in either form is irrelevant. So is the thesis of extrinsic/imposed teleology as it is a presupposition underpinning all techniques of selective breeding as well as biotechnology in the creation of biotic artefacts. However, biologists (other than genetic engineers), *qua* biologists, cannot dispense with the thesis of intrinsic/immanent teleology. They do not have to subscribe to Aristotle's four causes, but they do have to presuppose that the organism they are studying is organized and functions in certain ways in implementing its own *telos*. In other words, ironically, the definition given by Maturana and Varela of an autopoietic machine cited above serves to characterize, in part, even if not wholly, the *telos* of any one organism. The term *telos* may be replaced by "purpose," provided the latter is not confused with its cognate "purposeful" and its implications of being conscious and intentional. "Purpose" in the sense of being end-directed may be said to inform the activities, the behavior and the processes underpinning these, of an organism, without attributing to it consciousness and intention whatsoever.

In this sense of *telos*, an organism's *telos* is self-given and self-manifest. As such the term coined by Maturana and Varela—*autopoiesis*—is most apt. What is unfortunate is their use of it in conjunction with the term "machine." On this analysis, their resulting term, "autopoietic machine" amounts to a self-contradiction, while their other term "allopoietic machine," amounts to a tautology —a machine, as we have seen, is an entity or a system without an intrinsic/immanent *telos*, as it is an artefact embodying a human purpose designed into it. Such a human purpose is necessarily deliberate and intentional, involving full consciousness.

In other words, to labor an important point, there is need yet again to remind oneself of the two very different contexts in which the term *telos* may be used:

1. The self-given and self-manifest *telos* in the case of naturally-occurring living organisms is totally independent and autonomous of human activity and intervention. Such a *telos* is to be understood as "end-directedness" without reference to the conscious and the intentional on the part of the organisms themselves, as they are incapable of conscious intentional activity in the way we, humans, are.
2. The *telos* of an artefact, such as a machine, is a deliberate, intentional structure designed, created and imposed by humans. The end it serves is a conscious human one, as it has, by its very nature, no self-given ends.

Far from welcoming the language invoked by Maturana and Varela as an unproblematic ally, Deep Ecologists like Warwick Fox and Robyn Eckersley should, on the contrary, be extremely wary of it, as it embodies serious confusions about the natural and the artefactual categories of being. The language of "autopoietic machines" ironically sits far better with the program of biotechnology underpinned by the reductive sciences of molecular biology and molecular genetics than with the preoccupations of Deep Ecology. The reasons for saying so are as follows:

1. By removing *telos* understood as intrinsic/immanent teleology and purpose as end-directedness from talk about organisms, it becomes possible for Maturana and Varela to confine their account about the organizational unity of organisms to the processes, that is to say, the mechanisms involved in their ability to transform matter into themselves in such a way that the product of

their operation is their own organization.

2. Focusing on mechanisms makes it easy for the two theorists to slide into talk about machines (as standardly understood), for machines too operate through mechanisms to maintain their own organizational unity. In this way talk about organisms is subtly assimilated to talk about machines *via* the common notions of mechanisms and organizational unity.
3. In the hands of Maturana and Varela, the characterization of machines standardly understood is *pari passu* being restructured—*telos* in the sense of imposed, extrinsic teleology as human intention are removed, as we have seen, from an account of their organization.
4. The two theses just mentioned above then prepare the way for transforming the assimilation at the level of language into an assimilation of ontological kinds—organisms **are** machines. Let us call this use of the term '(MV)-machine'—'MV' stands for 'Maturana' and 'Varela.'
5. Any entity with an organizational unity may be said to be a system; (MV)-machines are systems which may now be divided into living and nonliving ones.
6. As the standard use of the term 'machine' connotes dynamism and as organisms exhibit dynamism, it is fitting that (MV)machines are said to be dynamic systems.[70]
7. The difference, then, between living dynamic systems and nonliving dynamic ones is to be marked by calling the former "autopoietic machines" and the latter "allopoietic machines." While the mechanisms involved in the former operate to produce the components that constitute their own organization, the mechanisms involved in the latter operate to produce something other than the components that constitute their own organization.
8. When the characterization of organisms has finally been transformed into that of autopoietic machines, when *telos* in the sense of intrinsic/immanent teleology has been expelled, this reductive transformation clears the way for further transforming organisms into (near-total) artefacts by biotechnological means, without touching their essential status as "living dynamic systems."
9. Far from worrying about such a prospect, Maturana and Varela positively welcome it. A long quotation from them on this point may be justified:

> Machines are generally viewed as human made artifacts with completely known deterministic properties which make them, at least conceptually perfectly predictable.[71] Contrariwise, living systems are *a priori* frequently viewed as autonomous, ultimately unpredictable systems, with purposeful behavior similar to ours. If living systems were machines, they could be made by man and, according to the view mentioned above, it seems unbelievable that man could manufacture a living system.[72] This view can be easily disqualified, because it either implies the belief that living systems cannot be understood because they are too complex for our meagre intellect and will remain so, or that the principles which generate them are intrinsically unknowable; either implication would have to be accepted *a priori* without proper demonstration. There

> seems to be an intimate fear that the awe with respect to life and the living would disappear if a living system could be not only reproduced, but designed by man. This is nonsense. The beauty of life is not a gift of its inaccessibility to our understanding.[73]

10. Indeed, it is the very understanding of such "living systems," given to us by molecular biology and molecular genetics, which enables biotechnology to design and create artefactual living systems. Transgenic organisms are precisely that. The objection mounted in this book is not that these cannot be made by man but that their fabrication substitutes an extrinsic/imposed *telos* for their own intrinsic/immanent *telos*.[74] This then amounts to an assault on their ontological integrity as naturally-occurring beings. As we have seen, *homo faber*, through biotechnology, has hijacked the mechanisms of such "living dynamic systems" to sustain and reproduce themselves for the purpose of performing chores for us, of fulfilling our intentions while eliminating their own ends. Maturana and Varela have indeed distorted and altered their ontological status as naturally-occurring beings, by characterizing them as (MV)machines, so that naturally-occurring biotic beings can then more readily be turned into biotic artefacts.

To conclude, while there may be something of relevance to environmental philosophy which can be salvaged from the notion of autopoiesis, it would, nevertheless, be a grave mistake to borrow it uncritically from Maturana and Varela as Warwick Fox and Robyn Eckersley seem to have done. Unfortunately, Maturana and Varela have embedded the notion in a discourse whose very language subverts the ontological integrity of organisms as naturally-occurring beings. Moreover, although they themselves are not aware of it, underpinning their account is the worldview of *homo faber* and of fabrication. The very explanatory framework they advocate as theoretical biologists for the understanding of living organisms is infused by an instrumental attitude to nature which, thereby, renders the framework itself suspect. As a result, it may be wise neither to use the term nor to invoke the notion, as it can be extremely difficult to dissociate it from its context of ontological misconstrual and betrayal at the hands of Maturana and Varela.

Notes

1. See especially Barry Commoner, *The Closing Circle: Nature, Man & Technology* (New York: Alfred A. Knopf, Inc., 1971). Although biotechnology is not an exception, as it is perceived to be a pollution risk, it will be argued in the next section that it poses a far more radical threat to the natural than mere pollution.

2. See Lee, *Social Philosophy and Ecological Scarcity* (London: Routledge, 1989).

3. See John Dryzek, *Rational Ecology* (Oxford: Blackwell, 1987).

4. An example in the near future concerns mining technology. At the annual meeting of the Australia and New Zealand Association for the Advancement of Science held in Geelong in late 1994, ten senior scientists presented papers as if they were writing them in the year 2020. One of them, Bruce Hobbs, outlined a technology which combines miniaturization (tiny robots) with designer molecules capable of selective 'harvesting' or scavenging of metals. Hobbs calls this system Miniature Robotic Function Chips. Molecules could be designed with the help of biotechnology to possess a certain structural arrangement, such as a hole to them, to attract whatever atom, such as gold, lead, zinc, that is being targeted. These would be organometallic molecules made up of carbon, nitrogen, hydrogen, oxygen, silicon and copper. They would be suspended in a polymeric liquid. The solution would then be soaked into rocks along cracks. Electrical stimulation would cause the molecules to expand, change shape and cause tiny explosions to shatter the rock. As the molecules contain copper capable of conducting electricity, they could double up as computers programmed to monitor the whole scavenging operation. Micrometer-sized robots could transport the mined mineral both underground and on the surface.

This coming revolution in mining would render quarries, opencast mines, and heavy-duty trains carrying ore and smelters obsolete. Infrastructure would be minimal and unobtrusive with the mineral collected in one spot and then dispatched by vertical takeoff aircraft. Coal could even be 'harvested' beneath a city such as Sydney, and environmentally sensitive areas, such as Antartica, could also be mined—see Ian Anderson, "Metal Guru Predicts a Future of Mini Miners," *New Scientist* (8 October 1994): 10.

5. Easterbrook's ecorealism may be considered to be the latest representative of this position.

6. J. A. Doeleman, "Environment and Technology: Speculating on the Long Run," in *Technology and the Environment (Research in Philosophy and Technology Series, Volume 12)*, Frederick Ferré, ed. (Connecticut and London: JAI Press, 1992), 13.

7. Certain crop plants are already being genetically engineered to become herbicide-resistant. Three justifications are given: (a) in cases of multiple planting without crop rotation, herbicide-resistant plants would be able to cope with the herbicide residue in the soil arising from the crop preceding it; (b) such plants allow for chemical control of weed at every stage in the crop's development; (c) they render tilling or turning over the soil unnecessary, thereby saving not only labor but also reducing soil erosion—see Sheldon Krimsky, *Biotechnics and Society*, 217-21.

8. Great excitement worldwide was instantly generated about the eventual possibility of human cloning by the publication of the success in cloning a sheep, called Dolly, from an adult cell taken from the donor's udder—see Ian Wilmut *et al.*, "Viable Offspring Derived from Fetal and Adult Mammalian Cells," *Nature* (27 February 1997): 810-13.

Initially, two caveats were entered about the Dolly experiment: (a) the experiment was not straightaway successfully repeated by other laboratories worldwide; (b) as a result, an unresolved matter remained, namely, whether the transferred nucleus was from a differentiated mammary cell or a stem cell of the donor sheep. See *Cloning Issues in Reproduction, Science and Medicine* (a consultation document issued in January 1998 by the Human Genetics Advisory Commission and Human Fertilisation & Embryology Authority, UK). However, in the light of further forensic DNA-testing, since July 1998, the scientific community has become convinced that Dolly is a real clone; furthermore, another successful cloning of mice has also been reported using a process called nuclear transfer—see Davor Solter, "Dolly *Is* a Clone — and No Longer Alone," *Nature* 394 (23 July 1998): 315-16.

9. See Cherfas, *Man Made Life*, 126-41; Krimsky, *Biotechnics and Society*.

10. See Peter Wheale and Ruth McNally, *Genetic Engineering: Catastrophe or Utopia?* (England and New York: Harvester, Wheatsheaf/St. Martin's Press, 1988) and *Animal Genetic Engineering: Of Pigs, Oncomice and Men*, eds. (London: Pluto Press, 1995). In the latter volume, see especially Joyce D'Silva's "Critical View of the Genetic Engineering of Farm Animals"; Ruth McNally's "Mad Dogs or Jackasses: The European Eradication Programme"; Sue Mayer's "Environmental Threats of Transgenic Technology."

See also Krimsky, *Biotechnics and Science*. But for another assessment, see *Bioscience⇌Society*, D. J. Roy *et al.*, eds. (Chichester and New York: John Wiley & Sons, 1991).

11. Mark Sagoff, "Process or Product? Ethical Priorities in Environmental Management," *Environmental Ethics* 8 (1986): 121-38; "Zuckerman's Dilemma: A Plea for Environmental Ethics," *Hastings Center Report* 21 (September-October 1991): 32-40.

12. Sagoff, "Zuckerman's Dilemma," 35.

13. Sagoff, "Zuckerman's Dilemma," 35.

14. Sagoff, "Zuckerman's Dilemma," 40.

15. However, to some, this last point may be politically naive. Given the deeply entrenched severe inequalities between nation-states in the world today and in the foreseeable future, it is not obvious that any new, powerful technology developed by the affluent economies would be quite so unproblematically transferred to the less developed ones. But here, for the purpose of exploring the issues considered to be crucial by this book, it would not be wise to be distracted by this real and grave problem.

16. On nanotechnology, see Drexler, *Engines of Creation*.

See Krimsky, *Biotechnics and Society*, 210-14. Krimsky points out that beginning in the 1970s, this new school emerged drawing upon the insights of writers as diverse as Lewis Mumford, E. F. Schumacher, Amory Lovins and Murray Bookchin.

17. Krimsky, *Biotechnics and Society*, 212.

18. Krimsky's technology assessment index does not make clear whether it is constructed from merely an anthropocentric point of view. Most of his index variables, like HE, EP, DJ, SN, MD, are overtly anthropocentric. But EI and ES could have a nonanthropocentric reference.

19. Benjamin Woolley, *Virtual Worlds: A Journey in Hype and Hyperreality* (Oxford, UK, and Cambridge, USA: Blackwell, 1992), 25.

20. Drexler *et al.*, *Unbounding the Future*, 6.

21. Barnet and Morse would, however, be silly and wrong if they were to think that any technology could overcome absolute scarcity as implied by the laws of thermodynamics. Nanotechnology does not claim to do that. Rather, it promises to overcome relative scarcity, but within the limits laid down by absolute scarcity, understood

thermodynamically, as we have seen.

22. J. Barnet and C. Morse, *Scarcity and Growth* (Baltimore: Johns Hopkins University Press, 1963), 11. Robert Solow of MIT, who was awarded the Nobel Prize in economics in 1987, may be said also to hold a similar view: "The world can, in effect, get along without natural resources, so exhaustion is just an event, not a catastrophe." This is cited by Vandana Shiva, *Staying Alive: Women, Ecology and Development* (London: Zed Press, 1990), 218. Of course, Solow could be interpreted as uttering the silly claim that technology transcends all dependence on nature. But what he has said is capable of a more plausible interpretation, namely, that natural kinds, such as oil and trees, as natural resources are subject only to relative, but not absolute, scarcity.

23. This concept is more fully discussed in the section that follows.

24. Ted Benton, "Marxism and Natural Limits: An Ecological Critique and Reconstruction," *New Left Review* 178 (November/December 1989): 73.

25. But to prevent misunderstanding, the following claims should be distinguished:

1. Nanotechnology embodies perfect mastery of nature by humans because the artefacts produced by it have been deliberately designed and consciously executed according to a human blueprint.
2. The more complete the mastery and control of the processes under which such artefacts are produced, the less unpredictable outcomes would occur.
3. Unpredicted outcomes would, nevertheless, occur, as these could never be totally eliminated—for instance, whether the environmental impact of such artefacts, both in the short and long term, could entirely be fully monitored and anticipated, as well as whether such unanticipated outcomes are positive or negative when judged by an agreed set of technology assessment variables, would remain an open matter.

Nanotechnology might score over extant technology as far as 1 and 2 are concerned but would, like all technologies, be subject to the imperfection and incompleteness under 3. The control of nature, in this sense, through the anticipation of the unintended consequences of human actions may be said only to be a regulative ideal. In practice, such an ideal may, at best, be approximated to. It is in its nature never to be be completely attained. (But for an alternative understanding, see William Leiss, *Under Technology's Thumb* Montreal and London: McGill-Queen's University Press, 1990, 80.)

The view that the more precise the control of nature made possible by science and technology, the less "negative externalities" it may lead to, may be illustrated by the following kind of example: in controlling malaria, a liberal spraying of DDT indiscriminately over wide areas and in all seasons would, undoubtedly, lead to greater "negative externalities" than a more discriminate and sparing use of the chemical which turns out to be just as effective in curbing the disease. This more encouraging result is achieved through a better understanding of the breeding habits of the mosquitoes which act as vectors in the spread of malaria, of the complex endogenous causal linkages between the use of the chemical and the evolution of the targeted organism, as well as, the equally complex causal links in the transmission of the chemical through the food chain itself.

26. This point will be explored in Chapter 5 section entitled **Resisting Humean Projectivism**.

27. It is beyond the remit of this book to discuss virtual reality and its contribution to the process of eventually rendering the natural redundant. But clearly, it could act as a powerful contributor to such a tendency.

On the state-of-the-art technology itself, see Woolley, *Virtual Worlds*.

28. For a detailed discussion, see Keekok Lee, "Beauty For Ever?" *Environmental Values* 4 (1995): 213-25.

29. On the role of the sublime, not merely the beautiful in Romanticism, see Mitcham, who cites Edmund Burke's definition of the former (in *A Philosophical Enquiry into the Origin of our Ideas of the Sublime and Beautiful*): "'Whatever is fitted in any sort to excite the ideas of pain, and danger, whatever is in any sort terrible, or is conversant about terrible objects, or operates in a manner analogous to terror, is a source of the sublime' " (Carl Mitcham, "Three Ways of Being-With Technology," in *From Artifact to Habitat: Studies in the Critical Engagement of Technology*, Gayle L. Ormiston, ed. London and Toronto: Associated University Press, 1990, 53). Mitcham says that Hiroshima and Chernobyl fit this definition admirably. He argues that the sublime plays a vital role in the attitude of Romanticism to technology and accounts for its ambivalence to technology itself. He writes: "The attractive and repulsive interest revealed by the sublime expresses perhaps better than any other the uniqueness of the romantic way of being-with technology" (Mitcham, "Three Ways of Being-With Technology," 53).

30. Ted Benton, "Ecology, Socialism and the Mastery of Nature: A Reply to Reiner Grundmann," *New Left Review* 194 (July/August 1992): 55-74.

31. Plato, *Republic*, 372d-373b.

32. For an excellent detailed discussion of the implications of Marx's distinction between the realms of freedom and necessity for environmental philosophy, see Robyn Eckersley, *Environmentalism and Political Theory: Toward an Ecocentric Approach* (London: University College Press, 1992), 75-94.

33. For a discussion of the concepts of sufficiency and abundance, see Nicolas Xenos, *Scarcity and Modernity* (London: Routledge, 1989).

34. However, Martha Nussbaum (*The Fragility of Goodness* (Cambridge: Cambridge University Press, 1986), 373-77) points out that while the Platonic Aristotle maintained this view, the Aristotelian Aristotle, so to speak, was not so single-minded in elevating the purely contemplative life as the supreme good.

35. But the Latin term itself appears only to have been introduced by Henri Bergson (*Creative Evolution* London: Macmillan and Co. Limited, 1911). According to Jean Leclercq ("Vers La Société Basée Sur Le Travail," *Revue du Travail* 51 1950: 3), it was Bergson who "threw the concept of *homo faber* into the circulation of ideas" (Hannah Arendt, *The Human Condition* New York: DoubleDay Anchor Books, 1958, 341).

Arendt (*The Human Condition*) takes a different view about *homo faber* from the account given in this book. She sees its central significance to lie in man's attempt to create a permanent world for himself out of fabrication, but at the expense of doing violence to nature. However, permanence is only secured if fabrication results in objects which, strictly speaking, have neither use nor exchange value. Hence, for Arendt, the paradigmatic product of fabrication is art, not craft—paintings, sculptures, books. *Homo faber* is engaged with doing, rather than contemplating, *vita activa* having displaced *vita contemplativa*. But this reversal of the classical hierarchy by *homo faber* is only one stage in the history of modernity. The next stage, according to Arendt, consists of the very defeat of *homo faber* himself by the principle of happiness, and through that, the final victory of the *animal laborans* in modernity.

36. As the earlier note shows, Marx did not coin the term itself. But commentators rightly use the term, after it has been introduced by Bergson, to characterize Marx's thoughts. But according to Arendt, this might not be the right interpretation. She says Marx

as good as substituted *animal laborans* for the more usual definition of man as *animal rationale*—see Arendt, *The Human Condition*, 325-26. She distinguishes labor from work and claims that central to Marx's thoughts is the concept of *animal laborans*, not *homo faber*. She says that there is a fundamental contradiction in the whole of his thoughts:

> While it was an "eternal necessity imposed by nature" and the most human and productive of man's activities, the revolution, according to Marx, has not the task of emancipating the laboring classes but of emancipating man from labor; only when labor is abolished can the "realm of freedom" supplant the "realm of necessity." ... [I]n all stages of his work he defines man as an *animal laborans* and then leads him into a society in which this greatest and most human power is no longer necessary. We are left with the rather distressing alternative between productive slavery and unproductive freedom. (*The Human Condition*, 90-91)

But perhaps the contradiction can be explained away. Marx moves from man as *animal laborans* to man as *homo faber*. Industrialism, relying on modern science and technology, enables man to transcend his status as *animal laborans* to become *homo faber*. For Marx, as we shall see, the process of humanizing nature through technology also naturalizes humanity. While the technology which preceded the modern era did not enable man to transcend his status as *animal laborans*, the ever more powerful technology of modernity could do precisely that. Under modernity, it would be fitting then to define man in terms of *homo faber* instead. When the need to labor is finally transcended, man would enter the new Garden of Eden, having exorcised the biblical curse which defined man's post lapsarian predicament. Man's regaining Paradise is to be accomplished by turning himself into *homo faber*, aided and abetted by modern science and technology. Salvation lies in following the path marked *homo faber*, science and technology. On this account, it follows that contrary to Arendt's contention, it is the notion of *homo faber* which ultimately is central to Marx's vision of the good society to come.

This view of Marx assumes that there is no real epistemological break between the early and the late Marx in spite of differences between the two periods—see, for instance, Norman Geras, *Marx and Human Nature: Refutation of a Legend* (London: Verso Books, 1983).

37. See Lee, *Social Philosophy*, 228-78.

38. Johann Gottlieb Fichte, *The Vocation of Man* (London: John Chapman, 1848), 136-38.

39. G. W. F. Hegel, *The Philosophy of Right* (London: Oxford University Press, 1942), 41.

40. Fichte, too, shares this sentiment. He writes:

> When men shall no longer be divided by selfish purposes, nor their powers exhausted in struggles with each other, nothing will remain for them but to direct their united strength against the one common enemy which still remains unsubdued — resisting, uncultivated nature. No longer estranged from each other by private ends, they will necessarily combine for this common object; and thus there arises a body, everywhere animated by the same spirit, and the same love. (*The Vocation of Man*, 147)

NeoMarxists today think in a similar vein. Marcuse, for instance, in spite of his attempt to develop an emancipatory science, actually ends up reaffirming the mastery of nature through science and technology. Like Fichte, he too is against "misery, violence, and cruelty." The elimination of such "disvalues," for him, takes place against the so-called "pacification of existence." This means "the development of man's struggle with man and with nature, under conditions where the competing needs, desires, and aspirations are no longer organized by vested interests in domination and scarcity—an organization which perpetuates the destructive forms of this struggle" (Herbert Marcuse, One Dimensional Man: Studies in the Ideology of Advanced Industrial Society Boston: Beacon Press, 1964, 16). In other words, he looks forward to the day when our science and technology have freed us from scarcity, and vested interests no longer use their power to enforce domination and prolong scarcity. He believes that this will "open a universe of qualitatively different relations between man and man, and man and nature." Yet, he has this to say about nature:

> Pacification presupposes the mastery of Nature, which is and remains the object opposed to the developing subject. But there are two kinds of mastery: a repressive and a liberating one. The latter involves the reduction of misery, violence, and cruelty. In Nature as well as in History, the struggle for existence is the token of scarcity, suffering, and want. They are the qualities of blind matter, of the immediacy in which life passively suffers its existence. This realm is gradually mediated in the course of the historical transformation of Nature; it becomes part of the human world, and to this extent, the qualities of Nature are historical qualities. In the process of civilization, Nature ceases to be mere Nature to the degree to which the struggle of blind forces is comprehended and mastered in the light of freedom.
>
> History is the negation of Nature. What is only natural is overcome and recreated by the power of Reason. The metaphysical notion that Nature comes to itself in history points to the unconquered limits of Reason. It claims them as historical limits—as a task yet to be accomplished, or rather yet to be undertaken. If Nature is in itself a rational, legitimate object of science, then it is the legitimate object not only of Reason as power but also of Reason as freedom; not only of domination but also of liberation. With the emergence of man as the *animale rationale*—capable of transforming Nature in accordance with the faculties of the mind and the capacities of matter—the merely natural, as the sub-rational, assumes negative status. It becomes a realm to be comprehended and organized by Reason. (*One Dimensional Man*, 236-37)

and

> All joy and happiness derive from the ability to transcend Nature—a transcendence in which the mastery of Nature is itself subordinated to liberation and pacification of existence. All tranquillity, all delight is the result of conscious mediation, of autonomy and contradiction. Glorification of the natural is part of the ideology which protects an unnatural society in its struggle against liberation. ... Civilization produces the means for freeing Nature from its own brutality, its own insufficiency, its own blindness, by virtue of the cognitive and transforming power of Reason. And Reason can fulfill this function only as post-technological rationality, in which technics is itself the instrumentality of pacification, organon of the "art of life." (*One Dimensional Man*, 236-38)

These passages may be said to indict him of holding the view that while domination of man by man could be eliminated, domination of nature by man is inevitable. Nonhuman nature, by its very essence, is inferior, awaiting liberation by human hands. But this, surely, is no true liberation but a confirmation of its lowly status—humans transform so-called brutish, inadequate and blind nature into something structured according to human design. It is as if nature in losing its status of being brutish and blind ends up losing its identity and freedom by becoming "civilized objects" or, indeed, slaves, in man's gilded cage. Nature, untouched by human hands, is an object for conquest, to be turned into "civilized" or "humanized" nature. In other words, for Marcuse, the "liberation" of nature consists of its being transformed into artefacts, a curious type of freedom, indeed. Nature as nature is necessarily unfree; it requires humans to "free" it. But upon achieving "freedom," it no longer is nature but has become a human artefact. Marcuse, for all his agonizing, has not gone beyond the deeply entrenched anthropocentric relationship between human and nature specified by Idealism as well as by the metaphysics of Scientific Naturalism. *Ipso facto*, the same holds true of Marx—see also C. Fred Alford, *Science and the Revenge of Nature: Marcuse and Habermas* (Florida: University Presses of Florida, 1985), 21-68; Andrew McLaughlin, *Regarding Nature: Industrialism and Deep Ecology* (New York: State University of New York Press, 1993), 118-20.

41. For a more detailed account of his contribution, see Lee, *Social Philosophy*, 161-68.

42. John Locke, *Second Treatise* (Section 40) in *Two Treatises of Government* (Cambridge: Cambridge University Press, 1988), 296.

43. Locke, *Second Treatise*: Section 43, 298.

44. On this point, see Geras, *Marx and Human Nature.*

45. Bergson, too, concurs with this view. The instrument, designed and executed by human intelligence

> reacts on the nature of the being that constructs it; for in calling on him to exercise a new function, it confers on him, so to speak, a richer organization, being an artificial organ by which the natural organism is extended. For every need that it satisfies, it creates a new need; and so, instead of closing, like an instinct, the round of action within which the animal tends to move automatically, it lays open to activity an unlimited field into which it is driven further and further, and made more and more free. (*Creative Evolution*, 148)

46. See David Concar, "The Organ Factory of the Future?" *New Scientist* (18 June 1994): 24-29.

47. See Drexler, *Engines of Creation*, 117-46.

48. Bergson, *Creative Evolution*, 146.

49. Bergson, *Creative Evolution*, 146.

However, to be fair to Bergson, when he talked about the ever expanding realm of manufacturing, he was, nevertheless, of the opinion that fabrication must be confined to abiotic and exbiotic nature only. In this, he was not far-seeing enough—the biotic too can be drawn into the domain of manufacturing.

50. The intelligence of instrumentation as the essence of modernity is once again reflected in a recent publication by Steven Weinberg entitled "Life in the Universe," *Scientific American* 271 (1994): 22-27. It contains a chart, captioned "The emergence of intelligence." Of the eighteen landmarks highlighted, spanning 1.9 million years B.P. to 1993, twelve are distinctly tools, instruments, machines of one kind or other. Of the

remaining six, four are inventions as techniques leading to the production of material artefacts, namely, rock engraving; writing (first on papyrus, then on paper); money (the first coins); the earliest farming which could refer to the practices of agriculture and domestication of plants and animals as well as to tools and implements to assist such practices. The remaining two refer to fundamental theoretical breakthroughs in scientific thought, namely, the theory of relativity in 1905 and the structure of the DNA in 1953. These may have been included, perhaps, because they obviously lead to radical technological innovations like the communications satellite in the case of the former (one of the landmarks mentioned) and biotechnology in the case of the latter (however, biotechnology, itself, does not figure in the list of highlights).

This little chart implies that human intelligence is a history of progress, by and large, in tool making, from the first simple stone tools of 1.9 million years B.P. to the extremely complex and sophisticated communications satellite of 1993. Furthermore, it betrays a predilection for modern devices—nine out of the twelve tools and machines highlighted come from the period 1440-1993. This seems to imply that while human intelligence was tardy and intermittent pre 1440, it suddenly assumed great leaps forward in the last five hundred years of human history. One may conclude from this attempt at deconstruction that the chart is a clear reflection of the spirit of modernity.

In Western thought, the earliest hint of the idea that instrumentation is a measure of human intelligence may be traced to Anaxagoras as reported by Aristotle. Anaxagoras is said to be of the view "that it is his possession of hands that makes man the wisest of living things" (G. S. Kirk and J. E. Raven, *PreSocratic Philosophers* Cambridge: Cambridge University Press, 1977, 393).

51. Bergson, *Creative Evolution*, 192-93.

52. Alternative definitions are: "a combination of rigid or resistant bodies, having definite motions and capable of performing useful work" (*MacGraw-Hill Dictionary of Science and Technology* and *MacGraw-Hill Encyclopedia of Science and Technology*); "a device consisting of two or more resistant, relatively constrained parts that may serve to transmit and modify force and motion in order to do work" (*New Encyclopaedia Britannica*, Macropaedia, 1975, 11: 231); or "a device for transfroming or transferring energy" (George H. Martin, *Kinematics and Dynamics of Machines* (New York: McGraw-Hill, 1969), 3). [These definitions are cited by Mitcham, *Thinking Through Technology*, 170, 327.]

Machines may be distinguished from mechanisms in mechanical engineering terms in the following way: a machine-like device, which does not perform useful work, is a mechanism. A clock, then, is a mechanism rather than a machine as it modifies motion (direction), not motion and force (amplification or reduction)—see Mitcham, *Thinking Through Technology*, 170.

53. See Drexler *et al.*, *Unbounding the Future*, 98.

54. This shows that nonhuman organisms may be said to be loci of intrinsic value. Chapter 5 section entitled **Resisting Humean Projectivism** sets out some arguments to challenge two key theses of anthropocentrism, namely, that humans are the source of all values and the sole locus of intrinsic value.

55. There is another objection—an ontological one—which will be spelled out in Chapter 5.

56. *Hutchinson Dictionary of Science* (1993), 643. Another account reads: "In physics, the term work refers to the transference of energy that occurs when a force is applied to a body that is moving in such a way that the force has a component in the direction of

the body's motion" (McGraw-Hill Concise Encyclopedia of Science and Technology 1984, 1891).

57. See Bud, *The Uses of Life*, 28-34.

58. An Edinburgh-based biotech company has of late patented Tracy the sheep, which has been genetically engineered to produce Alpha 1, an anti-trypsin drug used in treating people with emphysema. Herman the bull has had a human gene inserted into his germ-line so that his daughters may produce, in their milk, an anti-bacterial drug called lactoferrin. This would then make breast-milk substitutes more like human milk—see John Vidal and John Carver, "Lambs to the Gene Market," *Guardian*, 13 November 1994, 25.

59. Humberto R. Maturana and Francisco J. Varela, *The Tree of Knowledge: The Biological Roots of Human Understanding* (Boston: Shambhala, 1988). Especially in Chapter 2, they give a non-technical presentation of their theoretical ideas.

As for their work in theoretical biology, Maturana, Varela and Uribe have published an earlier version in English entitled "Autopoiesis: The Organization of Living Systems," in *Biosystems* 5 (1974): 187-96. Varela published *Principles of Biological Autonomy* in 1979. This was followed by Maturana and Varela, *Autopoiesis and Cognition: The Organization of the Living* (Dordrecht/Boston: D. Reidel Publishing Company, 1980), their influential major work.

Warwick Fox (*Toward a Transpersonal Ecology: Developing New Foundations for Environmentalism* Boston: Shambhala, 1990, 165-76) introduced the concept "autopoiesis" to environmental philosophy. See Eckersley (*Environmentalism*, 60-61, 70-71) for a direct endorsement, and Arran Gare (*Postmodernism and the Environmental Crisis* London and New York: Routledge, 1995: 129) for an oblique endorsement. But see Plumwood (*Feminism*, 210) for a more critical response. Freya Mathews while saying that her "idea of self-realizability matches up, in essential respects, with Maturana's notion of autopoiesis," nevertheless, recognizes that hers "differs from Maturana's in the following fundamental respect ... where Maturana considers that autopoiesis dissolves the apparent telos of living systems, I see the capacity for self-realization, understood in systems-theoretical terms, as definitive of telos" (*Ecological Self* London: Routledge, 1991, 173).

60. Erich Jantsch, *The Self-Organizing Universe: Scientific and Human Implications of the Emerging Paradigm of Evolution* (Oxford: Pergamon Press, 1980).

61. Ilya Prigogine and Isabelle Stenger, *Order Out of Chaos: Man's New Dialogue with Nature* (New York: Bantam, 1984).

62. Fritjof Capra, *The Turning Point* (New York: Simon and Schuster, 1982), 271-72.

63. Maturana and Varela, *Autopoiesis and Cognition*, 76.

64. Maturana and Varela, *Autopoiesis and Cognition*, 77.

65. Maturana and Varela, *Autopoiesis and Cognition*, 135.

66. See Maturana and Varela, *Autopoiesis and Cognition*, 79-80.

67. Maturana and Varela, *Autopoiesis and Cognition*, 135.

68. Maturana and Varela, *Autopoiesis and Cognition*, 84.

69. In ancient Greek thought, we have seen that *bios theoretikos* as the *summum bonum* informed its anthropocentrism. Moreover, as far as Aristotle was concerned, intrinsic/immanent teleology is logically prior to external teleology. In medieval thought, the glorification of God by man as the supreme goal of this worldy existence also acted as a constraint.

70. See Maturana and Varela, *Autopoiesis and Cognition*, 76.

71. The term here refers to the standard understanding.

72. These are (MV)machines, not machines, standardly understood.

73. Maturana and Varela, *Autopoiesis and Cognition*, 82-83.

74. For a criticism of biotechnology akin to this, see Michael Fox, "Transgenic Animals: Ethical and Animal Welfare Concerns," in *The Bio-Revolution: Cornucopia or Pandora's Box*? Peter Wheale and Ruth McNally, eds. (London: Pluto Press, 1990), 31-45 and H. Verhoog, "The Concept of Intrinsic Value and Transgenic Animals," *Journal of Agricultural and Environmental Ethics* 5 (1992): 147-60.

Chapter Five

Ontology and Axiology

The main thrust of the book so far has been (a) to examine the character of past, extant and near future technologies in order to work out their implications for a comprehensive environmental philosophy; (b) to show that the crucial underlying threat to nature, especially from recently established technology like biotechnology and rising technology like molecular nanotechnology, comes not so much from the pollution and the damage which extant technology, being ecologically insensitive on the whole, causes to the environment, but from its inherent tendency to transform the natural to become the artefactual; (c) to establish that the natural and the artefactual are two distinct categories; (d) to argue why and how we must defend the natural against a relentless assault mounted by modern technology in the last hundred years or so.

This chapter will strengthen the arguments for such resistance by showing that the main axiological argument for the defense of nature (as nature_{nh}, nature_{nk}, nature_{fa}), namely, that nature has intrinsic, not merely instrumental, value for humankind, is inextricably entwined with the ontological argument for the defense of the natural. In other words, it seeks to argue that nature as a distinct ontological category has value and, therefore, ought to be respected for itself. It would follow that the more fundamental and also the most fruitful way of establishing that nature is intrinsically valuable is *via* the ontological and not the direct axiological route more commonly pursued. And having established that, it would be obvious why this book considers it more appropriate to talk of nature possessing the value of independence rather than intrinsic value.[1]

The notion of independence as an ontological value is defended, as the conception of environmental philosophy advocated by this book is a comprehensive one, in two related senses of 'comprehensive'—that environmental philosophy includes within its remit not only biotic nature on Earth, but also abiotic nature on Earth as well as on other heavenly bodies (at least those in our solar system).

This, in turn, requires:

1. Introducing—in **Intrinsic Value, Trajectories and Independent Value**— (a) the concepts of the trajectory and of independent value, and (b) the distinction between two kinds of attribute an entity may be said to possess, the primary being ontological, while the secondary, axiological, in character.
2. Arguing for the notion of ontological dyadism but rejecting that of dualism to reinforce the concept of the trajectory—in **Ontological Elimination, Dualism and Dyadism.**
3. Reinforcing and further exploring the concept of independent value as being fundamentally ontological in character by examining the notion of ecosystem health in current environmental discussion, in order to show that ecosystem health (like complexity or intricacy) is an axiological characteristic, which both naturally-occurring and artefactual entities can be the bearers and, not an ontological attribute, like independence (autonomy), which only naturally-occurring beings and processes can be bearers—in **Ecosystem Health and the Human/Nonhuman Ontological Dyadism**.
4. Arguing for an ontology-led environmental philosophy, based on differences and the notion of independent value, rather than an axiology-led environmental ethic, based on similarities and the notion of intrinsic value—in **Environmental Ethics and Environmental Philosophy: Axiology and Ontology**.
5. Grasping the philosophical implications of the distinction between anthropogenic and nonanthropogenic for our attitude towards and policy-making about nature—in **Anthropogenic and Nonanthropogenic**.

The above themes form the core of the chapter. If these themes can be plausibly defended and established, they could be used to tie up with that pursued, especially in Chapter 4 and consistently throughout this book, namely, the threat posed to the natural by *homo faber*, relying on technology, which is increasingly nature-replacing rather than nature-saving. The last section —**Narcissism and *Homo Faber***—reinforces the claim that this threat is an ontological one *via* the notion of narcissism.

However, before the core themes can be properly set out, the ground has to be prepared in several important ways:

1. A critical examination is undertaken of one of the key obstacles to their defense, namely, the deeply entrenched view in modern Western philosophy that human consciousness is the source of all values. The first section —**Resisting Humean Projectivism**—seeks to undermine this claim by introducing the distinction between 'mutely-enacted' and 'recognized-articulated' values. But, in so doing, it succeeds in showing that, at best, only biotic nature (in the form of individual organisms) can be said to be intrinsically valuable in the absence of human consciousness and human valuers. However, such partial success is inadequate for establishing an environmental philosophy which claims to be comprehensive in scope.
2. Extant attempts to ground intrinsic value in abiotic nature will have to be critically assessed to see if help is ready to hand. However, it will be shown —in **Interests and the Intentional Stance**—that neither the more commonly canvassed notion of interests nor Plumwood's notion of the intentional stance could do the job.
3. This then calls for a fresh effort to establish that it makes sense to claim that abiotic nature (on Earth and on other heavenly bodies in our solar system) is morally considerable—**Abiotic Nature and Intrinsic Value** explores such a possibility and, among other strategies, introduces the distinction between beings which exist 'for themselves' and beings which exist 'by themselves.' The former is appropriate for characterizing biotic nature (as individual organisms); the latter for both biotic and abiotic nature. This deeper sense of 'by themselves' paves the way for introducing the notion of the trajectory and the substitution of independent for intrinsic value.

Resisting Humean Projectivism

In Chapter 1 we saw that Galileo, among others, made a major contribution toward the articulation of the new philosophy to support the new science then emerging. That contribution includes the idea of mathematizing and quantifying nature, the views that the secondary sensuous qualities are not real but mind-dependent, and most importantly from the standpoint of environmental philosophy, that nature itself, reduced only to the primary qualities which make up what Descartes called *res extensa*, as inert matter, is devoid of value. As we have also seen, this means that nature then only has instrumental value for humankind, thus leaving the way open for human agency (substituting for natural processes as the efficient cause) to control and manipulate nature through its science and technology in accordance with human purpose and design.

However, that nature is value-free is only one side of the coin—the other in the history of modern Western philosophy is completed by Humean projectivism. According to Hume, when someone claims a state of affairs or an entity to be good or bad, valuable or valueless, all attribution of values are simply the projections of human feelings, emotions or attitudes on to the world out there.[2] All values are, therefore, anthropogenically generated. Hume relied on this technique of projection quite generally; it may be helpful to see how he used it in another area of his philosophy.

Hume argued against those philosophers who considered the notion of cause to involve necessity and that necessity itself lay in nature. In other words, to hold that A causes B is to hold that there is something in the inner constitution of A such that it necessarily brings about B as its effect. A lump of arsenic ingested causes the death of the ingester. This means that there is something in the nature of arsenic *qua* arsenic which causes the death; clearly if the person had ingested a lump of chalk, it would not have killed him, unpleasant though the experience would have been. But Hume's empiricism as well as his skepticism led him to deny any such necessity. He gave his well-known analysis of cause as constant conjunction or uniform sequence, involving only variables which are overt and directly observable—to say that arsenic causes death in humans is to say no more than whenever anyone ingests (a certain quantity of) arsenic, s/he dies. However, Hume realized that the conceptual analysis he had given remained incomplete, as it could not account for the firm conviction which people have that arsenic is lethal, because of something it possesses, in virtue of the fact it is arsenic and not some other substance. So he argued that this conviction is really a trick played by our own psychology. Constant conjunction of the cause and effect variables has set up in the observer a habit of expectation such that when s/he sees the next instance of someone ingesting arsenic, s/he expects the person to die; but the observer then proceeds to project this expectation on to nature, leading to the claim, erroneous in reality, that the necessity of effect following cause lies in the nature of arsenic itself rather than in the observer's own psychology.

Hume extended his skepticism to attack not only the possibility of necessity in nature but also the possibility of values in nature. To this he added two related theses, (a) the so-called "is/ought" gap, and (b) severing the link between moral evaluation and reason. Regarding the first, in moral arguments, he noticed that the premises which constitute the evidence are usually factual matters but that the conclusion based on it is always evaluative or presciptive in character; furthermore, that accepting the factual premises to be correct but denying the conclusion would involve no contradiction. This logical gap between descriptive premises and evaluative/prescriptive conclusion ties in with his view that reason plays no role in moral evaluation, and is confined, at best, to choosing the most efficient means to achieve whatever end the human agent is predisposed to adopt, given her/his feelings and emotions. For Hume, reason is the slave of the passions; but as passions determine the ends we choose, Hume implied that ends are irrationally chosen.[3] It follows then that all values are mere projections of human emotions and feelings upon nature, and that in the absence of human consciousness, the world is devoid of all values.

One can see from this brief account of Hume's ideas that they provide part of the crucial groundwork for the deeply embedded anthropocentrism in modern Western philosophy, the view that human consciousness is the source of all values which is sometimes summed up by the dictum, "no values without (human) valuers," and its implicate that humankind is the sole locus of intrinsic value.[4]

To challenge this dictum is then in part at least to challenge Humean projectivism. This section will attempt to do so but in the context of biotic nature only (at least at the level of individual organisms),[5] leaving the sections to follow to complete the task when they look, among other things, at the possibility of value in abiotic nature.

The easiest point of entry is to take the well-established contrast in environmental philosophy between intrinsic and instrumental value. But as the former is less easy to clarify directly, it is best to do so obliquely through an initial clarification of the latter. Instrumental value is said to obtain whenever an agent A finds an object O of use to it as a means of satisfying whatever goal A might have. However, the first substantial step of the argumentation consists of showing that instrumental value can and does occur in the absence of humans. The important thing here to bear in mind is that A need not necessarily be a human agent. A human agent would find a rabbit of instrumental value in assuaging her/his hunger; but so would a bird find an insect of instrumental value in assuaging its hunger.

But, in evolutionary history, birds arrived on Earth before humans, and it is not inconceivable that they would survive beyond the extinction of *Homo sapiens* in the future. This then entails that instrumental value could exist in a world without human consciousness, as it is independent of such a consciousness. However, establishing this need not impress the upholders of the claim that human consciousness is the source of values and the sole locus of intrinsic values, unless one can go on to show that the existence of instrumental values entails the existence of intrinsic value.

This may be done in two steps. First by pointing out, as the Routleys ("Against the Inevitability of Human Chauvinism") have done, that in order to avoid an infinite regress, the agent A for which the object O has instrumental value must be presumed itself to be intrinsically valuable. The argument spelled out would run like this: if agent A is not presumed to be intrinsically valuable, then it itself is only of instrumental value to some other agent, say B, which in turn may be said to be only of instrumental value to yet another agent C, and so on, *ad infinitum*. As a matter of fact, anthropocentrism as defined above stops the infinite regress by postulating humans to be intrinsically valuable. The thesis of external teleology associated with Aristotle is said precisely to do this. Plants exist for the sake of animals and animals exist for the sake of humans. But this privileging of humans to break the chain of instrumental reasoning is unjustified, as it begs the question why agent A, which is nonhuman, could not be said to be intrinsically valuable.

To see why such a nonhuman agent could be said to be intrinsically valuable constitutes the second stage of the argument. In the example of the bird cited above, admittedly, the bird itself might in turn be of instrumental value to the cat, but this does not undermine the argument that the bird in eating the insect does so for no other end than that of sustaining its own functioning integrity.[6] In other words, the bird does not eat the insect in order to assuage the hunger of the cat but its own hunger. The fact that it could be of such instrumental value to the cat is only a happenstance, albeit an important one. This then shows that the bird possesses intrinsic value. It may be said to have a 'good of its own,' namely, striving to sustain its own functioning integrity. The same point may be made relying on the reconstruction of Aristotle's thought set out in the earlier chapters of this book, namely, that for Aristotle, it is his thesis of intrinsic/immanent teleology which is primary while that of external teleology is secondary, and that it distorts Aristotle's account of humankind's attitude to nature to invert the order between these two theses of teleology. For Aristotle, the bird exists, first and foremost, to unfold its own *telos* and only secondarily to be of use to cats, foxes or humans. It has not come into existence, nor does it continue to exist, to be of instrumental value to these other beings, nonhuman or human.

If the above argumentation is cogent, then it would have succeeded in challenging Humean projectivism in the following ways: that (a) instrumental value in nature is independent of human existence, (b) intrinsic value in nature (at least in individual organisms like birds) is independent of human existence, (c) humankind is not the only source of values, whether instrumental or intrinsic (d) humankind is not the only locus of values, whether instrumental or intrinsic.

However, anthropocentrists may still remain unconvinced by this challenge to Humean projectivism. At this stage their Humeanism may be joined by their Cartesianism and they may argue that human consciousness is indeed unique; and it is this which enables such a consciousness to endow the world with values, in a way which neither animals (which does not share the unique brand of human consciousness) nor plants (which have no consciousness whatsoever) can be said to project values on to nature.

To meet this objection, one must first distinguish between two senses of intrinsic value—(a) being valuable 'for itself' and (b) being valuable 'in itself.'[7] The sense established above is the first and is shared by both humans and nonhuman organisms alike. It is involved with their capability to sustain their own functioning integrity and their striving to do so. The second sense is confined solely to humans and is involved with their unique capability for language and to reason, particularly, in an abstract manner. Humans are indeed beings who not merely strive to maintain their own functioning integrity but also do so in a unique way, namely, by using their technology to ensure that they are more successful than other organisms in their respective strivings to maintain their functioning integrity.

As all individual organisms, human and nonhuman, each has a 'good of its own,' it remains true that the 'good of the human' goes beyond the 'good' of nonhumans—it includes a dimension which is related to the uniqueness of human consciousness. While each nonhuman organism merely goes about striving to sustain its own 'good,' it does so not being aware (in a self-conscious manner) that it is doing so. In contrast, humankind, while striving to sustain its own 'good,' is self-consciously aware that it is doing so. Furthermore, it can recognize that other organisms have each a 'good' of their own, while nonhuman organisms cannot recognize that humankind has a 'good' of its own or, for that matter, that other fellow organisms each have a 'good' of their own. While nonhuman organisms merely enact their own 'good,' humankind can, not only recognize, but also articulate, the 'good' of nonhuman organisms as well as its own 'good.'

This then shows the need to draw attention to another distinction, this time, between 'mutely-enacted' values and 'recognized-articulated' values. Nonhuman organisms, in unfolding each its own 'good,' in striving each to maintain its own functioning integrity and, thereby, exhibiting that each is intrinsically valuable 'for itself,' at the same time, embody and display mutely-enacted values. In contrast, humans, in being also intrinsically valuable 'in themselves,' embody and display not only mutely-enacted values but, also, are the bearers and generators of recognized-articulated values.

A world without human consciousness is a world indeed without recognized-articulated values, but is not a world without mutely-enacted values. The appearance of human consciousness on the evolutionary scene did add another dimension to the story of values in nature$_c$. For the first time, recognized-articulated values were generated. But although anthropogenically generated, these are not generated *carte blanche* and *au fond* by human consciousness. They are only anthropogenically generated in the more limited sense that human consciousness is capable of recognizing and articulating (through its linguistic and other powers) mutely-enacted values embodied in nonhuman (biotic) nature. In this sense, too, when humankind becomes extinct and no comparable consciousness were to take its place, then recognized-articulated values would disappear, although not necessarily mutely-enacted values.

In the light of the above clarification, one may conclude the following:

1. To say that human consciousness is the source of value does not have to be interpreted according to the metaphysics of Scientific Naturalism. Instead, it should be re-interpreted to mean that it is the source of recognized-articulated values only and not of mutely-enacted values. The recognition itself of the nonanthropogenic source of mutely-enacted values thereby renders it false to say that human consciousness is the source of all values, and misleading to say that it is the source of values *simpliciter*.
2. A world with human consciousness and nonhuman organisms in it is a world in which intrinsic value exists both in sense of 'for itself' and 'in itself.' While humans are intrinsically valuable in both senses, nonhuman organisms are intrinsically valuable only in the sense of 'for itself.' It follows that a world without human consciousness is a world where intrinsic value in that sense alone could be said to exist.

3. A world with human beings who are the unique bearers of intrinsic value in the sense of 'in itself' is also a world where moral values exist. It is the unique character of human consciousness which renders its bearers moral beings, capable of recognizing that other nonhuman organisms have 'goods' of their own and are generators of values in nature. This recognition in turn leads to the recognition that destroying or eliminating such values from nature is *prima facie* wrong. In other words, human consciousness can and ought to recognize that there are moral constraints on its goal to control and manipulate nature, and on its attempt to transform biotic nature to become biotic artefacts. The privileging of humankind over biotic nature does not necessarily, therefore, follow from its being the bearer of its own unique brand of consciousness. Rather, recognition of such uniqueness entails moral duties to nonhuman biotic others not to harm them as source and loci of intrinsic value understood in the sense of 'for itself.'

Interests and the Intentional Stance

In order to argue against the "Sole Value Assumption" (that humans are the sole locus of intrinsic value) and for intrinsic value in nonhuman entities, many environmental philosophers search for a concept which can intelligibly be attributed to such entities, and which is philosophically robust enough to sustain the weight put on it. One such concept is that of interests.[8] Those entities to which interests may meaningfully be attributed could be said to be the loci of intrinsic value. Animals, or at least the higher animals, may readily be seen to satisfy this criterion of moral considerability. If so, the class of beings with intrinsic value would certainly be wider than the class of humans. But what about plants? Can they meaningfully be said to have interests?

Within the individualist framework, those who answer 'yes' distinguish between two logically distinct senses of the notion 'interest,' which are reflected in the following two propositions:[9]

(p) Peter takes an interest in keeping fit and eating healthily.
(q) It is in Paul's interest to keep fit and to eat healthily.

Proposition p implies that Peter cares about his health, and he really wants to live healthily. Proposition q is true even if Paul is indifferent to the matter, if he remains in ignorance about it, and/or if he lives a lifestyle which is diametrically opposed to it. While p is mediated *via* conscious desires and beliefs, q is not. In the case of q, if conscious desires and beliefs are not understood as being linguistically expressed, then animals could be said to possess them;[10] but plants still would not qualify, as they lack consciousness altogether. However, when 'interest' is understood as in sense q, 'plants' could substitute for 'Paul' and the proposition would still be meaningful. Plants, then, may be said to possess interests and qualify, too, as loci of intrinsic value.[11] Furthermore, it makes sense to say that plants could be made better or worse off when they lack something which is instrumentally valuable to them. For instance, a plant is worse off in the absence of water, sunlight or a particular nutrient, and is clearly better off when such items are available to it. An entity which could be said to be better or worse off under certain determinate conditions could be said to possess interests.

Another way, perhaps, of making the same point, is to say that plants have needs which, if not satisfied, could result in their suffering harm. The notions of needs and interests overlap such that entities with needs are also entities with interests. However, this is precisely what Feinberg ("The Rights of Animals and Unborn Generations") denies. Although he is prepared to concede that plants have needs, he, nevertheless, argues that needs are only meaningfully attributed to plants, if and only if, human interests are frustrated when those needs are not met. But this is simply a bad argument. Plants may legitimately be said to have needs in their own right independent of whether human interests suffer when they are not met. Of course, if humans happen to have an interest in the apples from an apple tree, then the failure of the tree to meet its needs (and as a result fails to bear fruit, thus suffering harm) is frustrating to the humans concerned. But one could just as readily imagine a tree, like belladonna, whose fruit is poisonous to humans, failing to bear fruit, because its nutritional needs have not been properly met. Organisms, as we shall see (in **Abiotic Nature and Intrinsic Value)**, exist first and foremost 'by themselves' and not in order to satisfy human interests. In living 'by themselves,' they must strive to meet their own needs, and in that sense live 'for themselves.' If needs provide a legitimate basis for interests, then plants most certainly have needs and, therefore, interests, which have nothing to do with human interests whatsoever.

Assuming that Attfield, and not Feinberg, is correct about plant organisms, and that the notion of interest can legitimately underpin all life forms, can the notion of interest, nevertheless, be stretched to legitimize abiotic forms or entities and to endow them with moral considerability and intrinsic value? Let us explore a possible answer to this question. First of all, while the notion of needs and the corresponding notion of harm apply to organic entities, they do not seem to be so readily applicable to inorganic entities. Does a mountain have needs in the way an oak tree has needs? An oak, if persistently starved of water, would eventually perish. To persist as an oak and sustain itself as an oak, the tree must be successful in meeting its long-term need for water. Are there analogous circumstances in which a mountain would no longer persist as a mountain? There seems to be no obvious analogue. At best, one can say that a mountain will no longer be a mountain if an earthquake occurs to flatten half of it, or a quarrying company pulverizes most of it to gravel. But an earthquake is a teleomatic phenomenon, and so is the flattening of the mountain, which it causes in its wake. An oak, too, would be damaged and would not continue to exist if it were caught up in an earthquake. In this case, we do not say that the oak suffers harm because its needs have not been met. Similarly, we do not say that the needs of the mountain have not been met as a result of the earthquake occurring. In both cases, we simply speak of their destruction. It follows that at best, we can meaningfully talk about damage or destruction of abiotic forms, not harm to them. Similarly, in the case of the mountain being flattened by quarrying—the quarrying (which is not a teleomatic phenomenon but a human intention executed into action) destroys or damages the mountain such that it loses its identity as a mountain. To say that the mountain need not have been destroyed is something quite different from saying that the oak tree would no longer be a live oak tree but a dead one if its need for water systematically fails to be satisfied. The former simply is a shorthand way of saying that the quarrying company ought not to have been allowed to destroy the mountain, that the mountain should be protected from human desires. The latter, as we have seen, refers to genuine needs which the tree has, independent of any human interest in the matter of the tree meeting those needs.

Those who wish to argue that abiotic forms may be the loci of intrinsic value could lean on the notion of damage or destruction, though not on the notions of needs, harm and interests which are only appropriate in the case of individual organisms. But to see why the latter set of concepts is not available in the case of abiotic forms, and also to see how far the notion of damage or destruction can be helpful, the other related route to moral considerability must next be explored, namely, the so-called intentional stance,[12] to see if it could be meaningfully extended to cover abiotic forms.

But to do that, one must first distinguish between two types of logic—extensional (or truth functional) and intensional. Take the following two propositions:

A. Peter is convinced that there are leprechauns at the bottom of the garden.
B. There are leprechauns at the bottom of the garden.

Proposition B falls within extensional logic which, in its classical form, has two values—truth and falsity—such that it is straightforwardly either true or false. However, the truth or falsity of Proposition A is totally independent of the truth or falsity of Proposition B, which appears to be part of it. Proposition A can be true even if there are no leprechauns at the bottom of the garden. As such it falls within intensional logic—'Peter is convinced that' refers to a mental element which is absent in B. In this example, the mental element is about human psychology. But the mental element could extend beyond the human domain. For instance, it is not obviously unintelligible to assert the following:

C. The cat is convinced that there is a mouse in the cupboard.
D. The elephant is convinced that the man in question is the killer of her baby.

Moreover, intensional discourse covers matters much wider than those about psychology, whether human or (higher) animal. It includes the teleological (in the sense referred to by the thesis of intrinsic/immanent teleology)—ends, functions and directions. As such it covers agents which have no consciousness, such as plants. One can meanigfully say the following:

E. The honeysuckle is growing shoots to get around the large post which has been planted in front of it, blocking out the sunlight.
F. The dandelion has grown deeper roots than normal this year because of the unusual drought which has affected the moisture content of the soil.

Teleological (or teleonomic) talk in this sense has nothing to do with consciously entertained purposes and aims, whether linguistically expressed or not. It is this which Plumwood calls "the intentional stance" and in her view, the intentional is a subclass of intensional discourse.[13]

Plumwood and this author both agree that "the intentional stance" obtains in the case of living organisms. She writes: "We do not have to decree, for example, along with Dennett, that plants are 'degenerate, uninteresting and negligible Intentional systems' ("Conditions of Personhood," 180). To a more sensitive and less human-centred view, the plant world includes fully intentional others whose strivings, interactions and differences in life strategy are intricate, amazing and mysterious."[14] However, she goes further to say that it is even appropriate in the case of abiotic nature or inorganic entities: "[T]he glaciated valley can easily be conceived as such an intentional system, if considered as part of a directional, developmental process of the earth, and we might both hinder its journey and stop it telling its story by damming it, for example,"[15] and

> [d]rawing the moral boundary at living things (from the point of view of moral considerability) has the problematic consequence that the wild river, the forbidding mountain and the venerable glaciated landscape on which the story of the earth's history and power is inscribed, have value only for and in virtue of the living things they contain or entertain.[16]

However, although one may, in certain limited contexts, detect a direction to geological/chemical/physical processes in nature, it is not obvious that such teleomatic phenomena display 'direction' in quite the way biotic phenomena do. What is distinctive about the latter is that all living organisms strive to maintain their own functioning integrity and to realize their own respective *telos*. They keep entropic increase at bay through maintaining their structure and their functioning integrity to stay alive, and successfully to reproduce. Take the Emperor penguins in Antarctica. When the female has hatched its egg, both the female and male penguins strive to ensure that the egg is successfully transferred from the former to the latter—this operation must be completed within a very short period, as the egg would not survive the cold if exposed for more than two minutes. When the transfer has taken place, the male then protects and hatches it during the next four months, while his female partner travels about a hundred miles or so to get to the sea in order to feed and fatten herself. At the first stirring of the Antarctic spring, the egg is hatched. The hatching is timed to coincide with the return of the chick's mother from the sea. But in spite of the long separation, the partners successfully recognize and locate each other. At this point, another intricate transfer begins —the mother takes over and straightaway must feed the chick, as any delay would cause the chick to die of cold and hunger. When the offspring has been handed over, then it is the turn of the male partner to walk the hundred miles or so to the sea to feed himself—he has endured hunger and extreme cold for four months all through the depth of the Antarctic winter, protecting and hatching the egg.[17]

A plant, too, strives or 'takes steps' to ensure that its leaves grow in such a way as to be exposed to sunlight, to conserve moisture when the ambient temperature is too hot or too dry by varying its rate of evapotranspiration, to get water at lower levels of the soil by sending its roots to tap a deeper source of moisture, all in order to maintain its own functioning integrity. It is true that a plant is not mobile and cannot literally uproot and transport itself to another place when its environment becomes inhospitable, unlike an animal which can. When a water-hole dries up, or when thick snow on the ground prevents forage, an animal can migrate to another water-hole or to slightly warmer climates. But even a plant is not entirely helpless, as we have just mentioned.

But is there an analogue to striving in the case of abiotic (or teleomatic) phenonema? Take a mountain. When it is being eroded away by natural forces, whether physical, chemical or biological, does it strive to maintain its integrity in the way life forms do? Apparently not, as it 'sends out' no recognizable signals to another part of it to replace what is being eroded away; whereas, in the case of a plant, if its leaves are ripped off by the wind, it would 'send out' signals to replace them (and provided the damage is not too great or too persistent, the plant stands a good chance of recovery).[18] In the case of the mountain or rock, there is no known mechanism of self-repair or self-maintenance, whereas plants are known to be capable of tissue and cell renewal, even self defense.

Plumwood may think it unfair to use the example of a mountain. Would not a wild river be more relevant in making the point? Maybe not. The wild river, too, lacks any known mechanism for self-repair or self-maintenance when its integrity is being assailed, whether the assail comes from an anthropogenic source or not. Take the former; suppose humans dam it or divert its water in some way which tames it. It is not capable of self-rejuvenation. If it regains its former status, it is because the humans involved have constructed an inefficient dam which the weight of its water washes away. It is different with a plant. A gardener who considers it to be a weed may chop off its stem, but so long as its roots are intact, it could renew the stem. And if the gardener removes its roots as well, that is not necessarily the end of the matter, as even some remnants may be sufficient for regeneration in certain instances. Some plants are capable not only of renewing their own tissues and cells but also of reproducing themselves through cloning. But if a plant propagates by seeds, these could be disseminated by wind or animal; these will start to germinate and grow when suitable circumstances occur. Some could lie dormant for years until conditions are ripe. A river does not spawn another like itself—at best, it branches when it meets a particular set of circumstances.

Of course, one can agree with Plumwood up to a point. Take the damming of a wild river. It is true that human intervention here prevents the river from flowing freely, and in that way, undermines its integrity, which should have been taken into serious account in the first place, in deciding whether the river should be dammed to provide water for sprinkling the lawns of suburban homes miles away from it in some desert location. But the *raison d'être* for respecting its integrity may not necessarily lie in pursuing the intentional stance.

To see why not, one needs to explore further the distinction between anthropogenic causes of undermining its integrity as opposed to nonanthropoenic ones. While we may have qualms about engaging in the former, the situation is totally different in the latter context. Abiotic forms, be they rivers, mountains or lakes, change over time, so that eventually the river, the mountain, the lake would be no more and would have transformed themselves into some other geological forms. There are dynamic—physical/biogeochemical—processes at work, often slowly and imperceptibly, perhaps, to the naked human eye at any one moment in time, but, nevertheless, surely transforming extant geological formations. A river at one period of its history might be a fast flowing, raging torrent, but at another point of its history, it might be a lazy, meandering, sluggish thing. At one stage of its course, near the source, it may be a fast flowing current but further on, it becomes slow and sluggish. And at a still later stage, it may turn itself into a raging torrent yet again. What it turns out to be in its entire history and course depends on a variety of factors. Just to mention two key matters—how much water feeds into it at its source and further downstream, as well as the terrain it passes through.

Suppose we say that the wild, raging torrent captures the 'true spirit' of riverhood, that it is the 'real' thing while the sluggish version is a degenerate. This raises two nagging points. The preference for the wild, raging torrent is probably a matter of subjective aesthetics which is historically and culturally determined. At least eighteenth century English sensibility would not have endorsed it, while the Romantic sensibility, in turn, would find the sluggish river 'ignoble,' lacking 'passion.' In other words, there is no way in which we can say that the wild, raging torrent, and not the barely moving river, is the 'true spirit' of riverhood without imposing an extraneous anthropocentric, indeed, even anthropomorphic, scaffolding upon it, so to speak.

Furthermore, if the technology were available, on this view of the 'true spirit' of riverhood, would it follow that one restore the river to its wild state even if natural processes have turned it into a sluggish thing? But such restoration *per se* would be fundamentally flawed,[19] as it amounts to an imposition of both anthropogenic and anthropocentric demands upon natural processes.[20]

In conclusion, *contra* Plumwood, the intentional stance, while being appropriate in the case of biotic nature, is inappropriate in the case of abiotic nature. The section which follows introduces the distinction between beings which are valuable 'for themselves' and those which are valuable 'by themselves'; it will be argued that while individual organisms (animal or plant) are intrinsically valuable both 'by themselves' and 'for themselves,' abiotic nature, however, is only valuable in the sense of 'by itself.' The intentional stance is only conceptually relevant to beings which also embody the sense of 'for themselves,' as that sense is involved with the organisms' striving to sustain their own functioning integrity.

Abiotic Nature and Intrinsic Value

The main concern of most environmental philosophers who wish to argue that nature has intrinsic value concentrates on biotic nature. Such a route is both normal and reasonable given the boundaries presupposed by the subject called environmental ethics developed so far. Among the key problems which preoccupy theorists in the field, and others working in related areas, are those associated with pollution and its adverse effects on the functioning integrity of the biosphere as well as biodiversity itself. Given that technology, both past (in the last hundred and fifty years or so) and extant may be said to be ecologically insensitive and, hence, a significant source of pollution, given the quite alarming loss of biodiversity through human population increase as well as the demands made on natural habitats by the dominant economic system of production and consumption today,[21] it is not a wonder that the focus is on biotic nature. Furthermore, Earth is probably unique in the solar system as a planet containing life.[22] Given the inextricable contingent link between the biotic and the abiotic on Earth and the fact that there is hardly a location where life does not prevail, it seems less pressing, if not altogether unnecessary, to argue that abiotic nature itself, on its own, could be a locus of intrinsic value.[23]

On this commonsensical understanding, once biotic nature could be said to possess intrinsic value, then one could argue either that (a) the abiotic has instrumental value for the biotic, or (b) to undermine the one inevitably amounts to undermining the other—the living organism is part and parcel of its habitat, which contains other living organisms as well as abiotic matter, and these together with all their inter-relationships constitute an ecosystem. Biocentrism which is individualistically oriented would tend to endorse a, while ecocentrism tends to endorse b.[24] The former by-passes altogether the difficult matter of whether the abiotic could be said to be intrinsically valuable. The latter approaches the issue through a discussion of whether ecosystems and/or species could be said to be intrinsically valuable. Some holists agree with biocentrists that it is deeply problematic, if not impossible, to argue that abiotic nature could be intrinsically valuable.[25] For instance, Johnson, who grapples with the issue explicitly, is driven to the brink of reintroducing anthropocentrism by the back door: "Perhaps there are some things it is better for us that we value for their own sakes."[26] However, this does not mean that environmental philosophers, holists or otherwise, have totally ignored the matter. There are notable exceptions.[27]

But there is no escape from confronting the issue, as this book argues, first, that nature need not and should not be understood to refer only to what exists on this planet, Earth. Second, an environmental philosophy informed by features unique to Earth may be misleading in certain crucial ways, and prove inadequate as technology increasingly threatens to invade and colonize other planets in the solar system, so that a comprehensive environmental philosophy must encompass not only our attitude to Earth, but to other planets as well—in other words, it must not simply be an earthbound but virtually an astronomically bounded philosophy and ethic. Third, on Earth itself, as molecular nanotechnology in the near future could bypass natural abiotic kinds to create its own artefactual abiotic kinds, and combine with biotechnology and microcomputer technology (already in place) to transform natural to become artefactual kinds, resistance to the ontological elimination of abiotic nature as 'the Other' becomes urgent. But given these developments, it looks as if agonizing about the problem of abiotic nature becoming divorced totally from biotic nature, and/or whether the former, thus divorced, could be a locus of intrinsic value is no longer an optional extra.

But how can such a task begin? Let us first stay with Earth and see if one can outline a conception of intrinsic value for Earth without being too distracted, it is hoped, by the fact that organic life is indeed unique to Earth, and then extrapolate from it to another context, that of other lifeless planets in the solar system, to see if such a conception could also apply when suitably edited.

Such a conception of intrinsic value for an earthbound environmental philosophy would rest on three theses:

A. The 'No External Teleology' Thesis[28]

1. Earth did not come into existence and does not continue to exist to serve human purposes. In this sense, as we have seen, the thesis of external teleology is simply false, and should be distinguished from the thesis of intrinsic/immanent teleology which holds true in the case of organisms. An alternative language to make a roughly similar, though not identical, point may be used—for instance, the biologist Mayr distinguishes betweeen teleomatic processes at work (in abiotic nature) which simply follow physical laws, such as the law of gravity and the second law of thermodynamics, and teleonomic processes at work (in biotic nature), as a result of which, organisms display programed behavior, the program being the product of natural selection.[29] Neither teleonomic nor teleomatic processes and their products have come into existence or continue to exist to serve human purposes.
2. Humans, of course, find parts of nature useful as food, clothing, shelter, etc., just as nonhuman life forms find other parts of nature of use to them.

Plants (autotrophs) can make use of abiotic nature to sustain their own functioning integrity and in this sense, the carbon dioxide, minerals, water, heat and light from the sun, etc., have instrumental value for the plants. But it would be misleading to say that abiotic nature exists for the purpose or end of keeping plants alive. Similarly, the leaves of plants have instrumental value for insects but it would also not be correct to say that plants sustain their own functioning integrity in order to be of use to insects. As already argued earlier (see **Resisting Humean Projectivism**), neither can it be said that plants and animals exist for the purpose of keeping humans alive and flourishing although they, clearly, have instrumental value for humans.

3. From 1 and 2 above, it follows that just as nature does not exist for us humans, we humans do not exist for nature either. By this is meant the following: for instance, lice and microorganisms, no doubt, find human hair and human guts of instrumental value to them. But from this, we do not infer that the primary justification for human existence and its sole purpose is to serve lice and microorganisms in this instrumental fashion. Similarly, humans may find certain plants and animals useful as food, and caves useful as shelter, but from this one does not infer that the primary justification for their existence and their sole *raison d'être* is to serve us in this instrumental fashion.
4. The points above show that there is need to distinguish between two senses of 'for itself.' The standard sense is the one used so far and is tied up with individual organisms and the notion of their striving to maintain their own functioning integrity—see preceding sections. However, a different sense is used in this context here which is not involved with the notion of striving, and is much wider, as it includes abiotic entities to which it may be conceptually inappropriate to ascribe the notion of striving. This sense is to be marked by calling it 'by itself.' It is simply an entailment of the 'No External Teleology' thesis.[30] The more standard 'for itself' involves the thesis of intrinsic/immanent teleology. An organism exists, then, both 'by itself' (that is, it has not come into existence, nor does it continue to exist **in order to** serve human ends or purposes) as well as 'for itself' (that is, it strives to maintain its own functioning integrity). On the other hand, an abiotic item exists only 'by itself.'
5. If we were to consider humans as a locus of value because we are entities who exist 'by ourselves' (in the sense just characterized), then consistency should lead us to conclude that nature as a whole and the various items in it, too, are loci of value, for they, too, exist 'by themselves.' This, however, is not to deny what is obvious, that what has intrinsic value in this sense may have instrumental value, as a matter of fact, for another.

B. The Autonomy Thesis

1. The genesis of the universe and of earth are independent of humans. The Big Bang, which started the universe, is said to have happened 15 billion (or eons) years ago. Earth itself is said to be 4.5 billion years old. The genesis of life on Earth is also independent of humans. It happened at least 3.6 eons ago during the Archean period (4.5 to 2.5 eons ago) when the chemistry of the atmosphere was first dominated by oxygen. But the history of natural organic evolution is a very long one indeed. The lineage of anthropoid apes which led eventually to *Homo sapiens* emerged less than one-third of a million years ago.[31]
2. Earth and its biosphere would not be extinguished should humans themselves, for some reason, become extinct as a species. As far as the biosphere is concerned, the disappearance of the human species cannot be said to threaten it. Should human extinction happen, the niches formerly filled by humans will be taken over by other species. It would also, most probably, provide opportunities for new species to emerge. The continuing existence of Earth and its biosphere is clearly, in this fundamental sense, independent of humans. A simple thought experiment should establish this point.
3. Moreover, the ability of the biosphere to function integratively and well is also independent of humans.
4. In other words, Earth and its extremely complex biosphere are fully autonomous. 'Autonomy' is here used to mean no more and no less than its ability to exist, to function integratively and well without any reference to, assistance from, or reliance on humans.[32]
5. From the perspective of biospheric integrity, humans are, therefore, dispensable and could even be redundant.
6. It follows from the above that if an entity exists 'by itself,' and if its genesis, its continuing existence and survival, are independent of humans, then these are compelling reasons for us, humans, to recognize that it has a value independent of us. In turn, we ought then to recognize that we have a duty (in virtue of our ethical capability) not to undermine or destroy such a thing of value.

C. The Asymmetry Thesis

1. The above shows that there is a distinct asymmetry of causal dependence between humans and nature. While humans depend on nature, and cannot exist if it were absent, or if its functioning integrity were too drastically upset by humans, nature's own existence and functioning integrity is independent of human existence.[33]
2. Our total dependence on nature, but nature's independence of us, reinforces the Autonomy Thesis and re-emphasizes the view that nature has value which is entirely independent of us.

Can an analogous conception of intrinsic value (in the sense of 'by itself') for the other planets as well as the sun in our own solar system be constructed? It may be possible to do so as it is not immediately obvious that it cannot be done. An attempt would look as follows:

A. The 'No External Teleology' Thesis

To say that Mars (or any of the other planets) exists to serve human ends is more patently implausible than it is to claim that Earth does. Humans at least exist on Earth and have evolved to do so. But Mars has no humans. Instead, with our advanced technology we threaten to, and could probably, transform it into an object of use to ourselves. But this possibility of instrumental value does not undermine, in any way, the contention that its *raison d'être* has nothing to do with human ends. It exists 'by itself,' no more and no less. It is both an illusion and a fallacy to hold that whatever we humans happen to find useful in the light of our technology must have come into existence just for the purpose of serving our ends, or that its existence is to be justified solely in terms of such a purpose. Similarly, the undeniable fact that no life on Earth, including human life, is possible without the sun should not be understood to mean that the sun came into existence or continues to exist for the purpose of sustaining human life.

B. The Autonomy Thesis

The genesis of Mars, the other planets, the sun long antedated the appearance of humans on Earth. Their continuing existence has nothing to do with humans. What happened or happens to them, or on them, is totally independent of us. For instance, Mars might once have had water or indeed even bacterial life (as is being speculated), but today it is said to be waterless and lifeless. But neither state of affairs is caused by human effort or design.

C. The Asymmetry Thesis

Human survival and flourishing are dependent on the Earth's atmosphere and its biosphere, which depend on Earth's diurnal rotation around its own axis and its year-long revolution around the sun, which, in turn, depend on the sun staying still while the other planets rotate around it, exerting gravitational pull on one another in certain ways. So while the existence of humans depends on the existence of the sun and the other planets, the existence of the latter would not be affected should humans, as a species on Earth, become extinct. In other words, causally and cosmologically speaking, the extinction of *Homo sapiens* would not be significant.

The above set of arguments attempts to provide some reasons which may be considered to be compelling for us, humans, to recognize that abiotic nature as represented by the rest of the solar system (or on Earth when divorced from biotic nature) has a value which is independent of humans altogether, and, therefore, is not simply the result of mere human presence and the projection of human consciousness upon nature. It is, however, admitted—see **Resisting Humean Projectivism**—that the human presence does add another dimension to the matter. But at this stage of the argumentation, it suffices to point out that the account given so far may still be said to be both partial and interim, serving to pave the way for a fuller exploration of the notion of independent value and its intimately related concept of the trajectory in the section which follows.

Intrinsic Value, Trajectories, and Independent Value

An earlier section has shown that it is problematic to extend the intentional stance into the abiotic domain. A more appropriate notion to use may be that of trajectories.[34] A mountain, a river, a lake, a sand dune, a volcanic island, each displays its own respective trajectory. A trajectory, in this context, is simply the particular (or particular type of) history of the entity (or type of entity) in question including the processes that are the cause(s) of its origin to those leading to its final demise. It does not necessarily imply that the entity possesses an 'internal program' (analogous to the *telos* of the organism) controlling the various stages of its journey, although the trajectory of a biotic entity includes the interaction between its own *telos* and its external environment. The precise form of the trajectory of an abiotic entity is determined by a complex set of factors, some internal and others external. For instance, the precise trajectory of a rock, like Ayrs Rock, and the period of time taken to complete it, depend on its size, shape, make-up (whether it is igneous or sedimentary) as well as on its location, the wind that blows against it, the water or rain which falls on it, the animals which may trample over it, the plants which may grow in its cracks or around it, the movement of Earth's crust, etc.

Such trajectories would occur irrespective of whether humans are around or not. There was a time on Earth when there were no humans. Abiotic (and biotic) forms carried on with their trajectories. Outside Earth, on other planets in our solar system and elsewhere in astronomical space, abiotic forms undertake their trajectories (as far as we can ascertain) in the absence of life forms, both nonhuman and human. The main thing to grasp is that these trajectories, whether on Earth, in our own solar system, or elsewhere in outer space, occur totally autonomously and independently of us humans. As already argued, they owe neither their origins nor their continuing existence to humankind (although, admittedly, humankind may impact on some of them through its own activities, as the destruction of stratospheric ozone has amply shown).

It is important to grasp that this independence amounts to **ontological** independence, and should be differentiated from the usual sense, which is causal independence.[35] When we say that one entity is independent of another, there is no implication that the entities belong to different ontological categories. For instance, the penguin in Antarctica is totally causally independent of the polar bear in the

Arctic, and Ayrs Rock in Australia, of Jupiter's moons (although the Earth's tides are dependent on the behavior of its own moon). Similarly, the origin of Bill Clinton and his continuing existence are causally independent of Imran Khan's own origin and continuing existence. Clinton and Khan are both human beings but belonging to rather different cultural traditions. The penguin and the polar bear are both nonhuman animals but belonging to different species. However, in spite of their differences—chemical, physical, biochemical, biological or cultural as the case may be—and their respective independence from each other, no ontological divide exists respectively between the penguin and the polar bear, Clinton and Khan, Ayrs Rock and Jupiter's moon, or the polar bear and Ayrs Rock. But in the case of biotic and abiotic entities whose origins and continuing existence, whose trajectories are autonomous of human activity in the way delineated above, their autonomy constitutes ontological independence, for precisely the reason that these trajectories are initiated and sustained in the absence of humans, unaffected by human intentions, goals, purposes and actions, and regardless of human interests or desires.

By means of this key concept of ontological independence, one can distinguish between primary and secondary characteristics displayed by both the biotic and the abiotic. What is primary is their ontological independence. Their secondary characteristics are whatever they may display or come to possess once in existence in virtue of their membership of their respective natural kinds, such as their sentience, their mental and psychological activities (in the case of mammals or the higher animals), or their striving to stay alive in the particular way they do (in the case of all organisms), their being the bearers of interests, or of the intentional stance (which Plumwood argues both the biotic and the abiotic share but which has already been contested), or, indeed, even of complexity and intricacy (which the biotic as well as the abiotic may meaningfully be said to possess).[36]

With this recommended shift in focus to their primary ontological status, it is hoped that new light will be shed on an existing problem, namely, that in the main, theorists in environmental philosophy have not made clear the status of secondary characteristics when they concentrate on them to ground moral considerability.[37] One does not deny that naturally-occurring entities have the secondary characteristics attributed to them; however, one should go beyond secondary characteristics to acknowledge the more fundamental attribute that they exhibit, namely, that all secondary characteristics in any naturally-occurring entity are possessed and displayed (in principle) totally independently of humans.

According to the perspective pursued here, the central thrust of most environmental thinking is somewhat misleading as it ignores that the ontological status of naturally-occurring entities, rather than their secondary characteristics, provides the grounds for granting them moral consideration. While secondary characteristics ground intrinsic value, the ontological status of naturally-occurring entities grounds independent value and justifies their moral considerability in a more fundamental sense. The notion of independent value is an ontological one, that of intrinsic value, an axiological one. Ontology takes precedence over axiology.

To see more clearly the importance of distinguishing between the primary characteristic of independence and autonomy on the one hand, and secondary characteristics, like complexity, creativity, intricacy, sentience, beauty, or whatever, on the other, consider the situation examined earlier.[38] Our technology, informed by 'deep' science and growing more and more powerful, will enable us to design and construct entities which in every way may be as complex and intricate or as beautiful as natural kinds are. In terms of such characteristics, there may be nothing much to choose between them. Indeed, we cannot rule out in advance the possibility that technology in the distant future may enable us to devise something so ingenious as to surpass any complexity and intricacy displayed by naturally-occurring entities.

If these characteristics **alone** are taken to ground moral consideration,[39] it follows that any entity, whether naturally-occurring or an artefact, provided it displays complexity and intricacy (of an order agreed and specified in advance) should be accorded moral consideration. Hence, given that they warrant moral consideration, such artefacts could be substituted for naturally-occurring entities, and this would be entirely morally permissible. There may be some loss of value in the world when complex, intricate entities are destroyed but this loss could be compensated for so long as humans created and constructed complex, intricate artefacts at the same rate as they destroyed naturally-occurring entities.[40] To use Maturana's and Varela's arguments as an illustration—if they were accepted as valid, autopoeitic machines might be found today either in naturally-occurring living organisms or in those constructed by humans as biotic artefacts. The loss of biodiversity, on this view, is only worrying so long as humans cannot replace it with autopoeitic machines of their own design. Maturana and Varela, as we saw in the last chapter, positively welcome the prospect of humanly-designed autopoeitic machines. So, to be consistent, they should also welcome them as substitutes for naturally-occurring organisms. But this welcome is borne out of confusing the primary character of nature—nature$_{nh}$, nature$_{nk}$, nature$_{fa}$—as being autonomous and independent of humans with the secondary characteristics which it also displays.

In other words, if biotic forms display intricacy or if nature, in general, displays great complexity (such as the biosphere in its entirety), what is crucial to emphasize is that the intricacy or the complexity is embedded in a naturally-occurring context. What has independent value and, therefore, is entitled to moral consideration is the whole context, not simply the intricacy or the complexity divorced from that context. This point may become clearer if it is further pointed out that the complexity or intricacy may be understood as an aspect of nature's creativity. Human creativity is legitimate grounds for celebration. But so is nature's creativity. However, ontologically, they are two distinct types of creativity—human creativity is not a substitute for nature's creativity nor *vice versa*. The former should not crowd out the latter.

Let us return, for a moment, to the possibility of terraforming Mars. This book argues that such a project would turn the planet into an artefact. Mars, one day in the dim distant geological future, may or may not come to possess an atmosphere rather like Earth's from which life may or may not in turn evolve. Should it do so, that is an expression of Mars' own creativity. But what may be unacceptable is to deflect Mars from whatever trajectory it is on to become an *ersatz* Earth, constructed according to human design. If that were to happen, human creativity would be made a substitute for nature's own creativity.[41]

What this book is primarily arguing against is the attempt:

1. To turn all history into human history. According to the "1066 and all that" model of history, human history is no more than written accounts of what kings and great men did—lesser men and women, in general, are left out of the script. But social history, labor history, oral history and feminist history are all legitimate histories for celebration. Just as human history is not merely about what kings and great men did but also what ordinary laboring men and women in general did, so history confined to human history is too selective and distorting. If history is about past events and happenings, then there may be more events and happenings in the past than involved humans.
2. To undermine nature's future either by (needlessly or thoughtlessly) destroying parts of it (as through the loss of biodiversity) or by transforming it into the artefactual is to deny nature ontological space or *lebensraum*, to eliminate nature as 'the Other.'

In conclusion, the crucial point to grasp is that this section attempts to estabish the priority of the value of independence and autonomy embodied and exhibited by nature, which is ontological in character, over other values which are axiological in character. The ontological characteristics of independence and autonomy are primary while the axiological characteristics of being complex, intricate, sentient or whatever are secondary. Artefactual entities may amply exhibit the secondary axiological characteristics; *ex hypothesi*, they do not and cannot display the primary ontological characteristics which only naturally-occurring entities and processes possess.

Ontological Elimination, Dualism, and Dyadism

Some environmental philosophers including Deep Ecologists hold that dualistic thought which is part and parcel of mainstream modern Western philosophy since the seventeenth century is responsible ultimately for the serious environmental mess which the late twentieth century finds itself facing.[42] To get out of the mess, they argue, requires an outright repudiation of that metaphysical framework. This book agrees with Deep Ecology but only up to a point. It accepts in particular that the dualism between human and nonhuman or between culture and nature has to be rejected but, nevertheless, argues that the distinctions themselves must be retained not, however, as ontological dualism but as ontological dyadism.[43]

Plumwood has given a clear account of dualism; it involves what she calls "hyperseparation," which is more than a mere dichotomy or a plain distinction.[44] It systematically and pervasively construes the dualized other as inferior—humans (the sole locus of intrinsic value) and their culture lord over and denigrate what is nonhuman (which is of only instrumental worth to the privileged master category). As we shall see, ontological dyadism, far from denigrating the Other, recognizes that it has ontological worth.

Modern science and its technology are predicated upon nature as the dualized other. Its goal of controlling nature presupposes the inferior status of the dualized other. The successful execution of the modern scientific/technological program leads inexorably to the virtual extinction of that dualized other. On this view then, the dualism between human and nonhuman may finally liquidate itself, if science and technology can in principle systematically and at a deep level transform the natural to become the artefactual. We have seen that the artefactual is a human intentional structure, belonging to a different ontological category from the natural, and that to transform the natural to become the artefactual produces ontological impoverishment. Such impoverishment is thus an inevitable part of modern anthropocentrism.

Dualism—and hence the scientific/technological program based on it—is unacceptable because it denigrates 'the Other' and leads to its elimination both at the ontological and empirical levels. But one should not throw out the baby with the bath water. To prevent ontological impoverishment and to save the natural from being systematically transformed to become the artefactual, through the activities of *homo faber*, rightly requires throwing out dualism, but not the very distinction itself between the natural and the artefactual. As Plumwood has emphasized, differences should not be obliterated, distinctions not overlooked and respect for 'the Other' should be based on the recognition of relevant differences, not necessarily of similarities.[45]

Respect for nature in this deep sense requires two distinct ontological categories, the nonhuman and the human. Far from de-emphasizing the differences between the latter and the former, this argument requires that their differences be put center stage. After all, the present predicament arises primarily and precisely because humans as a species are so different from other species on planet Earth—humans, given their peculiar kind of consciousness, brain and other capabilities, have evolved in such a way as to possess, today, extremely powerful technologies with which they interact with the nonhuman environment. Moreover, as has been argued, their science and technology enable them systematically to transform the natural to become the artefactual, thereby imperiling the very existence of what is nonhuman.

Unpacked in greater detail, this ontological dichotomy includes the following overlapping distinctions:

The Nonhuman	**The Human**
nature	culture
$nature_{nh}$	
the natural	the artefactual
$nature_{fa}$/$nature_{nk}$	

Humans generate moral, social, legal and political codes to govern their behavior; and they regard themselves alone as morally responsible beings.[46] Culture is superimposed on whatever is natural in human behavior (that is, behavior as an organism, subject to the laws of biology and physiology). Even in such fundamental functions like eating, mating and defecating, each society lays down its own rules, determining what is the right thing and forbidding what is the wrong thing to do in each of these spheres of activity. However, sometimes, humans confuse nature with culture. They then transfer cultural norms which govern interpersonal conduct to nature. For example, in culture, it is morally right to help fellow humans in distress. Some people mistakenly extrapolate from this to determine the human attitude to nature.

One example should suffice to illustrate this confusion.[47] A bison bull in February 1983 fell through the ice in Yellowstone River and could not get himself out. The Park Authorities decided not to intervene on the grounds that this was a natural occurrence. A group of snowmobilers who came upon the scene was outraged by such a decision. About half of the party returned later, with a rope, to try to extricate the animal. However, they did not succeed, and when the night drew in, they abandoned their rescue. By the morning, the animal had frozen to death. (Coyotes and ravens soon lighted upon the carcass. In the following spring, after the snow melted, a grizzly bear was observed further downstream feeding on its remains.) One of the rescuers wrote to a radio commentator who said in denouncing the Park Service's cruelty: "The reason Jesus came to earth was to keep nature from taking its course." A curious interpretation of the divine mission perhaps, but the comment neatly encapsulates the confusion between nature and culture. If the snowmobilers and the radio commentator had paused to substitute, perhaps, a water rat for the bison, they might not have been so righteously outraged, and might not have so readily convinced themselves that it is the bounden duty of humans (in accordance with Jesus' will) "to keep nature from taking its course."

This book, by concentrating on the distinction between the natural and the artefactual, is affirming the necessity for ontological dyadism while rejecting dualism.[48] The main difference between 'ontological dyadism' and dualism is this.[49]

The latter involves denigrating and inferiorizing 'the Other,' while the former involves precisely the opposite as it recognizes that nature as 'the Other,' though very different from humans in many crucial ways, nevertheless is to be respected. It is born, not out of a 'gung ho' spirit of human superiority, but out of humility and awe for nature's achievements.[50] It follows that ontological space must be found for 'the Other' if one is to respect it.

However, respect for 'the Other'in interpersonal human conduct differs from respect for 'the Other' when 'the Other' is nature. Respect for another in the context of human culture could in some instances lead to complete or near complete identification with the other person(s). For instance, an anthropologist may go 'native' and may, indeed, be accepted by the very people s/he originally has come to study as one of them. But in the case of nature, no matter how hard one may strive to identify with it, one is not able, in the same way, to do so. The anthropologist could come to think, speak and act like, say, a Yanomami and see her- or himself as a Yanomami. S/he could shed her/his former identity as, say, a white North American social scientist and acquire a new identity as part of an indigenous people of the Amazonian forests. But the environmentally concerned citizen or the animal welfare activist cannot think (assuming that it makes sense to say that thought is not identical with the linguistic) and act like a chimpanzee. If s/he were to do so, s/he would be certified insane. But this fate would not await the anthropologist who has gone 'native,' although the data s/he produces might no longer be regarded as reliable—at least according to the standard canons of scientific methodology, the student should neither totally identify with the object of study, nor cause the object of study to identify totally with her- or himself.

To look for sameness between the human and the nonhuman in order to ground respect for the latter, perversely, may lead to the very trap of reductionism laid by the metaphysics of Scientific Naturalism itself. This postulates, as we have seen, that ultimately all the differences, observed or otherwise, between natural kinds or between individual specimens belonging to the same or different natural kinds are *au fond* merely different arrangements of atoms and their molecules. We have also seen that the program of molecular nanotechnology is predicated precisely upon that metaphysical foundation. The stuff of the universe is homogeneous and uniform. Ecology (at one level) and other sciences tell us that the biogeochemical and hydrological cycles recycle in the end the same atoms of carbon, oxygen, hydrogen, etc., although in different arrangements. It is true that an atom of carbon which once was part of the history of a particular tree in the Amazonian rain forest of millenia ago could conceivably now be part of the current me. In the same vein, when I die, an atom of carbon released from my decaying body could conceivably be part of a tree which would be in existence a hundred years after I am gone. In principle, if not in practice, it makes sense to trace the passage of an atom from one object to another object in nature$_c$. But it is not obvious that we, humans, ought to refrain from inferiorizing nature simply because both recycle the same atoms.

To argue so is to invite the kind of criticism which Marx and Engels made against the so-called "true socialists" in their second volume of *The German Ideology*.[51] The "true socialists" (which included Feuerbach) had argued that humans are themselves part of nature.[52] According to Marx and Engels, they posed the following questions: "Did not man too spring from the primeval world, is he not a child of nature, like all other creatures? Is he not formed of the same materials, is he not endowed with the same general energies and properties that animate all things?"[53] To these, Marx and Engels riposted: "The same general energies and properties which man has in common with 'all things' are cohesion, impenetrability, volume, gravity etc., which can be found set out in detail on the first page of any textbook of physics."[54]

While not by any means agreeing with Marx's and Engels's own endorsement of human chauvinism, their retort has a point. What has value is nature's own throwing up of different natural kinds, be these biotic or abiotic. An environmental philosophy must respect these natural kinds, as different natural kinds, and not be over-impressed by the homogeneity and uniformity which underlie these different kinds as nature$_r$. It is undeniably true that as nature$_r$, *Homo sapiens* is not so different *au fond* from an oak, a lion, or the water around us. The laws of physics and chemistry apply to all of these at such a basic level of organization. But at higher levels of organization, entities would display properties not displayed by those at relatively lower levels of organization. Again, this is not to deny that humans have evolved from other organisms. But it remains true that humans display a type of consciousness, as has already been emphasized, which has enabled them in the modern era to develop a type of science and technology which puts them in a unique position, namely, having the power systematically to eliminate other nonhuman entities around them, both at the empirical and philosophical levels.

That is why it is extremely important to uphold ontological **dyadism** as expressed in the distinction between the natural and the artefactual, while rejecting ontological **dualism**. It is required precisely for the following reasons: the natural and the artefactual belong to very different ontological categories; respect for the natural rests on recognizing its distinctive ontological status; such recognition is required in order to resist the systematic elimination of the natural *via* the application of science and technology.

Ecosystem Health and the Human/Nonhuman Ontological Dyadism

In the last few years, some environmental philosophers have started a trend toward a pragmatic consensus among practitioners, what Norton calls "the convergence hypothesis," thereby deflecting attention from the divisions at the theoretical level which have emerged within the discipline of environmental philosophy itself, as it has developed in the last twenty odd years.[55]

Two main sets of theoretical distinctions are generally recognized: anthropocentrism and nonanthropocentrism on the one hand, individualism and holism on the other. These in turn yield four distinct positions:

1. Anthropocentric individualism.
2. Nonanthropocentric individualism.
3. Anthropocentric holism.
4. Nonanthropocentric holism.

Position 1 is the dominant stance of modern Western philosophy—Kant or Descartes may be said to be examples, as they hold that human individuals alone have intrinsic value and everything else is merely of instrumental value to humans.[56] The other three all challenge the dominant Western tradition in one way or other, the most radical being represented by the last. Position 2 is exemplified by Singer's sentientism, Regan's "subject-of-a-life" philosophy or Taylor's biocentrism. Position 3 is best seen in Norton's work. Position 4 represents ecocentrism and may be illustrated by theorists as diverse as Deep Ecologists like Naess, and those who are not Deep Ecologists like Rolston or Callicott (or Leopold as a precursor of Callicott).[57]

The so-called pragmatic consensus is being forged between positions 3 and 4, or more specifically by Norton and Callicott who may be said to represent them respectively. They are both holists as each is prepared to argue that species, ecosystems and indeed, the biosphere are entities with properties and causal effects that cannot be reduced to those exhibited by organisms, whether plant or animal, as discrete and separate individuals. They both agree that it is not sufficient to preserve only individual organisms without preserving species and ecosystems. However, unlike Norton, Callicott argues that humans are not the sole locus of intrinsic value although the source of all values. But in view of this crucial difference, how could they begin to converge upon a pragmatic consensus? The mystery disappears when a distinction within contemporary anthropocentrism itself is distinguished—the extreme aggressive form as opposed to a less aggresssive form.[58]

Extreme aggressive anthropocentrism may be characterized in terms of the following theses:

1. The Sole Value Assumption—that humans are the sole locus of intrinsic value.
2. The Greater Value Assumption which is entailed by 1—that in a conflict of interests between humans and nonhumans, human ones always override nonhuman ones.
3. The Instrumental Value Assumption which is entailed by 1 and 2, namely, that nonhumans are merely of instrumental value to humans.
4. The All-permissive Assumption which is entailed by 1, 2, and 3, namely, that there are no moral constraints, but only technological ones, to govern how

humans use nonhumans to serve their own ends.

Less aggressive anthropocentrism (at least as advocated by Norton) may be characterized in the following terms:

1. It accepts the Sole Value as well as the Instrumental Value Assumptions.[59]
2. But it rejects the All-permissive Assumption, while ignoring (without actually ditching) the Greater Value Assumption, as it is not called upon to do any real work. This is because it recognizes that technological optimism has its limits and that there are ecological constraints to the way in which humans may use nonhuman nature to serve their own ends. For instance, Norton (*Why Preserve Natural Variety?*) has argued that given the complexities in the relationships between species and their ecosystems, there are very convincing ecological arguments against the rate and scale of anthropogenic extinction of species that is occurring today. Norton may not mourn the loss of species *per se*, but he argues that the vicious downward spiral of species extinction has unfortunate implications for the human species. In other words, to ensure the survival and flourishing of humans requires, at the same time, ensuring the survival and flourishing of other nonhuman species and their habitats. The ecological constraints mean that prudence at least, if not morality, dictates that there are contexts in which the satisfaction of human desires and interests, simply in virtue of their being human desires and interests, should not take precedence over the survival and flourishing of other nonhuman species and ecosystems.[60] This view then focuses on ecosystem health and integrity as a fundamental value, albeit on anthropocentric grounds.

However, this more subdued anthropocentrism can be successfully reconciled with nonanthropocentrism, at least on the level of decision-making in matters of crucial concern to both sides. In this, Norton may well be right, for Callicott ("La Nature Est Morte" and "The Role of Technology"), a noted nonanthropocentrist, nevertheless, agrees that the preservation of nature *via* the maintenance of ecosystem health and integrity is the main goal of environmental management. This particular pragmatic consensus is possible as it takes place within a fundamental theoretical agreement, namely, that they are both holists. For them, given the complexities of ecological relationships, the survival and flourishing of a particular species or ecosystem is more than simply the survival and flourishing of some individual organisms, or of certain individual items which constitute their habitat, and the survival and flourishing of the biosphere goes beyond the survival and flourishing of certain particular ecosystems.

But no analogous pragmatic consensus could emerge between anthropocentric and nonanthropocentric individualists—a Kantian, for instance, would not be able to see eye to eye with, say, a biocentrist like Taylor, in arriving at fundamental goals in environmental policy-making. Nor would the possibility of such agreement obtain between anthropocentric holists and nonanthropocentric individualists—for instance, while Norton is deeply concerned with the threatened extinction of species and ecosystems, Taylor would see the saving of individual organisms as the only basis for saving the species, as these alone are morally considerable, and not the species and the ecosystems of which they are a part.[61] The same would apply in the case of anthropocentric individualists and nonanthropocentric holists—the former would simply adhere to the Sole Value Assumption and its derivatives, while the latter would, in principle, endorse (though not, obviously, in all contexts of conflict) preserving species and ecosystems at the expense of human interests.

The above discussion serves to provide evidence for asserting the following:

1. Holism leads to a broader, and therefore, more comprehensive environmental philosophy than individualism. The saving of individual organisms does not amount to the saving of the species or ecosystems to which the organisms belong or are a part. On the other hand, the preservation of actual species and ecosystems at the same time also preserve the natural processes at work underlying the generation of species and their ecosystems. This account is more in keeping with our understanding of ecological relationships, of the natural evolution of species and the evolution in general of life on Earth.
2. While a crucial theoretical gap remains between anthropocentric and nonanthropocentric holists, a pragmatic consensus is possible between them, which means that both sides can agree that maintaining ecosystem health and integrity is the fundamental goal of environmental policy-making. They can set aside the theoretical dispute about whether nonhuman individual organisms, species or ecosystems may be said to be morally considerable or the loci of intrinsic value.
3. Anthropocentric holists are able to endorse a much weaker instrumentalist attitude to nature (as resource preservation) which does recognize severe ecological constraints to the economic use of natural resources entailed by aggressive anthropocentrism and the strong instrumentalism (as resource conservation) it entails.
4. As part of this *rapprochement*, Callicott, as we have seen, appears to endorse only one sense of 'nature,' namely, 'nature$_{c}$.' (Chapter 3, in **The Natural: Different Senses of 'Nature,'** points out that such a sense is too wide.) Furthermore, he is also keen to discard what he calls the idea of nature which involves nature as 'the Other.' The distinction between humans (as being apart from the nonhuman) and nonhuman nature is abandoned—humans and nonhumans are all part of nature (that is, nature$_{c}$), as attested, according to Callicott, by the theories of evolution and ecology in biology, of relativity and quantum in physics. Behind the pragmatic consensus, therefore, stand certain radical theoretical adjustments on the part of the nonanthropocentrist-*cum*-holist.

5. The consensus assumes that (a) certain agreed, objective, scientific indicators for assessing ecosystem health are available or could be developed,[62] (b) these are applicable to so-called natural ecosystems (even when these are not instances of nature$_p$) and artefactual ecosystems (such as agro-ecosystems) alike, (c) social and cultural values—which undoubtedly prevail—would not affect these criteria of ecosystem health and integrity to such an extent as to undermine and even destroy the consensus itself.[63]

However, laudable and plausible as this pragmatic consensus may be, unfortunately, behind its back, so to speak, certain technological developments are taking place which would force one to focus attention on a crucial divide which could develop, namely, between those who are prepared to endorse ecosystem health, but at the expense of transforming nature$_{fa}$ and nature$_{nk}$ into artefactual kinds, and those who may not be so prepared. It is conceivable that artefactual biotic kinds, in particular, could be successfully managed as part of a 'healthy' ecosystem—for instance, an ecosystem of specially biogenetically engineered trees could satisfy most, if not all, the criteria of ecosystemic health in terms of its capability for nutrient recycling and for resisting disease and 'pests,' for withstanding toxic contamnination, stability and so on.[64]

One criterion which an artefactual ecosystem might not, at first sight, satisfy is that of species diversity. An agro-ecosystem of biogenetically engineered trees for commercial timber might not be much of a host for other flora and fauna, and would create something of a species-poor environment. But artefactual ecosystems which are not so severely dedicated to mono-cultivation might fare better in terms of this index of ecosystem health. For instance, several tree species could be genetically engineered to absorb more carbon (as a contribution to solving the global-warming problem) than natural$_{fa}$ trees, and these could be grown in such a way as to constitute a mixed forest capable of playing host to more species than its counterpart devoted to mono-cultivation.

The indices of ecosystem health that are now being developed are admittedly based on natural$_{fa}$ ecosystems: two such indices are the diversity of species and the proportion of native to non-native species in the ecosystem. For the more stressed an ecosystem is by external factors, the greater the decline, in general, in species diversity and, in particular, in native species, as those unable to adapt to the stresses disappear and their places are taken by more 'opportunistic' species from outside (usually fewer in number). However, once these indices have been developed and are in place, there is no reason why ecosystem management could not deploy them as guidelines in designing, structuring and maintaining artefactual ecosystems to ensure that these, too, would be judged healthy by such yardsticks.[65] Clearly, those who hold that the fundamental goal of ecosystem management is the maintenance of ecosystem health intend the indices to apply, irrespective of whether the ecosystems are natural$_{fa}$ or artefactual.[66] For them the distinction between the natural$_{fa}$ and the artefactual is immaterial. Indeed, Callicott, given his new mood of ecumenical postmodernism, is of the view that the sooner we give up "the idea of nature" which could be said to recognize such a distinction, the better off everyone would be, humans and nonhumans alike.[67]

But from the perspective of this book, it clearly matters that the new goal of ecosystem health should not mask the ontological difference between the $natural_{fa}$ and the artefactual. If all that really counts is ecosystem health, then does it matter if $natural_{fa}$ ecosystems are replaced by artefactual ecosystems? The answer to this rhetorical question is, presumably, no for Norton and Callicott. But from the point of view of ontological impoverishment, it matters greatly if healthy $natural_{fa}$ ecosystems, for one reason or other, come to be replaced by equally healthy artefactual ecosystems. If secondary characteristics, like ecosystem health, become entrenched as the fundamental goal of environmental policy-making, $natural_{fa}$ ecosystems would become greatly endangered. So would natural evolution itself. Human processes of selection and control in the designing and maintaining of healthy artefactual ecosystems would displace the processes of natural selection and evolution themselves. Indeed, postmodernism, at least in recent environmental thought, in upholding $nature_{c}$ as the only sense of nature, even boasts of ushering in a postevolutionary world.[68]

As has been argued, such replacement amounts to the triumph of the modern philosophy of *homo faber*, of fabricaton and of the humanization of $nature_{fa}$. Such a philosophy behind modern science and its technology would ultimately destroy $nature_{fa}$ as 'the Other,' both as concept and as reality. Furthermore, the elimination of $nature_{fa}$ as 'the Other' not only diminishes the world both ontologically and physically; it also diminishes us, humans, who are the agents of such impoverishment, as Chapter 4 has shown.[69] The philosophy of *homo faber* reduces the 'essence' of human nature to its ability to fashion tools, to make things—the intelligence of fabrication is all the intelligence that matters and, therefore, that there is in humans. This marginalizes all other aspects of human intelligence as well as other human characteristics and potentials. Resistance against this kind of single-minded determination on the part of modernity or, indeed, of postmodernity, both to undermine $nature_{nh}$ ($nature_{nk}$, $nature_{fa}$) and to simplify and diminish human nature, is called for.

Environmental Ethics and Environmental Philosophy: Axiology and Ontology

About twenty years ago when philosophers became concerned about environmental matters, they concentrated, in the main, on developing an ethic which was axiology led.[70] The discipline was to be called 'environmental ethics.' Theorists who are not too comfortable with this agenda, for one reason or other, call it 'environmental philosophy.' For this author, the reason for following such a tendency is the one already raised, namely, that ontology precedes axiology in determining our attitude to nature. Ethics and axiology appear to go hand in hand. Ontology, however, is another main branch of philosophy. To draw attention to its contribution, it seems apt to opt for the label 'environmental philosophy' for the new discipline.

In the light of the points raised in the preceding sections, it may also be clearer to use the term 'theories of intrinsic value' to characterize those theories which concentrate on secondary characteristics, in developing their respective environmental ethics. In contrast, this theory is a theory of moral considerability grounded in the primary ontological characteristic of independence from humankind. Such an environmental philosophy is led by ontology rather than axiology. Intrinsic value complements this understanding of moral considerability, but is not itself a fundamental criterion of moral considerability.

An axiology-led environmental ethic, focusing on intrinsic value, is an ethic which relies on the notion of similarity rather than difference between entities. Similarity, in turn, relies on the notion of consistency. The reasoning goes somewhat as follows: one ought to treat entity A in way X because A possesses attribute P; one ought, therefore, to treat any other entity, like B, in way X in virtue of B possessing attribute P. For instance, one ought to accord rights of a certain kind to A (a human being) because A is a being capable of making choices; one ought, therefore, to accord B (another human being) rights of a certain kind because B, too, is a being capable of making conscious choices. Or, one ought not, *ceteris paribus*, to confine A (a human being) to such a small space that A cannot even turn round or lie down because A is a being capable of suffering pain and frustration; one ought not, therefore, to confine B (a chicken) to such a small space that B cannot even turn round or flap its wings because B, too, is a being capable of suffering pain and frustration. On this line of reasoning, what is considered to be morally relevant is the similarity between A and B, namely, in the first instance, that A and B are beings capable of making conscious choices, and in the second instance, that A and B are beings capable of suffering pain and frustration. What is (are) considered to be morally irrelevant is (are) the difference(s) between A and B, namely, that, in the first instance, A has a white skin and B a black one, and in the second instance, A is a human and B, a chicken. Rationality as consistency bids one to overlook the difference(s) and concentrate on the similarity. In that sense, one can say that universalization is built into this mode of ethical reasoning.[71] It is the standard mode, irrespective of whether the ethicist in question is concerned with trying to accord intrinsic value only to humans, or to nonhumans as well. It follows that in the latter context, a so-called 'new' environmental ethic may not be so new after all, as it continues to rely on the same mode of ethical reasoning which obtains in traditional inter-personal ethics.

To get to a distinctly new perspective, one has to leave environmental ethics behind for an environmental philosophy which is ontology led. An ontology-led environmental philosophy relies on the notion of difference, not similarity. As such, it cannot straightforwardly invoke rationality as consistency in the way the standard mode of ethical reasoning does.[72] On this view, what is morally relevant is precisely that naturally-occurring beings and artefactual beings are, in terms of their ontological status, **different** categories of being. What matters primarily is this **difference**. What is less crucial are the similarities which could well exist between members of these two ontologically distinct categories of being.

The difference between an axiology-led environmental ethic and an ontology-led environmental philosophy may be appreciated more readily by looking again at biotic artefacts. The Beltsville pig, though not a transgenic organism, nevertheless, is a biotic artefact; it shares a characteristic with ordinary domestic pigs and wild pigs, indeed even with humans, namely, the capability to suffer pain and frustration when treated in certain ways. An axiology-led environmental ethic—like that held implicitly, or explicitly, by nearly all animal welfare groups—would deplore the suffering of the Beltsville pig on the grounds of consistency. If it is morally unjustifiable to inflict such pain on humans, it is equally morally unjustifiable to do so to the Beltsville pig, or, indeed, to any other factory-farmed pig liable to suffer pain in an industrial system of keeping animals. However, by the same reasoning, as Richard Ryder consistently maintains, if animals were genetically engineered so that they no longer could feel pain and frustration, then moral censure would no longer be appropriate. The attribute which grounds their intrinsic value (and in turn their moral considerability on this view) has been removed. They, then, no longer share the characteristic which other morally sensitive beings have, the capability to feel pain and frustration.

Those who find such consistency hard to take are groping to get out of this kind of ethical straitjacket. In trying to grope for a way out, some of them may be vaguely hinting at an ontology-led environmental philosophy instead. For instance, some argue that it is unnatural or against the *telos* of a pig that it should grow so large as not to be able to hoist itself up on its four legs, or suffer from anoestrus (lack of breeding period or sexual urge). To invoke the *telos* of an organism is to imply that without specific human intervention, the individual organism would not exhibit the deplored characteristic in question at all, or, to the extent, it does. It is, therefore, to hint obliquely that the individual is not a naturally-occurring being but an artefactual one. And that even if the pig in question were genetically engineered to feel no pain or not to mind being unable to get up properly and walk with co-ordinated gait, it would still be wrong to do so. In other words, what is really wrong, morally speaking, is that the individual organism has been designed to suit our human ends and intentions rather than to pursue its own *telos*, its own projects, or in the terminology advocated here, to get on with its own trajectory. Its altered ontological status is the really worrying thing.

The biotic artefacts cited above have negative features in their altered ontological state—being unable to get up and stand on all fours as a pig is distinctly a negative attribute. But such negativity is not crucial to the argument; instances where the features found in beings in their altered ontological states are positive have also been mentioned. We have seen that healthy ecosystems may include genetic diversity, ability to resist pests, etc., all positive features. Yet an artefactual ecosystem may also display them. The same holds for other positive characteristics like complexity or intricacy which may obtain in both naturally-occurring and artefactual entities and systems.

This, then, brings us back to the distinction between the primary characteristic of ontological independence and the secondary characteristics which are axiologically relevant. A perspective which argues for the primacy of ontology over axiology must ultimately rest on the notion of ontological difference, not axiological similarity. It, therefore, departs radically from the standard mode of ethical reasoning. However, an analogue exists within inter-personal ethics. Equality, understood as a distributive notion, is about recognizing the existence of **different** and **separate** persons, each pursuing her/his own projects. The projects themselves are secondary and have to be considered within the context of their source—in other words, it matters **whose** projects they are, **who** initiates and controls them, **who** has access to, and to how much of the resources which render projects possible and realizable. A society in which 10 percent own 90 percent of the resources is an affront to equality. The denial to the rest of 90 percent of the resources to pursue their projects would not, from this point of view, be morally compensated for even if the well-resourced 10 percent were frenetically and endlessly initiating exciting projects of one kind or another of their own. As one philosopher has succinctly observed some time ago: "to give two treats to one twin is not the same as to give one treat each to both."[73] No matter how alike the twins are in all respects—physically, temperamentally, psychologically—they are not substitutable, one for the other, for they are different and separate persons.

The difference between separate persons, however, is not an ontological one, unlike the difference between naturally-occurring and artefactual beings. All the same, it makes sense even in such a context to discern an analogous distinction between primary and secondary characteristics—the former pertains to the fact that twin A and twin B are each the respective site of her/his own trajectory, the latter to whatever other characteristics they may share, such as their height, color of eyes, enjoyment of the same sort of jokes, or music, or liking the same type of personality in seeking out a partner in life, etc. If twin A were to die in an accident, the world would be minus that one individual person. We do not say that there has been no loss on the ground that the surviving twin will be leading the same sort of life that the deceased sibling would lead anyway, if s/he were still alive.

However, it could be argued that the focus on difference in the analogue above is only an apparent one. After all, the twins do have something fundamental in common, namely, that each is the site of self-consciousness. Self-consciousness, at least in modern European philosophy, is *par excellence* what makes humans intrinsically valuable. This then seems to bring the argument back to the standard mode of ethical reasoning, namely, relying on the notions of similarity in intrinsic value and consistency.

However, the detour is not without methodological merits. An analogy, *ex hypothesi*, is not an exact, perfect match. Its dissimilarity simply reinforces the point already made, namely, that the standard mode of ethical reasoning cannot accommodate a context where the difference is an ontological one, and that to argue for ontological priority is indeed a radical departure from such a mode.[74]

Claiming radical departure is one thing. Trying to make such radical departure intelligible is another. The second task is obviously more difficult than the first. This book has tried to offer several strands of argumentation, the first of which has to be a clarification of what may be called an ontology-led environmental philosophy. This should not lead one to deny that axiology has a role to play in such a philosophical perspective. The ontological independence of naturally-occurring beings enables such beings to display, in their trajectories, diverse intrinsically valuable features. But it is their primary ontological characteristic which gives a different order of value—independent value—to their secondary axiological characteristics, enabling them to be distinguished from the artefactual which may also display similar axiological secondary characteristics.

Such independent value in nature$_{nh}$ is mutely-enacted value. However, humans, given their peculiar type of consciousness, can linguistically express such a value as recognized-articulated value. A world without human consciousness is a world without independent value as recognized-articulated value, but it is still a world with independent value mutely enacted. Independent value is then not simply humanly conferred, although it can be humanly recognized and articulated.

But critics may not be impressed by the distinction between independent value as mutely-enacted and recognized-articulated value. They wish to raise the more fundamental criticism: why is ontological independence, thus construed, said to be valuable at all?

One short answer is simply to say that the question may be asked of any characteristic whatever, not merely the primary quality of ontological independence—whatever P (a characteristic of any entity, whether a primary ontological or a secondary axiological one) may be, it makes sense to ask 'why is P valuable?' and to reject any possible justification as logically uncompelling. For instance, take the Kantian secondary axiological characteristic of self-consciousness. The question can be posed: 'Why is an entity possessing self-consciousness (intrinsically) valuable?' Other similar questions may also be posed: 'Why is an entity able to feel pain (intrinsically) valuable?' or 'Why is an entity which is the bearer of interests (intrinsically) valuable?' No logically compelling answer can be given to any of these questions if being logically compelling amounts to logical derivability or deducibility.

A longer route undertaken (so far) may briefly be summarized here by tying together some of the points raised in preceding chapters and sections of this chapter as well as anticipating a key point in the section to follow—**Narcissism and *Homo Faber***—which argues that collective narcissism, expressed through the relentless thrust of *homo faber* and its fabrication, we shall see, is that condition which is blind to nature$_{fa}$ as 'the Other.' Such blindness is worrying because:

1. It ignores the fact that Nature$_{nh}$ exists first and foremost 'by itself' and is only

secondarily or incidentally to be of instrumental value to humans. This sense of existence is a logical implicate of the 'No External Teleology' thesis. This then in turn entails that the Instrumental Value Assumption is simply false.

2. We have seen that awe in the presence of human artefacts is not the same as awe in the presence of $\text{nature}_{\text{fa}}$. Complexity, intricacy, ingenuity, or creativity in artefacts is not a substitute for complexity, intricacy, ingenuity, creativity in $\text{nature}_{\text{fa}}$. The awe induced by artefacts is *au fond* narcissistic in character; that induced by $\text{nature}_{\text{fa}}$ is not.
3. Moral blindness is not simply confined to an inability to recognize that similar others exist, but also includes an inability to recognize that **different** others exist. In other words, moral blindness is not only about blindness to axiological similarities and disregarding them, but also about blindness to ontological difference and disregarding it.
4. Ontological blindness to nature and elimination of nature as 'the Other' is ontologically deflationist. Modern Western philosophy has, by and large, dispensed with the ontological category of transcendant entities, like God. *Homo faber* in modern Western philosophy increasingly threatens to dispense also with the ontological category of $\text{nature}_{\text{fa}}$. So minimalist is such a world that only one kind of ontological being exists and matters, namely, the human being and its extended self through its fabrication. In such a world, humans have eliminated both God and $\text{nature}_{\text{fa}}$ as 'the Other.' Ontologically, the world without 'the Other' is a lonesome world. Wherever one turns, one only sees images of oneself; whenever one shouts, one only hears one's own echoes. Humankind is then imprisoned within an existential or ontological solipsism of its own making.
5. To eliminate $\text{nature}_{\text{fa}}$ as 'the Other' amounts to a refusal to consider ontological independence as a value. This in turn involves driving a wedge between ontological and axiological values. The latter are regarded as unproblematic, except to a moral skeptic. But those who do not wish to go down the skeptical path cannot reject the ontological value of independence offhand without giving reasons for the rejection. But the reasons proffered could well turn out to commit *petitio principii*—they will simply be a re-assertion in various guises of the view that humans, in virtue of their unique type of consciousness or self-consciousness, and/or their unique ability for symboling, deserve moral consideration while all other naturally-occurring beings are only of instrumental value to them. But this assumption is precisely the one that is at issue. To assert that humans alone are worthy of moral consideration is to beg the question, not to provide for a defense against the charge of denying moral considerability to beings which embody the ontological value of independence.

Anthropogenic and Nonanthropogenic

This section will look at a distinction, namely, between anthropogenic and nonanthropogenic, which runs throughout this book. Transforming biotic and abiotic beings to become artefactual ones is clearly both an anthropogenic and anthropocentric enterprise, as it involves humankind, attempting to embody its intentions in material terms.[75] Environmental philosophy in general has focused on anthropocentrism and its polarity, nonanthropocentrism. But the distinction between anthropogenic and its antonym, nonanthropogenic, has not attracted the same amount of scrutiny. Earlier on, this book has touched on the distinction, but some further discussion, even if it is brief, is clearly called for and its philosophical significance brought out into the open.

Increasingly in environmental debate, attention is being drawn to the fact that nature on Earth, in its history of four and a half billion years, has been producing effects no different in kind from anthropogenic ones, but on an infinitely grander scale. For instance, take carbon dioxide, a main greenhouse effect gas. Anthropogenic emission mainly through power generation, car use and the burning of forests is estimated at about seven billion tons per annum, while nonanthropogenic emission, mainly through volcanic eruptions and plant decay, runs to about 200 billions tons. One environmental writer, Gregg Easterbrook, has recently worked out that:

> Carbon dioxide constitutes roughly one percent of the full greenhouse effect, with the human-caused component of the carbon dioxide cycle at roughly four percent and the rate of artificial carbon dioxide increase around one percent annually. This works out to the human impact on the greenhouse effect roughly 0.04 percent of the total annual effect. That is, 99.96 percent of global warming is caused by nature, 0.04 percent is caused by people. The present rate of increase in human-caused greenhouse forcing, meanwhile, works out to be about 0.004 percent per annum of the total effect.[76]

Or take species extinction. Since life first appeared on Earth three billion years ago, there have been five major global extinctions as well as several relatively minor ones in the last 500 million years—the Ordovician (440 million years ago), the Devonian (365 million years ago), the Permian (245 million years ago), the Triassic (210 million years ago) and the latest, the Cretaceous (65 million years ago). The first four are attributed to climate cooling, but as yet there is no consensus as to what caused the cooling in the first instance, although continental drift has been speculated. The last, the so-called K-T extinction, attracts three competing hypotheses to explain it, in terms of the aftermath of volcanic eruption, the impact of an asteroid hitting Earth or a combination of the two accounts. But after each catastrophic loss, biodiversity recovered to the level before the crisis; admittedly, the time for full recovery varied from 100 million (after the Permian and the Triassic extinctions taken together, as they were close in time) to 20 million years (after the Cretaceous episode). Of all the species which had ever lived, 99 percent had become extinct.

The number of animal and plant species known today is about 1.5 million. The total number of species estimated varies from 10 million at the conservative, to at least 30 million, at the more generous end. In the absence of humankind, it is estimated that one species becomes extinct per 1000 years. Between 1600 and 1960, over 400 years, more than 1000 species have become extinct. And since 1960, as many as 1000 are said to be lost per year. Some experts predict that by the end of the century one will be lost each hour; however, an authority like E. O. Wilson regards that to be on the low side and predicts three per hour. However, this kind of estimation is built upon a raft of assumptions, empirical and theoretical, which may be open to challenge. But supposing that the figures turn out to have a realistic foundation, the following response may still be forthcoming: loss of biodiversity owing to anthropogenic causes is no different really from loss of biodiversity owing to nonanthropogenic causes. And we have seen how Earth had recovered from such eco-catastrophes before the emergence of humankind. Is there any ground for believing that anthropogenically-caused extinctions are specially alarming or pernicious as, after all, extinctions are extinctions, no matter the cause? On the contrary, in terms of scale alone, alleged anthropogenically-caused contemporary biodiversity loss measured against non-anthropogenically-caused extinctions in the geological past is small.

This is not the place to answer the question in empirical terms, that is, whether Earth, at this moment in time, unlike Earth in earlier times, could or could not recover eventually from biodiversity loss and how long such recovery might take. Instead, it will concentrate on the philosophical issues that underpin the anthropogenic as opposed to the nonanthropogenic sources of biodiversity loss.

The central thing to bear in mind is a point which has been made throughout this chapter, but especially in **Resisting Humean Projectivism**, namely, the uniqueness of human consciousness itself and the particular kind of intelligence which flows from it. The uniqueness of human intelligence lies in its ability systematically to find out about the world, to ascertain, foresee (to a greater or lesser extent) and monitor the consequences of one's acts, to use information and knowledge deliberately in pursuit of a given goal.[77] Furthermore, as the epistemic exploration of the world is mediated by the human capability for symboling, that knowledge may be highly abstract in character.[78] For instance, while lions know instinctively that humans or gazelles make satisfactory dinners when they are hungry, they do not know and cannot know that their victims make desirable dinners because they contain protein, without which lions would no longer be effective lions or be lions at all. Humans (although not necessarily any one named particular individual human) know this as well as that for their own metabolism to function properly, they need protein among other nutrients, and that this may be found in meat, fish or parts of some types of plant like lentils and soya beans.

Human language enables human consciousness to deal not only with the abstract, but also with the absent as part of the abstract. Animal intelligence, however, can only cope with the present.[79] Sultan, the clever chimpanzee, was clearly brilliant in solving the particular problem he was faced with, namely, how to get the food which was just lying beyond his reach outside his cage. First he tried getting it with a pole which was not long enough, upon which he gave up and started to play with two sticks in another part of the cage. He accidentally slotted them together, making one single stick. Then Köhler observed Sultan rushing back immediately to the food which he could now reach with the aid of the longer stick.[80] Köhler argued that Sultan must have "realized" the relevance of his accidental discovery to the problem he faced about getting the food. This then is a clear demonstration of intelligence. However, Sultan's eventual solution to the problem was made possible only in the physical presence of the food and the sticks. In the absence of these physical objects in Sultan's immediate environment and his hunger, it would not be conceivable for Sultan to entertain the problem and solve it in the abstract and purely theoretically. But human intelligence is distinctive and more powerful precisely because it can grasp what is absent as part of what is abstract. It is this capability which allows humankind to build successful spaceships to get to places and traverse over space never traversed before. Some birds fly great distances and some sea animals travel many nautical miles. But they do so as the evolved beings that they are, unaided by the equivalent of artefacts like spaceships or submarines. Humans, on the other hand, use the intelligence they possess as the evolved beings that they are, to construct artefacts to enable them to do things they would otherwise be unable to do; these artefacts are the embodiments of their abstract intelligence and intentionality.

The future, too, is an aspect of the absent. We have already said that human intelligence could be directed toward finding out what the future would be like either without human intervention or in the light of human intervention. This means it could entertain alternative futures or possible worlds. This ability in turn makes choosing between alternative futures intelligible. Choice and (moral) freedom go hand in hand.[81]

Human intelligence, in the way just outlined, is a pre-condition of morality, as it makes it possible for us to grasp the difference between good and evil, to entertain conception(s) of right and wrong, and to appreciate what it is intentionally or knowingly to do what may destroy or damage the good, thereby promoting what may be evil or wrong. This means that humans are morally responsible beings.[82] Some nonhuman beings, like the higher mammals, may be intelligent in the sense of being able to solve problems as we have just seen, use tools, or even, as some people have more controversially claimed, acquire and use human language to some extent, or exhibit altruistic behavior. However, no nonhuman beings, other than humans, could agonize about what counts as being moral and what immoral, what constitutes good or bad behavior, what is virtue or vice, what is right or wrong, not only about matters happening right now, but also about things which are likely to happen in the future as a result of our intervention. Humans could be reprimanded, criticized and punished for intentionally or even knowingly doing wrong. But the modern mind finds the idea of holding other nonhuman beings, like animals, morally responsible for the damage they may cause to others, nonhuman or human, a bizarre one.[83]

However, this capability for moral discrimination and responsibility based on human intelligence should not be read as providing support for the dominant paradigm of human superiority in modern Western thought derived from the ancient Greek or Cartesian tradition. On the contrary, it forms the basis of moral duties and obligations toward nonhuman others regarded as the sources and loci of independent value which, given the peculiarity of our type of consciousness and intelligence, we alone can recognize, even while the bearers themselves of such value are incapable of recognizing and articulating it.[84]

Natural forces, which are purely geophysical like climate cooling or warming, and volcanic eruptions or forces, which are biochemical like those involved in the decay of organisms, cannot be praised or blamed either for what they are or do. Nor does it make sense to hold animals causing pain to other animals (as in the predator-prey relationship) as culpable—the pain that happens in nature, as earlier argued, is not to be confused with the pain that happens within culture. To remove pain which is nonanthropogenically produced in nature is an attempt on the part of humankind to alter and control how nature operates, that is to say, to humanize nature so that it would only embody our way of doing things, not nature's way, our values, not nature's own.

Moral responsibility is part of moral agency, and may be said to obtain if and only if a being is capable of moral discrimination, of appreciating what damage or harm could be done to others (as well as to the agent her/himself) by acting in certain ways. Thus defined, only human beings can qualify to be moral agents. But one needs straightaway to consider a possible objection, namely, that some humans do not satisfy the conditions of moral agency, such as the very young, those in a coma, etc. But while this observation is undoubtedly correct, it does not amount to a valid criticism of the position outlined. Two theses must be distinguished: that (a) not all humans (at any one moment of their lives) qualify to be moral agents, and (b) all moral agents are humans. Thesis b is the one which is advanced above; thesis a does not contradict it. Thesis b implies that no nonhuman beings are moral agents; it does not imply that all human beings are moral agents.

While the notion of moral agency is central to the moral vocabulary, it is not the only one. Other notions include, for example, being a moral patient.[85] For a being to count as a moral patient, it is enough that the being could be harmed, damaged in some ways or be the bearer of a value which is independent and autonomous of the articulator of that value. Guardians articulate on behalf of the unborn, the very young or the very senile what harm or damage could be inflicted on them, or what value in them could be violated if certain things were to be done to them. The same could be said on behalf of both biotic and abiotic entities which, too, could be either harmed or damaged by human activities.

In other words, moral patients are fit objects of moral concern on the part of moral agents. Moral agents have duties and responsibilities to moral patients. However, moral patients themselves, *ex hypothesi*, cannot be said to have moral duties to moral agents or to one another. Some moral patients which are human may sooner or later acquire the status of being moral agents—the very young will grow out of their immaturity but the very senile would not recover their former status of being moral agents. But nonhuman moral patients are entities which lack the capabilities for the ascription of moral agency and could not be said to acquire them at any stage of their existence.

In the light of the discussion above, let us return to the distinction between anthropogenic and nonanthropogenic sources of environmental damage or change. Take the following two examples cited by Rolston.[86] The first involved a rancher in Wyoming who erected a fence twenty eight miles long and five feet high to keep out the wild antelope from his grazing land. The winter of 1983-84 turned out to be both early and severe, covering the ground with snow and ice. The antelope could not migrate to the milder Red Rim area where they could forage because the fence blocked their path. As a result about 1,500 antelope faced starvation. The same severe winter in southern Wyoming caused 85 percent of an antelope herd (that is, 7,200 out of 8,400) to perish whereas the normal winter death rate is 15 to 25 percent. Many in the herd tried to migrate south into the western slope of Colorado where other animals were already starving. In the first instance, the court was persuaded by the Game and Fish Department and the Wyoming Wildlife Federation to order the rancher to remove the fence and to save the animals. In the second instance, the Colorado Division of Wildlife resisted a call from animal welfare groups to feed the antelope.

In the first case, the death would have been anthropogenically caused, admittedly not deliberately, but knowingly, as the consequences concerning the fate of the antelope of erecting and maintaining such a fence in a severe winter were foreseen. So it was morally proper and desirable that the fence be removed. But in the second case where the death had nothing to do with anthropogenic causes, humans as moral agents could not be said to have a moral duty to save the animals; on the contrary, to do so would be to substitute human intervention and control for the forces of natural selection at work. Nature follows its own trajectories; to thwart them each time we humans are minded and able to do so amounts ultimately to eliminating nature as 'the Other,' both empirically and ontologically. Our duties to nature are not identical to our duties to fellow humans. In the case of fellow humans, humanitarianism requires that efforts be made to save them should they be starving in a sevre winter or drought, whether anthropogenically or nonanthropogenically caused. But humanitarianism should have its limits and these should be drawn at anthropogenically caused damage or changes to nonhumans. Otherwise, one runs into the distinct danger of humankind, out of misguided humanitarianism, assuming the role of controlling and manipulating nature, thereby undermining its ontological status of independence and autonomy from humankind.[87]

Go back to the examples earlier cited which appear to have impressed Easterbrook so much, such as carbon dioxide emission and biodiversity loss. He calls his own position ecorealism; part of that remit is to tell us to stay cool and not to overreact to the so-called ecological crisis which has gripped many people in the last thirty years or so. Earth is not that fragile. It has been massively bashed about by its own natural forces in its long history. By comparison, whatever damage humankind has allegedly done is pitiably puny. We magnify ourselves surely in magnifying the damage we claim we are capable of inflicting upon Earth. Nature recovers, sometimes even faster than we assume it can.

While this voice against over-reaction sounds sane and is, perhaps, called for as a counter-balance to certain exaggerated claims which have been made by some environmental thinkers and activists, Easterbrook's ecorealism fails totally to grasp and appreciate two fundamental matters: the distinction between anthropogenic and nonanthropogenic, and its relevance to the notion of nature as the ontological other. The anxiety over the loss of biodiversity or the emission of carbon dioxide which is human-caused is not alleviated by taking note of the fact that such loss is either much less spectacular than nature's own loss in previous eco-catastrophes, that nature eventually recovered from such losses, or that carbon dioxide is carbon dioxide whatever its source of emission. One may even be able to agree with him on this; but the anxiety remains, as it is not what Easterbrook has made it out to be. The anxiety is really centered on the respective ontological status of humankind and of nature. Humans and their artefacts belong to a different ontological category from nonhumans. It makes sense to say that humans have a *prima facie* moral duty not to eliminate naturally-occurring beings or entities as well as the processes underpinning their continuing existence which determine when they alter, how they alter and the rate at which they alter.

This book shares a common assumption with Easterbrook, namely, that pollution may in the longer term recede as a threat to the integrity of the environment. Easterbrook has indeed made a reasonable case for saying that at least in the affluent developed North, more ecologically sensitive measures and practices have been or are being put in place to abate the various types of pollution identified, which have been worrying many people in the last thirty years or so. Given the general success in this direction, he feels that a degree of environmental optimism which he detects over the horizon is justified. While this book acknowledges that the trend toward pollution abatement (at least in the North, if not globally) is to be welcomed, environmental optimism may not be justified.

The implicit theoretical presuppositions of Easterbrook's book are not the same as this. He has simply embraced the dominant paradigm of (axiological) anthropocentrism and, as shown above, has failed to appreciate the philosophical significance of the difference between anthropogenic and nonanthropogenic causes of environmental damage or change. The oversight of such significance, whether witting or unwitting, is congenial to the project of managing and controlling nature—any environmental state or change, irrespective of how it has come about, is regarded as within the remit of managerial control provided the intervention is judged to serve some worthwhile human goal, and the technology is available to do so. His book amounts, in short, to an unqualified celebration of *homo faber* whose destiny, with the help of modern science and technology, is systematically to transform the natural to become the artefactual, thereby eliminating nature both empirically and as ontological other. This, for Easterbrook—although he does not articulate it as such—constitutes the coming age of environmental optimism. But from the standpoint of this book, such transformation constitutes ontological impoverishment in the world. At the same time, it also impoverishes humankind, as such elimination amounts to the entrenchment of a completely narcissistic civilization which destroys the source of true wonderment. True wonderment is appropriate only in the presence of beings and processes which have come into existence and continue to exist, as well as go out of existence at a rythm and in ways that have nothing to do with human ingenuity, design or intervention. But when the natural has been transformed to become the artefactual, humankind, in expressing awe and wonder at its environment, is simply admiring its own handiwork and projects, not nature's own projects and trajectories. One of the aims of this book is precisely to argue that the new technologies, induced by the theoretical discoveries of so-called basic science at the end of this millennium and the beginning of the next, will enable *homo faber* to advance, even more deeply than before, its totalizing project of humanizing nature. While some transformation is inevitable if humankind were to live and flourish, nevertheless, humankind, given its unique intelligence, is also capable of recognizing that there are moral constraints to the project of humanizing nature. Those who uphold the anthropocentric paradigm, like Easterbrook, have failed to grasp this. Furthermore, they and most environmental theorists and writers have also failed to grasp that, in the final analysis, the fundamental threat posed by humankind to nature is ontological in character.

Narcissism and *Homo Faber*

Narcissism in a person may considered to be a pathology or, at the very least, a form of immaturity.[88] Analogously, a civilization dedicated so obsessively to celebrating the philosophy of *homo faber,* as modernity appears to do, may be said to be a narcissistic civilization, and therefore, either pathological or immature. A healthy, mature civilization, like a healthy, mature person, recognizes different others and respects them.[89] Such a civilization would then recognize $nature_{nh}$ ($nature_{nk}$, $nature_{fa}$) as significant Other and would not wish to eliminate it by systematically transforming it through science and technology to become a near-total, if not a total, human artefact. Narcissistic individuals feel ontologically secure and emotionally satisfied only if they see themselves, and no other, reflected in every mirror they look at. Likewise, a narcissistic civilization is ontologically secure and emotionally satisfied only when everything in it has become transformed into embodiments and structures of human intentionality, acting as mirrors, to reflect human consciousness and its projects. Unless scientific/technological determinism obtains, we can and should stop our civilization from going down the total narcissistic road and reset it, instead, on to a path to maturity. In so doing, we save human nature from impoverishment and nature$_{nh}$ from being systematically undermined or eliminated. In this sense, ontological dyadism, not dualism, underpins axiology and provides a philosophical framework within which both humans (provided their numbers are not excessive) and nonhumans (biotic and nonbiotic) may co-exist, with sufficient ontological space left by humans for nonhuman others to pursue or carry on their own varied trajectories.

Narcissism is egomania. Let us first take the standard case in which the ego involved is that of a particular individual person. Any assertion attributing egomania to someone may be said, logically, to be Janus-faced. It is at once descriptive and evaluative. It is descriptive because it refers to a state of affairs which can be checked in much the same way as any other descriptive assertion, such as, 'X is HIV positive' may be checked. It is evaluative because it refers to a state of affairs which is disapproved, rather than approved of.[90] The very term 'mania' means 'fixation' or 'obsession.' While an obsession with something is not necessarily unhealthy, it is dangerously near to being so. Indeed, an obsession, when it becomes total, may be ethically suspect. Total obsessions necessarily ignore other aspects of one's personality, as a result of which that personality becomes lop-sided. Furthermore, legitimate ethical demands from others would go unsatisfied. For instance, writers who shut themselves up in the loft or basement, while remaining deaf to the claims of their children on their affection and attention, may be said to be less than virtuous even if the products of their intellectual labor were to turn out to be historically momentous.[91] But when the object of the fixation is primarily not with an external cause but with one's ego *per se*, whether that ego is in pursuit of money, political power or prestige, then such an obsession is doubly ethically unwholesome. In such pursuit, the legitimate ethical demands of others are usually not simply ignored but are positively undermined and denied.

As already observed, our present modern civilization celebrates humanity primarily as *homo faber*, and appears to be obsessed with the control of nature through its science and technology. Its ostensible object of fixation is the promotion of human well-being primarily as human comfort.[92] However, beneath that seemingly benign humanistic ideal, is another, perhaps less ethically attractive, enterprise which amounts to the elimination of nature, empirically as well as conceptually. The collective project of totally humanizing nature by transforming the natural to become the artefactual is analogous to the absolute pursuit of money, political power, or prestige in the case of an individual person. Just as such an individual project necessarily involves trampling upon the legitimate interests and ethical demands of other persons, the collective project, on behalf of humanity to control nature, also necessarily involves trampling upon the legitimate ethical demands of nonhuman others. To say, then, that such a civilization exhibits a form of collective egomania is at once to say something which is both descriptively true and evaluatively negative.

Human collective egomania is revealed behind yet another seemingly virtuous and noble goal, namely, that of self-realization with the help of science/ technology. With every new radical technological innovation, humankind can realize projects not realizable before. These innovations often result in complex and intricate objects and forms of activity which embody great human skills, ingenuity, and creativity. Yet, as we have seen, when we admire them, we are in reality admiring ourselves—the products simply are mirrors held up to reflect ourselves. We ought not, as we saw in an earlier section in this chapter, to appreciate creativity, ingenuity, complexity, intricacy *simpliciter*. In this context, we are, in reality, admiring **our** creativity, ingenuity, complexity, intricacy, and congratulating ourselves for the feats, against all the odds, which we have performed. It is, therefore, ontologically and morally **not** the same thing as admiring the creativity, ingenuity, complexity, intricacy expressed by nature through the trajectories of its entities and their relationships with one another. In the context of naturally-occurring creativity, etc., when we admire it, we are **not** admiring ourselves—we are admiring nature as 'the Other.' However, as *homo faber* appears relentless in its efforts at fabrication, through its science/technology, to transform systematically the natural world to become an artefactual one, the fabricated world which results would, undoubtedly, indeed, display great ingenuity, complexity, intricacy, creativity.[93] The price paid for such an achievement would be nature's own ingenuity, complexity, intricacy, creativity and its numerous trajectories in which these characteristics are expressed. Humankind would no longer feel awe and wonder at nature's handiworks, for the simple reason that such handiworks would be few and far between, even if not totally non-existent (at least on Earth itself). It will still be able to feel awe but only in the presence of its own handiworks. And that response, we have said, is of a totally different ontological and moral order from the awe elicited in the presence of nature's own creativity and ingenuity.[94]

Notes

1. This sense of independent value will be explored later in the section entitled **Intrinsic Value, Trajectories, and Independent Value**.

2. Strictly speaking, Hume spoke in terms of passions, not of feelings, emotions or attitudes. But contemporary philosophy, in the spirit of Hume, prefers to talk in such terms, especially attitudes, rather than passions.

3. See the discussion in Appendix 1 (**Nature as Work of Art**), on the related thesis about means/end rationality as opposed to the irrationality of ends themselves.

4. The Humean version of the thesis, "no values without valuers," differs somewhat from the Cartesian one. At first the distance between them may appear much greater than it really is. Descartes emphasized human self-consciousness, capability for language, more generally, the possession of mind. Hume did not focus on the intellectual side of human consciousness but on the emotional side. So this might lead one initially to think that Hume

could have extended, at least to some animals, the possibility of being the generators of values on the grounds that these animals, too, are capable of emotions. However, this is not Hume's route, as it remains true that for him, human emotions are distinctive, precisely because they are mediated *via* the unique kind of consciousness and self-consciousness which humankind possesses.

The implication of the thesis that there can be no values in the absence of valuers may be spelled out as follows:

1. Human consciousness as self-consciousness linguistically expressed is unique.
2. This attribute is the source of all values.
3. This attribute also grounds moral considerability.
4. Moral considerability in turn grounds intrinsic value.
5. Therefore, human consciousness is the sole locus of intrinsic value.

But this conclusion, at best, only has limited validity—this point will be made clear as the argument develops in what follows in this section.

5. The arguments used here are based on some of those the author has set out in a slightly different context—see Keekok Lee, "The Source and Locus of Intrinsic Value: A Reexamination," *Environmental Ethics* 18 (1996): 297-309, which addresses the controversy between Callicott and Rolston about the source/locus distinction itself.

6. This example cites only individual animals. But the argument holds also in the case of individual plants—for details, see section which follows, **Interests and the Intentional Stance**, as well as Lee, "The Source and Locus of Intrinsic Value."

7. For a further distinction between this sense and 'by itself,' see section: **Abiotic Nature and Intrinsic Value**.

A gloss is called for regarding the terms—'for itself' and 'in itself'— used in this book. Confusingly, they are the opposite of Sartre's use of *pour soi* (for itself) and *en soi* (in itself)—for Sartre, the former, not the latter, applies to humans. (Furthermore, the term—'by itself'—to be introduced later should not be identified with Sartre's *en soi.*) But as this book is an attempt to elucidate issues in environmental philosophy, it follows the well-established usage in environmental philosophical literature, which is the opposite of the Sartrean use—see, for instance, James B. Callicott, "On the Intrinsic Value of Nonhuman Species," in *The Preservation of Species: The Value of Biological Diversity*, Byran G. Norton, ed. (Princeton, N.J.: Princeton University Press, 1986), 160.

8. An example from the theoretical perspective of individualism is Peter Singer. But as he relies on sentience to generate interests, his account is restricted in its applicability only to certain animals; obviously, it is not relevant to the preoccupation here. An example from that of holism or ecocentrism is Lawrence Johnson. But this chapter, and indeed the whole book, will also have to ignore Johnson's theory. This is because ecosystems are precisely those entities where biotic and abiotic nature are inextricably entwined. So even if Johnson's arguments are successful, this would not establish that abiotic nature, on its own, could be said to possess interests and, hence, be the locus of intrinsic/independent value.

9. See for instance Robin Attfield ("The Good of Trees," *The Journal of Value Inquiry* 15 1981: 35-54), but Joel Feinberg ("The Rights of Animals and Unborn Generations," in *Philosophy and Environmental Crisis*, William T. Blackstone, ed. Athens, Georgia: University of Georgia Press, 1974) stands for the opposite view. He argues that only

humans and animals may be said to possess interests.

10. Those who argue otherwise include Raymond Frey (*Interests and Rights: The Case Against Animals* Oxford: The Clarendon Press, 1980) and Peter Carruthers (*The Animals Issue: Moral Theory in Practice* Cambridge: Cambridge University Press, 1992). (However, Frey has since conceded that some animals can have interests—see, "The Significance of Agency and Marginal Cases," *Philosophica* 39 1987: 39-46.)

This may also be a convenient point to dispose of one red herring which is traced to Frey. Frey argues that if 'interest' in sense q obtains in the case of animals (if not plants), this

> yields the counter-intuitive result that manmade/manufactured objects and even things have interests, and, therefore, on the interest thesis, have or at least are candidates for having moral rights. ...[I]f tractors do have interests, then on the interest thesis they have or can have moral rights, and this is a counter-intuitive result. ("Rights, Interests, Desires and Beliefs," *American Philosophical Quarterly* 16 1979: 233-39)

The crucial point that Frey appears to have failed to grasp is that an artefact—especially an abiotic one like a tractor—has no intrinsic *telos* but is the material embodiment of human intentionality. In other words, it exemplies extrinsic/imposed teleology. An artefact does not exist 'by itself'—see following section, **Abiotic Nature and Intrinsic Value**, for an explication of this sense—as it has come into existence and continues to exist in order to serve a particular human purpose. Insofar as it has a 'good of its own,' that good is determined and informed by human interests, as it does not have interests of its own independent of those designed into it by its human creator. In short, it belongs to a different ontological category from that of naturally-occurring beings. The section to follow later—**Intrinsic Value, Trajectories, and Independent Value**—argues that it is crucial to distinguish between primary and secondary characteristics. Frey may be said to have confused the two—the former is ontological, the latter axiological. An artefact differs in ontological status from a naturally-occurring entity no matter how similar their (secondary) characteristics may be. Both may, for example, share the secondary characteristic of interest-bearing; but in the case of the artefact (like the tractor), as we have seen, the only interests it can bear and does bear are human ones, whereas the interests, the naturally-occurring entity bears are its own, which are entirely independent and autonomous of human interests.

11. For an account which confines the notion of interest to animals only with sentient life (thereby also excluding plants), see David DeGrazia, *Taking Animals Seriously: Mental Life and Moral Status* (Cambridge and New York: Cambridge University Press, 1996).

12. See Plumwood, *Feminism*, 131 40.

To prevent confusion, it is best here to clarify how Plumwood uses the following terms and sees the relationships between them:

1. "The intentional," following Brentano and others in the phenomenological tradition, is the hallmark of the mental (in humans).
2. "The intensional" is contrasted with 'the extensional' or 'the truth-functional' in logic.
3. The intentional is a sub-class of the intensional.
4. Both classes may contain mind-like properties.
5. "The intentional stance" is applicable to all Earth Others, including abiotic Earth

Others. As such it is an attempt to re-introduce agency to nonhuman nature.

This section, however, queries the extension of the intentional stance to abiotic nature, and argues that it should be confined to humans and biotic nature only—see text which follows.

13. For a more 'orthodox' understanding of intentionality see John Searle ("What is an Intentional State?", 74-92 and "Intentionality and the Use of Language," in *Meaning and Use*, A. Margulit, ed. (Dordrecht, Holland: D. Reidel Publishing Co., 1979)). For a critique of Searle, see C. B. Martin and Karl Pfeifer ("Intentionality and the Non-Psychological," *Philosophy and Phenomenological Research* XLVI, (1986): 536-56) which Plumwood (*Feminism*, 132, 209) has cited.

14. Plumwood, *Feminism*, 134-35.

15. Plumwood, *Feminism*, 138.

16. Plumwood, *Feminism*, 210.

17. This oversimplified account is based on a film by David Attenborough made for the BBC *Life in the Freezer* series called "The Big Freeze." Admittedly, we are ignorant of the precise mechanisms by which such highly co-ordinated activities take place. But the fact that they lack consciousness, like human consciousness, is neither here nor there—their behavior is distinctly purposive, that is, end-directed, co-ordinated, non-random.

18. See Andy Coghlan, "Sensitive Flowers," *New Scientist* (26 September 1998): 24-28, for the latest outcome of scientific research in understanding the mechanisms by which plants may be said to 'see,' 'smell,' 'taste,' etc.

19. But to prevent misunderstanding, perhaps one should add that this argument is intended to show that abiotic forms and processes in nature have a value which is independent of human valuation of them. It is not meant to argue that in management policy terms one ought never to tamper with such forms and processes. It argues that environmental policy-making should not be based on the presupposition that they are devoid of human-independent value.

20. But, surely, restoring the river to its wild, raging state illustrates only anthropogenic, not necessarily anthropocentric, considerations? This, however, is not correct, as the conception of the 'real spirit' of the river being embodied in its wild, raging torrents is, as earlier observed, simply a disguise for a particular type of aesthetic sensibility. The river is just as much a river whether in a hyperactive or sluggish mode.

21. See, for instance, Bryan Norton (*Why Preserve Natural Variety?* Princeton, New Jersey: Princeton University Press, 1987); E. O. Wilson (*The Diversity of Life*). Also in his address to the American Association for the Advancement of Science in February 1995, Wilson said that based on current knowledge and even a very conservative estimate—that is, assuming that Earth is home to only 10 million species and that many of them have a wide geographic range—27,000 species each year would become extinct, or 74 per day or 3 per hour.

22. In August 1996, NASA caused a stir by claiming that a Martian meteorite actually contains signs of (bacterial) life. The Open University group of scientists in the UK followed this in October with a similar claim about a second Martian meteorite. But the claim is controversial. See, for instance, the editorial—"Martians Have Landed Again"—*New Scientist* 152 (9 November 1996): 3, and W. Wyat Gibbs and Corey S. Powell, "Bugs in the Data?: The Controversy over Martian Life Is Just Beginning," *Scientific American* 275 (October 1996): 12-13. In *Nature*, 4 December 1997, in its Scientific Correspondence section, p. 454 ("No 'Nanofossils' in Martian Meteorite"), John

Bradley of the Georgia Institute of Technology and two colleagues (R. P. Harvey and H. Y. McSween Jr.) dispute the claim that meteorite ALH84001 contains nanofossils; they hold that the evidence suggests a non-biological origin. (This communication is followed by a reply from the authors—D. S. McKay *et al.*—of the original paper making the nanofossil claim.) On 2 December 1996, NASA launched the Mars Rover on the Pathfinder mission (expected to land on the planet in July 1997, which it did successfully), the first lander since the Viking missions in 1976. It would collect and sample data, transmitting them to Earth for 30 days. It would look for the presence of water, both in the past and now; if there (was) is liquid water, there could be life. See Radford, "A Clean Pair of Wheels"; Jeffrey S. Kargel and Robert G. Strom, "Global Climatic Change in Mars," *Scientific American* 275 (November 1996): 60-68.

See also Carl Sagan, "The Search for Extraterrestrial Life," *Scientific American* 271 (1994): 70-77.

23. Organisms thrive even in the craters of volcanoes, in hot springs and in volcanic vents in the deep ocean. In Antarctica—on the land and the frozen surface of its ocean —there are some life forms; beneath the ice of the latter are even more. Inside the exposed but porous rocks of one part of Antartica, a lichen manages to find a niche for itself.

24. See Paul Taylor, *Respect for Nature* (Princeton, New Jersey: Princeton University Press, 1986); Robin Attfield, *The Ethics of Environmental Concern* (Athens and London: The University of Georgia Press, 1991).

25. The belief that abiotic nature is beyond moral consideration and intrinsic value is part of a general consensus which obtains, not only among many environmental theorists, but also among those who are actively engaged in policy-making or in influencing such policy-making. Just one example will suffice to illustrate the point—see John Vidal, "Eco Soundings," *Guardian*, 8 February 1995, Society Section: 4. In 1995, a quarrying firm called Redland Aggregates wanted to remove 600 million tonnes of rock from the Isle of Harris off the north of Scotland (the Outer Hebrides) which amounts, no more and no less, to destroying and removing the Roinebhal Mountain from the island. The company itself had commissioned an ecological survey of the mountain which identified 11 biophyte species (mosses and liverworts), and concluded that it is of little botanical interest. Scottish Natural Heritage (SNH), the official government's conservation agency, conducted its own ecological assessment and came up with 164 biophyte species, including some rare ones nationally. In the light of this, SNH argued that Roinebhal more than satisfies top British conservation rating. Indeed, it should even be acknowledged to be internationally significant. By arguing thus, SNH and other conservation bodies hoped to impress the inquiry which had been set up to look into the matter of whether to grant permission to the quarrying company to reduce Roinebhal to rubble for export. This shows that all sides were agreed that abiotic nature itself has no intrinsic value apart from it playing host to life forms (or significant life forms). An alternative explanation could be that even those parties—if such existed—which held that the abiotic could be intrinsically valuable, had seen fit not to present such an argument to the planning inquiry on the grounds that it would be counterproductive to do so, given the generally accepted consensus that the abiotic is beyond the pale of moral consideration or intrinsic value. In other words, while it may be intellectually and politically acceptable to plead that the biotic be respected, it is probably intellectually, and certainly politically, not acceptable to plead that the abiotic on its own be respected. In the Roinebhal inquiry, besides the appeal to biotic values, other evidence referred to anthropocentric values, such as its status as a historic and prominent landmark in the area. But as far as one can determine, there was no reference to its moral standing as

abiotic nature.

26. Lawrence E. Johnson, *A Morally Deep World* (Cambridge: The University of Cambridge Press, 1991), 282.

27. Among others, Andrew Brennan, "Moral Standing of Natural Objects," 35-56; contributions in the volumne edited by Eugene Hargrove, *Beyond Spaceship Earth: Environmental Ethics and the Solar System* (San Francisco: Sierra Club Books, 1986), of which see specially Rolston, "The Preservation of Natural Value in the Solar System"; Stone, *Earth and Other Ethics*; Plumwood, *Feminism*; Thomas H. Birch, "Moral Considerability and Moral Consideration," *Environmental Ethics* 15 (1993): 313-32.

Birch's view is different from the others mentioned. He rejects altogether the notions of moral considerability and moral *consideranda*. Instead, he argues that seeking a criterion or criteria of moral considerability belongs to the imperial modernist project of domination. He puts, in its stead, the notion of universal consideration which applies to all things, whether biotic or abiotic.

28. Some of the points made in this section are based on parts of an earlier paper by the author, "Awe and Humility: Intrinsic Value in Nature. Beyond an Earthbound Environmental Ethics," in *Philosophy and the Natural Environment*, Robin Attfield and Andrew Belsey, eds. (Cambridge: Cambridge University Press, 1994 with some amendments. But some finer distinctions have since been made; it would be more accurate to call this 'the No External Teleology Thesis' rather than simply 'the No Teleology Thesis' as in the original. But the major amendment involves introducing the distinction between 'for itself' and 'by itself.'

29. Ernst Mayr, *Toward a New Philosophy of Biology* (Cambridge, Massachusetts, and London: Belknap Press of Harvard University Press, 1988).

30. This sense of 'by itself,' as an implication of the No External Teleology thesis, is reinforced by the empirical claim that there are entities and processes which have come into existence, continue to exist and, eventually, go out of existence entirely independently of humankind, its intentions and its activities; such entities and processes may be said to exist 'by themselves.'

31. According to Mayr, if the age of Earth (4.5 billion years) is made equivalent to a calendar year, then the dates of origin of organisms are as follows:

Origin of Earth = 1 January
Life (Prokaryotes) = 27 February
Eukaryotes = 4 September
Chordates = 17 November
Vertebrates = 21 November
Mammals = 12 December
Primates = 26 December
Anthropoids = 31 December, 01:00 a.m.
Hominid line = 31 December, 10:00 a.m.
Homo sapiens = 31 Decmeber, 11.56½ p.m.
(= 3½ minutes before year's end)
(Mayr, *Toward a New Philosophy of Biology*, 69)

Prokaryotes, the first form of life, did not appear until 3.5 billion years ago. Eukaryotes then arrived some 1.5 billion years ago when a symbiosis occurred between two (or more) kinds of prokaryotes, one supplying cytoplasmic organelles, the other the nucleus to form a totally new organism. Within several 100 million years after that event, four new kingdoms evolved—the protists (one-celled eukaryotic animals and plants), fungi, plants and animals. Of the latter, in the Cambrian period, there evolved at least 50 different phyla, most of which became extinct. But the chordates survived, from which, eventually, mammalian life descended. However, the vertebrates (which include fish, birds and reptiles) enjoyed a long period of ascendancy in the Paleozoic period before the mammals became dominant. Among mammals, it was from the anthropoid apes, who are primates, that *Homo sapiens* finally emerged.

32. This sense of autonomy has nothing to do with that of Kantian autonomy— the Kantian sense obtains only in the context of humans and their capabilities for reason and freedom in the moral choices they make.

33. Two caveats should be entered about this thesis:

1. It is not claiming that **no** nonhuman species is ever dependent upon the survival of the human species. For instance, it might be true that there are species of bacteria living in the human guts which are entirely non-viable outside such a specialized environment. It is also true, for example, that the San in the Kalahari, through their fire regime, had, unwittingly, suppressed thornbushes but favoured grasses. The grasses in turn provided food for the antelope and other grazers, resulting in a diverse biotic community. Without the anthropogenic fires, certain species in the Kalahari might not have prospered in quite the way they did.
2. Neither is it claiming that the human species would not survive if any one or more nonhuman species were to become extinct. For instance, were the wild boar, the wild pigeon or the cultivated strains of wheat to disappear, this would not necessarily bring about the demise of humans. But as the thought experiment bears out, while one can readily imagine Earth minus humans carrying on in the way it normally does, given our understanding of evolutionary and ecological processes at work, one cannot imagine humans surviving in the absence of Earth's biosphere and atmosphere. The only way of imagining it is *via* a technosphere simulating the biosphere and atmosphere, reproducing precisely those conditions, in the absence of which humans cannot survive and flourish.

In this light, the principle of interdependency which underpins ecological understanding may have to be qualified. It is true, for instance, that bees, to survive, depend on the flowers for their nectar, while the plants bearing the flowers, in turn, depend on the bees for cross-pollination. The destruction of the one will lead to the destruction of the other. But in the case of humans and their relationship to nature, while it is true to say that the disappearance of a species here and there, or the localized erosion of top soil would not affect human survival as a whole, it is not true that humans can systematically survive without the survival of the biosphere as a whole, as has just been mentioned above. It follows that an asymmetry exists notwithstanding an acknowledgment made above of the dependence of some species on the human species. The asymmetry inclines one, then, to a principle of ecological dependency on the whole rather than interdependency as far as humans and their reliance on nature is concerned. This point, it appears, was appreciated by the Lakota who "regarded nonhuman life-forms as, if anything, superior to themselves. Indeed, animals and plants, in permitting themselves to be taken for legitimate human needs, are said to 'pity' people and to voluntarily sacrifice themselves for the sake of their younger siblings, the human beings" (J. Baird Callicott, *Earth's Insights: A Survey of*

Ecological Ethics from the Mediterranean Basin to the Australian Outback Berkeley and London: University of California Press, 1994, 124). On the other hand, Callicott himself could only maintain the symmetry and the interdependence perspective in very strained terms indeed. He says: "We depend on plants and animals for goods and services, and they depend on us for their very existence—since without our respecting them and their needs ... many of them will be driven to extinction" (*Earth's Insights*, 131). Our dependence on them is physical and material while their dependence on us is cultural and moral. One side of this curious equation is upheld by ecological relationships but the other side is upheld by morality or prudence.

34. The term 'trajectories' is intended to cover nonhuman biotic beings as well as abiotic entities. Each elephant living today has its own trajectory; and so does every river or lake existing now. But for exegetical clarity, only the trajectories of abiotic forms will be emphasized. While biotic forms possess *tele*, abiotic forms do not; while biotic forms exist both 'for themselves' and 'by themselves,' abiotic forms exist only 'by themselves.' But both have trajectories.

35. The sense of independent value introduced here is, up to a point, compatible with Marietta's definition of the term as follows:

> What is meant by "*x* has independent value" is "*x* has value independently of being valued by some *s*." If *x* has independent value, speaking of the value of *x* is not a shorthand way to talk about someone valuing *x* or may come to value *x*, and the value of *x* is not derived from the fact that some person values *x* or may come to value *x*. This would mean that forests and rivers, a diversity of species, and some of the other things we value in nature have their value apart from our thinking them valuable. (Don E. Marietta, *People and The Planet: Holism and Humanism in Environmental Ethics* Philadelphia: Temple University Press, 1994, 124)

An earlier section of this chapter—**Resisting Humean Projectivism**—has distinguished between mutely-enacted and recognized-articulated value. 'Independent value' as used by this author clearly refers to mutely-enacted value in the context of individual organisms whereas Marietta does not have the conception of mutely-enacted value. But it is wider than the class of mutely-enacted values, as it is meant to obtain not only in the case of beings which exist 'for themselves' but also of beings which exist 'by themselves.' For both authors, it appears, that the source and the locus of independent value is not the human valuer or human consciousness. However, for this author, the human valuer could come to recognize and articulate such a value. We shall also see that Appendix 2 (**Is Nature a Mere Social Construct?**) makes it clear that what is anthropogenic does not entail what is anthropocentric and is, therefore, not necessarily incompatible with what is nonanthropocentric. Recognized-articulated values are necessarily anthropogenic but not necessarily anthropocentric. Independent value when recognized and articulated is necessarily anthropogenic as well as necessarily nonanthropocentric.

There is another crucial difference in the use of the term 'independent value' between the two authors—in this book, the term is meant to carry ontological weight, whereas Marietta appears not to have such an intention when he invokes the term. Instead, for him, things identified as possessing independent value are simply those which play a valuable functional role in a system which is itself valuable. (For an account which also emphasizes the ontological dimension of independent value, see Leena Vilkka, *The Intrinsic Value of Nature* Amsterdam and Atlanta, GA: Editions Rodopi B.V., 1997).

36. For the sake of simplicity, the category of the exbiotic is left out. It is usually assimilated to the abiotic. Sometimes, it is difficult to separate out the exbiotic from the abiotic as they have become inextricably entwined—some fossils are embedded or impressed into the rock itself.

The biotic and the abiotic both exhibit great complexity and intricacy. But while animals and plants manifest such characteristics not only in their structure, but also in their function, the crystal, as an abiotic entity, is intrincate and complex only in structure, as it has no function in the sense that organisms have structurally-related functions.

37. Rolston may be exempt from this charge. Although he does not quite distinguish between the natural and the artefactual in the way this book does, his work can be said to celebrate the fecundity and creativity of nature generating, what he calls, "systemic value." (Of course, theorists who raise, in one way or other, the distinction between the natural and the artefactual, like Katz, are clearly not liable to this charge. But see the note after the next for comments on Elliot.)

38. Two comments are appropriate here.

First, from the perspective pursued here, animal welfare and animal rights groups who, in the main, protest against genetically engineered or modified animals , both farm and laboratory, on the grounds that they suffer pain or even frustration, are missing a very important point. To say this, however, is not to deny that such groups are right in protesting against such evil. But it is to say that they appear to be only concerned with a secondary characteristic, namely, the animal's capability to feel pain. Such a focus raises two issues:

1. If these animals were engineered to suffer no pain, would not these protagonists have to endorse such GMOs? For instance, Ryder ("Animal Genetic Engineering and Human Progress," in *Animal Genetic Engineering*, Wheale and McNally, eds., 12) would have to say 'yes' in response to Attfield ("Genetic Engineering: Can Unnatural Kinds be Wronged?" in *Animal Genetic Engineering*, Wheale and McNally, eds., 201-08), who raises precisely this point. (Ryder was, in the past, chairperson of the Council of the Royal Society for the Prevention of Cruelty to Animals and the Liberal Animal Welfare Group in the UK. He was/is director of the Political Animal Lobby Limited, UK.)
2. Such a focus cannot distinguish between suffering and pain in wild animals, on the one hand, and domesticated animals, like pets, on the other—see Callicott, "Animal Liberation."

(Some spokespersons of such groups do mention other objections, such as D'Silva in "Critical View," 107. D'Silva is Director of Compassion in World Farming and of Compassion in World Farming Trust in the UK.)

Second, compare Robert Elliot's position with the one argued here. Elliot ("Extinction, Restoration, Naturalness," *Environmental Ethics* 16 1994: 135-44) seems to understand 'the otherness' of nature to be something distinct from the absence of (human) intentional design or purposive intervention. He also understands intentional design to include divine design. The argument here confines itself solely to the lack of human intentional design and regards that lack as constitutive of 'the otherness' of nature. The aesthetic worth of nature is considered under two headings in terms of subjective aesthetics, with nature acting as a mere trigger of human aesthetic experiences or as objective aesthetics—see Chapter 4—where the aesthetic response is grounded on secondary characteristics like complexity or intricacy. Elliot, however, regards naturalness on the one hand and complexity and the like on the other as all value-adding properties—see Elliot, "Intrinsic Value, Environmental Obligation and Naturalness," *Monist* 75 (1992): 138-59. In other words, he does not distinguish between naturalness as a primary ontological characteristic and beauty or complexity, etc., as secondary characteristics.

39. The operative word in this clause is 'alone'. The point made in this context is the failure (generally in environmental philosophy) to ground moral consideration ontologically which is more fundamental than doing so axiologically. However, this should not be misunderstood to mean that secondary characteristics like complexity, intricacy, or whatever cannot be a basis for ascribing intrinsic value to bearers of such characteristics.

40. However, it is admitted that this argument works best in the case of non-sentient entities, whether biotic or abiotic. In the case of conscious/sentient beings, whether animal or human, additional factors obtain to undermine this conclusion. This is precisely because intrinsic value is assigned to such beings, not so much on grounds of their complexity or intricacy but that, indeed, they are conscious, are capable of feeling pain as in the case of the higher animals, and are even self-conscious as in the case of humans. Certainly, as far as humans are concerned, it would not be morally permissible to kill extant 'natural' humans ('natural' because humans are not naturally-occurring beings within the definition of the term used in the book) with their clones on grounds that they both share the same characteristic of being self-conscious or whatever.

41. However, the perspective of this book does not necessarily rule out a particular expression of human creativity in space—if humans are determined to live outside Earth, they could build platforms in space and terraform these should they so wish. As nanotechnology anticipates, an artefactual material could be manufactured, infinitely stronger and lighter than any known material in nature which could be mined and quarried, to construct such a platform and other objects on it. On it, too, humans could build a biosphere (like the one they are experimenting with in the Arizonian desert today), within which may live specially genetically engineered organisms. To add to the excitement, the humans who opt to dwell on the platform could also be specially genetically engineered to live more comfortably in such a habitat. Such genetically engineered humans, if isolated from Earth's humans for a sufficiently long period of time, may eventually no longer be able to inter-breed (without the help of technology itself) with Earth's humans. They could be the beginning of a new species.

42. See, for instance, Arne Naess, "The Shallow and Deep, Long-Range Ecology Movement: A Summary," *Inquiry* 16 (1973): 95-100, "Identification as a Source of Deep Ecological Attitudes," in *Deep Ecology*, M. Tobias, ed. (San Diego: Avant Books, 1985), *Ecology, Community and Lifestyle: Outline of an Ecosophy* (Cambridge: Cambridge University Press, 1988), "Self-realization: An Ecological Approach to Being in the World," in *Thinking Like a Mountain*, John Seed *et al.*, eds. (Philadelphia: New Society Publishers, 1988); Bill Devall and George Session, *Deep Ecology: Living as if Nature Mattered* (Salt Lake City, Utah: Peregrine Smith Books, 1985); Mathews, *Ecological Self.*

43. According to Plumwood, Deep Ecologists seem to think that to transcend dualist thought requires collapsing the "nonhuman half" of the dualism into the "human half," by expanding the old "human half" into one called "the ecological self" or the "Self" of "Self-realization." Ironically, this is tantamount to eliminating nonhuman nature as 'the Other.' (See Plumwood, *Feminism*, 173-82.)

Plumwood, too, resists such elimination, though not for the same reasons as this book does.

44. By hyperseparation or radical exclusion, Plumwood means the following:

> The relation of radical exclusion has special characteristics. For distinctness, for non-identity or otherness, there need be only a single characteristic which is different, possessed by the one but not the other, in order to guarantee distinctions according to the usual treatment of identity (e.g. in Leibniz's Law). Where items are constructed or construed according to dualistic relationships, however, the master tries to magnify, to emphasise and to maximise the number and importance of differences and to eliminate or treat as inessential shared qualities, and hence to achieve a maximum separation. ... Denial or minimisation of continuity is important in eliminating identification and sympathy between members of the dominating class and the dominated, and in eliminating possible confusion between powerful and powerless. ... A major aim of dualistic construction is polarisation, to maximise distance or separation between the dualised spheres and to prevent their being seen as continuous or contiguous. ... A further important feature of dualistically construed opposition is that the underside of a dualistically conceived pair is defined in relation to the upperside as a lack, a negativity. (*Feminism*, 49-52)

and:

> The upperside is an end in itself, but the underside has no such intrinsic value, is not for-itself but merely useful, a resource. The identity of the underside is constructed instrumentally ... the underside is not part of the sphere to be considered morally, but is either judged by a separate instrumental standard ... or seen as outside morality altogether. (*Feminism*, 53)

45. This thesis will be explored more fully later in the section, **Environmental Ethics and Environmental Philosophy: Axiology and Ontology**.

46. Modernity, at least, so regards humans. But in pre-modern times in European cultures, animals (not only horses and bulls but also rats and fleas) were held as responsible beings too, and could be tried in court for the damage they caused to humans. They were liable for capital punishment, for instance, in the case of pigs which had devoured human infants. Indeed, even as recent as the late nineteenth century, writs were issued against vermin like rats and fleas—see Edward P. Evans, *Criminal Prosecution and Capital Punishment of Animals: The Lost History of Europe's Animal Trials* (London: William Heinemann, 1906); Luc Ferry, *The New Ecological Order* (Chicago and London: Chicago University Press, 1995), ix-xvi. Of course, to our contemporary mind, this is incomprehensible and bizarre.

In this chapter, see section—**Resisting Humean Projectivism**—which argues that this unique human attribute, of being morally responsible, does not entail human arrogance as it is normally taken to do, as well as section—**Anthropogenic and Nonanthropogenic**—for further discussion about human responsibility and moral agency.

47. Cited by Christopher D. Stone, *Earth and Other Ethics: The Case for Moral Pluralism* (New York: Harper and Row Publishers, 1988), 155-56, and Holmes Rolston, *Conserving Natural Values* (New York: Columbia University Press, 1994), 110-11.

Another more recent instance which appeared in the British press concerned a three-year-old English bull terrier called Lucy which chased and killed a cat called Fluffy, belonging to a neighbor of her owner. Lucy was arrested by the police and detained, facing a possible death sentence under the Dangerous Dogs Act, on a charge of having been "dangerous and out of control." Lucy's arrest as well as her potential death sentence set a precedent—up to then, the Dangerous Dogs Act had been interpreted as applying only to dogs behaving aggressively to humans. (See Edward Pilkington, "Fur Starts to Fly as Police Arrest Bull Terrier Which Killed a Cat," *Guardian*, 21 August 1995, 1). But as the story eventually turned out, Lucy was released after five days of detention, upon the successful application, lodged by the solicitor of Lucy's owner, against the police, for unlawful holding of the animal—see Nick Wright, "Lucy, the Cat Killer, Released," *Guardian*, 24 August 1995, 7.

A dog is a domesticated animal, and so a biotic artefact to a great extent. Bull terriers, as a breed, have been specially selected for their hunting ability. As they are, by and large, humanly created and designed, it would stand to reason, that their creators and owners should be held liable for their vicious characteristics, and not the dogs themselves. The police seemed to be hopelessly muddled and confused. First, they failed to appreciate that according to modern thinking it makes no conceptual sense to hold animals, even domesticated ones, to be responsible for their behavior. Second, that while there is a justification for putting down dogs which kill humans, one would be hard put to find any justification for dogs which kill other animals. Third, if a justification were to be found, then by the same logic, cats which kill birds would also have to be put down. This kind of moral policing among animals smacks of misplaced zeal, the main source of which may be traced to the transposition of culture on to nature, albeit not wild nature, as in the bison case. However, all the same, the bull terrier, though a biotic artefact, nevertheless, still has a residual *telos* of its own, that is to say, given its biological origin, being descended from wolves, and being the kind of biotic artefact that it has become, it is "in its nature" to kill anything that runs away from it. More generally speaking, dogs and cats have been known since the dawn of their domestication to fight and, occasionally, to kill each other.

48. The term 'ontological dyadism' has been coined by this author for want of a better term.

Ontological dyadism is akin to an argument advanced by Peter Reed, although he arrived at it by a rather different route. He leaned upon the existentialist theology of Rudolph Otto and Martin Buber. He wrote: "When we modify or destroy natural systems or objects, we destroy something that did not take shape under our hands or spring from our brow. We destroy something that is essentially other" ("Man Apart: An Alternative to the Self-Realization Approach," *Environmental Ethics* 11 1989: 53-69. 63). Neil Evernden (*The Social Creation of Nature* Baltimore and London: The Johns Hopkins University Press, 1992, Chapter 7) cites Reed, among others, pursuing a similar line of thought.

Reed also criticized the Self-realization approach of Naess' ecophilosophy. See Arne Naess, "Man Apart and Deep Ecology: A Reply to Reed," *Environmental Ethics* 12 (1990): 185-92.

49. Plumwood and this author agree that dualism embodies a power difference between the privileged master class and the inferiorized other. Distinctions and dichotomies *per se* are not necessarily infected by this unequal power relationship. They merely focus on differences between things which have to be taken into account if they are to be adequately and properly assessed and appreciated. However, there is an alternative view which holds that all dichotomies and polarities are implicitly dualistic *a priori*—see James Cullen, *A Systems Approach to Environmental Values: Systems Process and the Bifurcation of Nature*, MA diss. (University of Lancaster, 1994), 30 (note 2).

50. This point will be pursued in a later section.

51. Karl Marx and Friedrich Engels, *Collected Works*, vol. 5 (London and Moscow: Lawrence and Wishart & Progress Publishers, 1976).

52. In other words, the insight of postmodernism in challenging the human/nature dualism in Western European thought is not new. The "true socialists" around 1840 had already articulated it, except that what they said was not called "postmodernism." They formulated three theses which are similar, if not identical, to some of the theses of Deep Ecology: (a) humans are themselves part of nature, (b) nature forms a unity of life, (c) humans should live in harmony with nature, like all the other forms of life, which are part of the whole that is nature. (For a more detailed discussion of "true socialism," see Lee, *Social Philosophy*, 259-61.)

53. Marx and Engels, *Collected Works*, vol. 5, 472.

54. Marx and Engles, *Collected Works*, vol. 5, 472.

55. See Norton, *Why Preserve Natural Variety?*, "Epistemology and Environmental Values," *Monist* 75 (1992): 209, and "Where Do We Go from Here?"; Callicott, "La Nature Est Morte, Vive La Nature!" and "The Role of Technology."

56. In one sense, this position could be said to fall outside the orbit of environmental ethics/philosophy. Along the shallow/deep spectrum, it is so shallow as to fall off the edge. The only way in which it can cling on as the shallowest of the shallow positions is, as in the case of Kant, *via* indirect duties to animals or indirect harm to humans.

57. Ecocentrism, in this sense, excludes holism when paired with anthropocentrism. However, it must also be borne in mind that there are may forms of ecocentrism. For instance, Rolston's and Callicott's views are very different from each other. For a start, Callicott's account of nonanthropocentrism consists of only challenging the Sole Value Assumption understood as the thesis that humans are the sole loci of intrinsic value while adhering to the view that human consciousness is the source of all values; Rolston's account challenges both.

58. Chapter 2 has earlier distinguished between two forms of anthropocentrism. The distinction there is made in the context of contrasting the passive anthropocentrism of Aristotle and the Ancient Greek world on the one hand, with aggressive modern anthropocentrism, on the other. But here the distinction is made within modern anthropocentrism itself, between a more aggressive and a less aggressive variety.

59. However, regarding the Instrumental Value Assumption, it has some qualifications—see the argument as it develops.

60. See Bryan Norton, "Sustainability, Human Welfare and Ecosystem Health," *Environmental Values* 1 (1992): 97-111.

61. See Bryan Norton, "Environmental Ethics and Weak Anthropocentrism," *Environmental Ethics* 6 (1984): 131-48 and Why Preserve Natural Variety?

In so far as Taylor appears to endorse the protection of species through, for instance, his principle of restitution, then he is being inconsistent in doing so.

62. On this point, see David J. Rapport, "What Constitutes Ecosystem Health?" *Perspectives in Biology and Medicine* 33 (1989): 120-32; Robert Costanza *et al.*, *Ecosystem Health: New Goals for Environmental Management* (Washington, D.C.: Island Press, 1992); see also J. Baird Callicott, "Benevolent Symbiosis: The Philosophy of Conservation Reconsidered," in *Earth Summit Ethics*, J. Baird Callicott and Fernando J. R. da Rocha, eds. (Albany: State University of New York Press, 1996). For a handy collection of contributions on the subject of ecosystem health—both approving and skeptical—by authors including Rapport and Callicott, see *Environmental Values* 4, No. 4, 1995.

63. Rapport ("What Constitutes Ecosystem Health?") mentions one possible clash—a forest of commercial timber could be managed as a sustainable ecosystem of economic production. However, judged by the criterion of species diversity within an ecosystem, such a forest could not be said to be as healthy as a nonagro-ecosystem, as it supports far fewer species of flora and fauna than the other.

64. In May 1995, the Institute of Physics in the UK organized its first major conference on technology foresight. One of the speakers, Professor J. R. Hillman of the Scottish Crop Research, was reported as holding the view that biotechnology would be

> used to reprogramme trees not only to resist blights but to produce different standards of pulp, timber or veneer for industry, while cutting wasteful processes and environmental pollution. He sees tree-breeders as cotton-growers did, developing different cultivars for different markets. ... He also sees genetic engineering as necessary for tomorrow's foresters. Trees today just aren't good enough. "All the very best parental material has already been chopped down and built into boats or used in furniture. You might argue that in many parts of the developed world there is no parental material existing. The prize trees have all been utilised and we have ended up with inferior material. The thing people often forget about biotechnology is that it can help generate new variation." (Tim Radford, "Futurology: A Pastime with a Rosy Future?", *Guardian*, 11 May 1995, Online Section: 9)

65. One assumes that there would be no insuperable empirical difficulties to wreck such a project. An analogue to native species in natural ecosystems may be constructed in the case of artefactual ecosystems. For instance, 'native species' could be defined as those which are deemed to be the most important or the most desirable species, whatever the criteria of 'importance' or 'desirable' may be.

66. For a critique similar to that pursued here, see Eric Katz, "The Big Lie: Human Restoration of Nature," in *Technology and the Environment*, Frederick Ferré, ed. (Connecticut and London: JAI Press Inc., 1992) and "Artefacts and Functions."

67. For a recent view of Callicott on the subject of ecosystem health, see J. Baird Callicott and Karen Mumford, "Ecological Sustainability as a Conservation Concept," *Conservation Biology* 11 (February 1997): 32-40. Here Callicott and his co-author argue that sustaining ecosystem health cannot be the primary but only a secondary goal of environmental policy-making.

On the term 'postmodernism,' Gare writes: "*The Modern-day Dictionary of Received Ideas* says of 'postmodernism': 'This word has no meaning. Use it as often as possible'" (*Postmodernism*, 4). Gare understands postmodernism as the rejection of grand narratives and of progress—"The loss of credibility of grand narratives is essentially a loss of belief in 'progress'" (*Postmodernism*, 4). However, Callicott, who presents himself as a postmodernist thinker in environmental philosophy, does not appear to see postmodernism in the same way as Gare. But whatever the term may mean in other contexts, in this discussion, one takes it that Callicott is challenging, among other things, the human/nature divide, and by implication, the distinction between the artefactual and the natural. (Callicott, of late, and da Rocha have distinguished between two types of postmodernism. The brand they approve of is called "constructive or revisionary postmodernism" which does not entail relativism or nihilism. See *Earth Summit Ethics*, xi.)

68. In such a world, neoDarwinism has no place. For Callicott, this ought to pose a problem, as his environmental philosophy (at least in his earlier works) pays due respect to Darwinian evolution. Postmodernism as presented by Callicott in current environmental

philosophy may harbor two conflicting tendencies—that aspect which incorporates Darwinian thinking and the other which, in rejecting the human/nature dualism, opts for nature$_c$ as crucial to environmental philosophy. From the standpoint of this book, however, postmodernism in the Callicott/Norton form is, paradoxically, *au fond*, an affirmation of Cartesian homocentrism as it celebrates human control of biotic nature.

69. This view should not be confused with another, namely, that 'the good of humans' or human flourishing requires nonhumans to exist and to flourish. According to such a view, nonhuman flourishing has simply been co-opted as part of human flourishing. But according to the view argued in this book, nonhuman flourishing has a value independent of human existence or valuation, and it is precisely because nonhumans have such a value that their elimination by humans constitutes an impoverishment of humans themselves. This conception of human impoverishment rests on a prior recognition by humans that nonhumans are the source and locus of independent value, and that they constitute a distinctly different ontological category from that of human agency and its artefacts.

70. It is, therefore, apt that the leading journal of the new discipline is entitled *Environmental Ethics*. However, to be fair, its contents do go beyond axiological to ontological concerns. The point made here is simply this: that the title captures succinctly the focus on axiology rather than ontology as the key preoccupation of the new discipline.

71. This is not what Hare is talking about when he advocates universalizability and presciptivism as a joint strategy for generating moral rules—see R. M. Hare, *Freedom and Reason* (Oxford: The Clarendon Press, 1963).

72. This, however, should not be misunderstood to mean that the new perspective has no use for the notion of rationality as consistency. It only means that the notion also applies not merely in the axiological but also ontological context.

73. D. H. Munro, *Empiricism and Ethics* (Cambridge: Cambridge University Press, 1970), 200.

74. A critic could retort that traditional morality has no difficulty in accommodating God, an ontologically different kind of being from human beings. But that theology also says that God—who, at least in the great monotheist traditions arising from the Middle East, is a being defined as all-good and all-powerful—commands us to do all sorts of things, including to care for and respect the natural world, His creation. However, for secularists, this invocation has no resonance. Instead, the secularist recognizes that nature is self-created, not God's creation. The point at issue is precisely this: that nature's self-creation is ground for respecting it.

75. In the sense of axiological as opposed to existential anthropocentrism—see Appendix 2 (**Is Nature a Mere Social Construct?**)

76. Easterbrook, *A Moment on the Earth*, 2-3.

77. Of course, philosophical skepticism denies the possibility of obtaining knowledge of any kind whatsoever. But it would be beyond the remit of this book to refute such a claim. For the purpose of this discussion, the skeptical claim that we can never foresee all the consequences of one's actions is irrelevant. Admittedly, as human intelligence is not omniscient, it is necessarily limited.

78. The uniqueness of human consciousness and intelligence is claimed as a species characteristic. This means that the unique attribute may fail to obtain in the case of any one particular individual human, but this would not jeopardize the thesis as such. In other words, one needs to distinguish between (p), that only humans possess that kind of intelligence capable of grasping what is abstract, and (q), that not all humans possess it (as some humans may suffer severe brain damage). p is undermined by 'some or all non-

humans possess it,' not by q; p and q are logically compatible.

79. Of course, this is not to deny that a dog, for instance, knows that its mistress is absent and pines for her. Some animals do entertain absence of this kind but, however, not of the kind which is being discussed here.

80. W. Köhler, *The Mentality of Apes* (London: Routledge and Kegan Paul, 1925).

81. The metaphysical issue of freewill and determinism may be ignored for the purpose of this discussion. But see the note which follows for a further comment.

82. One biologist (a geneticist) has argued that the conditions for ethical behavior have evolutionary roots, and that ethical behavior itself is predicated upon the advanced intellectual abilities characteristic of *Homo sapiens*. Francisco J. Ayala—"The Difference of Being Human," in *Biology, Ethics, and the Origins of Life*, Holmes Rolston, III., ed. (Boston and London: Jones and Bartlett Publishers, 1995)—mentions three such conditions: the ability to anticipate the consequences of one's action, to make value judgments, and to choose between alternative courses of action. The first is related to the ability to establish a link between means and ends which sets humans on the developmental path as *homo faber*. Bipedalism freed the front limbs of our ancestors for the manipulation and construction of objects as tools. Tool using and tool constructing led to further enhancement of human intelligence. The ability to make value judgments is linked with the first ability and also depends in turn on the capability for abstraction, to see objects and actions as belonging to general classes, some of which are considered more desirable than others. And as for the third ability, even if freewill turns out to be illusory, it remains the case that even the illusion that we can choose between alternative courses of action presupposes we have the intelligence and the foresight to ascertain and construe these alternative futures.

It is important to point out that Ayala does not argue that ethical behavior is itself adaptive, only intelligence is. But all the same, it is the adaptive development of intelligence which renders ethical behavior possible. Indeed, for Ayala, the three abilities mentioned are each necessary but jointly sufficient conditions for ethical behavior. This book makes a more restricted claim, namely, that the uniqueness of human intelligence is a pre-condition for moral agency and responsibility.

83. See Evans, *The Criminal Prosecution and Capital Punishment of Animals*, and Ferry, *New Ecological Order*, cited earlier. But of late, some philosophers have tried to reclaim the notion of responsibility on behalf of (some) animals, at least in some limited contexts—for instance, see DeGrazia, *Taking Animals Seriously*.

84. Environmental philosophers, on the whole, are eager to deny either that humankind is unique in any way or that its unique characteristic is relevant to environmental ethics. This is because they believe that admission of human uniqueness entails human chauvinism. But they are wrong; as pointed out by Rolston, admission about the uniqueness of human intelligence is compatible with the claim that humans, alone, may be said to be morally responsible agents, and with the related one that they have direct duties or obligations, not only to fellow humans, but also to nonhuman others.

85. In **Appendix 2**, a further category, namely, moral subject, will be introduced to refine the discussion.

86. Rolston, *Environmental Ethics: Duties to and Values in the Natural World* (Philadelphia: Temple University Press, 1988), 54-55.

87. For instance, Easterbrook eagerly anticipates "the New Nature" which he says is within the grasp of humankind to engineer in the near future, when predation by animals against other animals would be abolished. In his view, carnivores are a flaw in nature's 'design,' and we have a duty to remove that flaw—see *A Moment on the Earth*, 671, 681-

82.

88. In this context, it is not meant to carry any load from the psychoanalytic tradition. It has in mind only the Greek myth itself.

89. To prevent misunderstanding, it may be wise to point out straightaway that this remark and others in the section should not be read as an exercise in virtue ethics, a so-called third way of doing substantive philosophical ethics. (The other two are the Kantian and consequentialist ethics which are well-established in modern Western ethical thought.) Virtue ethics has its roots in ancient Greek philosophy and is especially associated, in particular, with Aristotle. Of late, it has been enjoying a revival.

However, it is not easy to determine what precisely virtue ethics is, beyond the briefest outline of a few general features, as there appear to be very different, tentative accounts of it in recent writings, an observation made by a prominent advocate—see Michael Slote, "Virtue Ethics," in *Three Methods of Ethics: A Debate*, by Marcia W. Baron, Philip Pettit and Michael Slote (Oxford and Malden, Massachusetts: Blackwell, 1997), 233.

The main features gleaned from Slote's contribution cited above include: (a) virtue ethics is aretaic rather than deontic in character, (b) it is agent-based (at least, this is Slote's own preferred slant), (c) it focuses on morality as inner strength of the agent. According to Slote, the aretaic is fundamental, and the deontic (what is right) as well as the consequentialist (what is right based on what is good) may be derived from it. For instance, from the aretaic evaluation that to act in a certain way would be a stupid, ignoble or terrible thing to do, one could derive the deontic judgment that it ought not to be done; from the aretaic evaluation that someone is a caring daughter and that she is in possession of the relevant knowledge about the medical state of her elderly, ill mother, one can infer the utilitarian/consequentialist judgment that the best and right course of action to pursue is not to insist on doctors taking heroic measures to save her mother. Thus, from agent-based characteristics and considerations, one could derive judgments about what is right or wrong, either in deontic or utilitarian/consequentialist terms. This, in turn, means that moral judgments and decisions flow directly from, or can be traced directly back to, the inner strength of the agent's character.

At first sight, it may look that the argument presented in this section is amenable to being cast within the virtue ethics mold. That argument may be formulated schematically (which the author acknowledges with thanks to John Nolt but also with apologies for having modified it somewhat) as follows:

1. If we do not respect nature, we will eventually transform it into artefacts.
2. This transformation would be narcissistic (analogous to the absolute pursuit of money, political power, or prestige in the case of an individual person).
3. Narcissistic pursuits are morally defective or wrong.

Therefore,

4. If we do not respect nature, we will be guilty of a moral defect or committing a moral wrong.

So it appears that one is arguing from the aretaic evaluation that narcissism is a moral defect of character to the deontic judgment that one ought to respect nature.

However, this is not the line of inquiry and argumentation pursued in this section and in the entire book. That specifically challenges the two entrenched normative ethical theories, but has nothing to say about the merits or otherwise of virtue ethics. More to the point, virtue ethics assumes that the aretaic is fundamental, as we have seen. But, according to this book's orientation, that would be to put the cart before the horse—a narcissistic civilization is morally defective because it fails to recognize the ontological value of independence in nature and, therefore, seeks systematically to eliminate (both philosophically and empirically) nature (in its biotic and abiotic forms) as ontological Other. This is the very opposite of arguing, following virtue ethics, that one ought to respect nature or ought not to transform it (systematically), because to do so is to suffer from a defect of moral character, namely, narcissism. The moral censure against (collective) narcissism as a defect in humankind only makes sense if nature manifests or exhibits the ontological value of independence.

90. 'Egomania,' as a concept, is like 'murder' but yet unlike it in a crucial respect. 'Murder,' too, is logically Janus-faced. Its descriptive content refers to the killing of another person, but its negative evaluative overtone is conveyed through its meaning as morally unjustified killing. But one may envisage circumstances when killing another, no matter how unfortunate, is considered to be morally justified, in which case, one no longer uses the term 'murder,' strictly speaking, to characterize the act. For instance, the act of killing or assisting in killing another to terminate excruciating pain from which there is no hope of recovery is not called murder, but euthanasia. Of course, euthanasia itself does not command universal moral approval. But whether it does or not, there is felt to be a need to introduce the term in order to focus attention on the crucially relevant moral differences between the two respective acts of killing another, differences which even those who do not approve of euthanasia are capable of appreciating. (However, 'murder' as a legal concept is different—because contemporary legal systems are, at best, deeply ambivalent about legalizing euthanasia, and at worst, deeply hostile to doing so, it makes sense to say that someone has committed a murder, that is, unlawfully killed another, even when the motive is entirely to assist the person to end intense pain from which there is no hope of recovery.) In the case of narcissism or egomania, however, there appears to be no analogous set of circumstances one could envisage, which might lead one to condone egomania as being morally permissible or justified. Hence there is no term analogous to 'euthanasia' to demarcate between two potentially morally different types of egomania. No obvious moral championship of egomania springs to mind. This is because egomania appears not to be morally defensible under any description.

Egomania is not to be confused with either psychogical egoism or ethical egoism. Psychological egoism is either a false empirical thesis or a vacuous, trivially true one. Ethical egoism (including rational egoism) is a normative thesis which is compatible with understanding self-interest broadly to include the interests of others *via* enlightened self-interest. Egomania is a condition which amounts to being morally blind to the existence of others, and incapable of appreciating that others could make legitimate moral demands upon oneself. Analogously, when a civilization, such as ours at present, is said to be a narcissistic one, the collective egomania it embodies is a condition which amounts to being morally blind to the existence of nonhuman others, being incapable of appreciating that nonhuman others can make legitimate moral demands upon us.

91. It is important to grasp that the word 'virtuous' here is not understood in the same sense as it would be within the context of so-called virtue ethics. As earlier pointed out, virtue ethics (at least as understood by a prominent advocate today) argues that the aretaic is fundamental, and the deontic as well as the consequentialist to be derivative from it. But according to the stance taken by this section and throughout this book, what is virtuous presupposes other notions. In the case of the individual who is an egomaniac, to say that one is not virtuous is to say, among other things, that such a person is morally blind to the

needs or rights of other persons which ought to act as constraints upon one's conduct. In the case of a civilization which suffers from egomania, it is to say that such a civilization is morally blind to the status of (nonhuman) nature as ontological Other and to the constraints which such a status imposes on the conduct of humans as morally responsible agents. In other words, the term 'virtuous' may be used and understood in more ways than one; virtue ethics has no monopoly over it.

92. According to Scitovsky, capitalist economies in general, but the American economy in particular—as paradigmatic embodiment of the advanced mature economy—promote comfort almost to the exclusion of all else, even of culture. Comfort, Scitovsky says, amounts to

> relieving or forestalling discomfort. That includes behaviour which satisfies various bodily and mental needs and so lowers arousal that is too high; it also includes behaviour which combats boredom and so raises arousal that is too low. Though somewhat different and of opposite sign, the two kinds of behaviour are alike in that both aim at securing a negative good; freedom from pain, unpleasantness or discomfort. The positive good is pleasure, and it is very different from comfort. (Tibor Scitovsky, *The Joyless Economy: An Inquiry into Human Satisfaction and Consumer Dissatisfaction* London: Oxford University Press, 1976, 59)

93. In the Introduction, it is pointed out that this book, admittedly, is only concerned with the impact of recent and near future deep technologies upon nature. The application of such technologies, so far, is by and large, confined to Earth; terraformation as conceived today, for instance, concerns only some of the other planets in our solar system. As such, the rest of 'nature, out there' is beyond the scope of the kind of deep technologies so far generated by deep science. It follows that the whole of nature would not be humanized even if these technologies were in the end successful; there is plenty of 'nature, out there' which will continue to embody 'the ontological Other.' However, while this conclusion is, strictly speaking, correct, it does not follow that the main thrust of the arguments laid out in this book has been undermined. The claim that humankind is absolutely determined and shows itself eminently capable of transforming the natural to become the artefactual remains untouched. One cannot *a priori* rule out certain possibilities as 'technological impossibilities,' even though they clearly are here and now. It remains an open question what science and technology beyond the next few decades may discover and, in turn, transform and fabricate. Furthermore, what is crucially worrying, philosophically, is the spirit of narcissism which is embodied in and which defines this kind of civilization, a spirit which appears to seek, unceasingly and relentlessly, to eliminate the 'ontological Other' by embedding the intentionality of the human self in nature, as an extension of itself.

94. Leiss cites two instances of machines which made people feel that they were in the presence of something sublime, inducing both awe and dread in them. The first was a Corliss steam engine, with a weight of 680 tons and a height of 39 feet, displayed at the Philadelphia Centennial Exposition in 1876. According to Leiss:

> [O]ne guidebook ... offered its readers a lesson in aesthetic judgment. Whereas traditionally poets located the experience of the sublime in our reactions to wild nature or powerful human passions, the guidebook claimed that the modern age recognized the sublime in the design and operation of its great machines. And a newspaper reported that in the presence of the Corliss engine 'strong men were moved to tears of joy.' (*Under Technology's Thumb*, 36)

The second instance cited by Leiss was an attempt in the 1970s, at the Museum of Modern Art in New York, by the French neo-Dadaist artist, Jean Tinguely, to mount a spectacular auto-da-fé through the self-destroying machine he had constructed. The machine was 23 feet long and 27 feet high.

> [I]ts main distinguishable components were a piano, an old Addresso-graph machine, eighty bicycle wheels, steel tubing, a meteorological balloon, a huge klaxon on wheels, a wide assortment of small mechanized devices, and various chemicals—smoke, flash powders, and foul-smelling substances. When the main motor was switched on, the piano keys were struck, wheels turned, klaxons sounded, a radio blared, clouds of smoke billowed forth; a number of small constructions broke free and wheeled about; and small objects were hurled through the air. Then the piano caught fire, the steel tubing supports began to give way, and the terrified museum authorities ordered in firemen with axes and extinguishers to finish off the machine. ... The late machine was described as both a beautiful and a terrible thing, and it was reported that at the end some spectators had wept. (*Under Technology's Thumb*, 37)

Conclusion

This book, in exploring a cluster of related themes, is an attempt to show that the crucial issue in environmental philosophy is an ontological one, namely, the systematic supersession of the natural by the artefactual. The ontological threat to the natural occurs both at the empirical and the philosophical levels.

Empirically, the source of the threat comes from modern science and its technologies, extant but also near future. The history and the philosophy of science, since the mid nineteenth century, show that as science yields 'deeper' and 'deeper' theories, such theories induce more and more powerful and radical technologies to manipulate nature, eventually, directly at the molecular and atomic levels, thereby producing artefacts with a greater degree of artefacticity than previous ones. Biotechnology, in place for only about two decades, shows what it can do in manipulating biotic nature by converting it into biotic artefacts. Although it has by no means solved all the problems which lie in its path, nevertheless, it is immensely fruitful as a research program and promises a good deal more than it has so far already delivered. Molecular nanotechnology is increasingly being taken seriously, promising to do to abiotic natural kinds what biotechnology has done to biotic natural kinds.[1] Microcomputer technology, which is already well established, is expected to join forces with biotechnology and nanotechnology. The synergistic effects of such a partnership would lead to an, as yet, unsurpassed attempt to transform nature, both biotic and abiotic, into artefacts.

Philosophically, two sources of the threat have been identified. The first is linked to the metaphysics which underpins modern science and technology. The metaphysics of Scientific Naturalism lays down the framework within which science seeks to predict, explain and control natural phenomena. The beginning of the modern Scientific Project in the seventeenth century was accompanied by the emergence of a new philosophy to back it. Five crucial strands, at least, can be identified to transform nature itself, philosophically speaking, into a fitting object of human control and manipulation. Descartes put in place the human/nature dualism which in turn is derived from his more fundamental mind/body dualism. If it is human consciousness—*res cogitans*—which renders humans uniquely superior and privileged, then not only is the human body inferiorized, the rest of nature, which is devoid of such consciousness, is also inferiorized as *res extensa*. The second strand consisted of the Galilean project of mathematizing and quantifying nature. The third, also stemming from Galileo, was his distinction between primary and secondary qualities which Locke later systematically incorporated into his philosophy of empiricism. Only the primary qualities of shape, size and motion were real and resided in an object; the secondary sensuous qualities like color, though ultimately based on the primary qualities, were not real and were only conjured into existence by a human perceiver or mind. The fourth emptied nature of all values; what may be called the thesis of Humean projectivism, namely, that values are the mere projections of human emotions or what Hume called passions upon nature, is part of it. The fifth, in reinforcing the fourth strand, purged from the explanation of natural phenomena two out of Aristotle's four causes, namely, the final and formal, retaining only the material and efficient causes. Together, these provided the philosophical foundation which launched modern science on its project of controlling and manipulating nature as well as for the anthropocentrism found deeply embedded in modern Western thought itself, namely, the theses that human consciousness is the source of all values, that it is the sole locus of intrinsic value, and that nature is only instrumentally valuable (for humankind).

Nature, the object of scientific study as *res extensa*, devoid of consciousness, value, final and formal causes as well as the sensuous qualities, becomes reduced to uniform, homogeneous, inert matter whose real qualities could then be quantified and mathematized. But that is the first stage of the Scientific Project. To comprehend the next stage, beginning in the mid nineteenth century, one has to look more closely at the retention of the remaining two Aristotelian causes in modern philosophy. This stage, which consists of developing technologies induced by fundamental scientific discoveries in physics, chemistry, and later, genetics, has the effect of replacing the efficient cause of natural phenomena with human agency executing its intentions and ends upon the material cause understood in terms of molecules of the natural elements—in this phase, synthetic urea and synthetic rubber, for instance, replaced natural urea and natural rubber. Technology in this phase also permits the creation of new kinds of artefactual substances, like plastic, out of natural substances like oil. In sum, artefactual kinds replace natural kinds. But science and technology are dynamic in character. As the twentieth century draws to a close, they promise to deliver further artefactual kinds even at a deeper level of manipulation than that involved in the production of plastic—for instance, nanotechnology works at the level of manipulating atoms to create new artefactual kinds, not merely deriving new ones from existing natural kinds. This, if successful, will constitute yet a further stage in modernity's triumphant march of progress; *homo faber* in serving either the mundane goal of improving the material well-being of humankind, or the more noble one of increasing human freedom through providing expanding opportunities for the expression of the human spirit, systematically transforms the natural to become the artefactual.

However, *homo faber* is not the only threat, powerful as it is, in the late twentieth century to the survival of the natural. It is also assisted by the influence of postmodernist thought within environmental philosophy itself, mounting an attack on the very distinctions between human and nonhuman, nature and culture on the grounds that these make no sense given the deep penetration of the nonhuman (the natural) by the human (the cultural). At first sight, this so-called postmodernist attack might even appear to be in agreement with the critique of modernity pursued by this book. But upon closer scrutiny, it turns out not to be so. Although postmodernism, too, denies dualism, its denial is committed to the elimination of nature as 'the Other.' At this fundamental level, ironically then, the postmodernist assault mounted within environmental philosophy is in agreement with the very thesis it criticizes, namely, modernity itself. But this book has made clear that the artefactual is the material embodiment of human intentionality, a category very different ontologically from that of the natural; hence, it argues for the necessity of upholding the distinction between the natural and the artefactual as an ontological one.

One can readily see that the systematic transformation of the natural to become the artefactual through technology has prepared the way for the postmodernist undermining of the human/nonhuman distinction. They mutually reinforce the role of humankind as total manager and controller of Earth. For instance, the postmodernist tendency in environmental thinking has recently generated a pragmatic consensus (about ecosystem health as the policy goal of environmental management), as endorsed by James Baird Callicott—the non-anthropocentric holist wing of environmental philosophy—and Byran Norton—the anthropocentrist holist wing. While the pragmatic consensus is not philosophically worrying to Norton and the position he represents, it should be worrying to Callicott and the position he represents. By claiming that whatever humans do is natural, Callicott and Norton may think that they have banished the ghost of the human/nature dualism, but the price for that which Callicott, at least, has to pay is to admit that humans are in ultimate charge of manipulating nature as they see fit. If ecosystem health is the ultimate goal of environmental management and manipulation, then, in principle, there is nothing to prevent environmental managers from sustituting artefactual ecosystems for natural ones, provided these ecosystems satisfy the *desiderata* of 'being healthy.'

Such a predicament shows the need not to throw the baby out with the bath water. Modern (Western) thought has indeed privileged humans and inferiorized nature through maintaining the human/nature dualism. But while such dualism is unjustified, it does not mean that the distinction between what is human and what is natural is itself unjustified. If ontological elimination of nature as 'the Other' is to be avoided, then the distinction must be retained, **not** in the form of dualism but of ontological dyadism. Unless ontological space is reserved for nature, nature will be eliminated ontologically and empirically. When nature becomes thoroughly humanized, when the natural has been transformed to become the artefactual, our civilization correspondingly would have transformed itself into a totally narcissistic one. Such a transformation could not be checked unless one argues for the priority of ontological values over axiological ones—the latter involves secondary characteristics like complexity, intricacy, creativity while the former involves the more fundamental values like being independent and autonomous (of humankind). While artefacts and naturally-occurring entities may share the same secondary characteristics, nevertheless, they do not share the same fundamental ontological status.

To appreciate 'the ontological distance' that humankind has traveled since the inception of the modern Scientific Project in the seventeenth century in Western Europe, one must go back again to Aristotle. At the beginning of this book, readers were reminded that Aristotle maintained that in explaining all phenomena, the four causes must be invoked and distinguished, even in the case of naturally-occurring beings like individual organisms. However, in reality, he admitted it was very difficult to separate out these causes where biotic nature was involved, as the individual organism *qua* organism exhibited them all but in a way which rendered them inextricably entwined with one another. Aristotle realized that it was only with artefacts that the four causes could be externally identified and assigned each to its source. In the case of a building, one could easily point to wood as its material cause, to the mason who constructed it as its efficient cause, to the plan either drawn on a sheet of paper or held in the mason's head as its formal cause, and to the purpose which the owner had intended for it as the final cause. But in the case of individual organisms, although explanatory cogency and intelligibility demanded that one identified these four causes, the distinctions between them could only be made intellectually rather than empirically.

Artefacts in Aristotle's time were paradigmatically abiotic or exbiotic in character. However, by the late twentieth century, humankind through biotechnology, as we have seen, is able to create artefacts in the form of living organisms whose degree of artefacticity goes well beyond that achieved by the earlier technology of selective breeding or even by the Mendelian-induced technology of hybridization—the immanent *telos* of a naturally-occurring organism under which the four causes are inextricably entwined is replaced by an external *telos* imposed by humankind, under which its four causes, as an artefact, can be externally identified and assembled. Under biotechnology, the material cause of the artefactual organism can be introduced from another species, be that plant or animal; its efficient cause is, thereby, anthropogenic; its formal cause is the external *telos* imposed on it by its anthropogenic efficient cause, such as the desire for a genetically engineered sheep to produce milk containing human rather than sheep hormones; and its final cause is to serve the end of its human creator, such as to produce human hormones which, when harvested from the organism, could be used to improve human health. Such an individual organism, in the language of this book (see Chapter 5), no longer exists 'by itself' although it does exist 'for itself.' A being which exists 'by itself' is a being which has not come into existence or continues to exist in order to serve human ends. A biotic artefact has lost that status. But a biotic artefact continues to exist 'for itself,' that is to say, it continues to strive to maintain its own functioning integrity just as a naturally-occurring organism does. Humankind has learned to use its science and its technology, especially in the last twenty odd years, precisely to capture the status of a naturally-occurring organism as a being capable of existing 'for itself' to turn it into a being which is no longer capable of existing 'by itself.' Just as the four causes in a naturally-occurring organism are so inextricably entwined that they can only be distinguished intellectually, in the same way, a naturallly-occurring organism in existing 'for itself' is at one and the same time inextricably also existing 'by itself.' But modern science and its technology has succeeded spectacularly in prizing apart these two inextricably entwined modes of existence. This transformation of the biotic as a naturally-occurring being to become an artefactual being has taken more than two thousand years to reach this most recent deep level of artefacticity. And we have just seen that such a transformation involves a revolution at once at two levels, conceptual and technological—it now makes sense both empirically and conceptually to separate out the four Aristotelian causes assigning them to external anthropogenic sources and to destroy the being which exists 'by itself' without destroying also the being which exists 'for itself.' This achievement enables theoretical biologists like Varela and Maturana to conceptualize even naturally-occurring organisms as mere machines. Such a revolution is awesome but ontologically worrying.

We should not delude ourselves that the humanization of nature will stop at biotic nature or indeed be confined only to planet Earth. Other planets in our solar system, too, may eventually be humanized; given the technological possibility of doing so, the temptation to do so appears difficult to resist on the part of those always on the lookout for new challenges and new excitement. To resist the ontological elimination of nature as 'the Other,' environmental philosophy must not merely be earthbound but, also, astronomically bounded (at least to the extent of our own solar system). We should bear in mind that while there may be little pristine nature left on Earth, this does not mean that nature is not pristine elsewhere in other planets. We should also be mindful that while other planets may not have life on them, this does not necessarily render them only of instrumental value to us. Above all, we should, therefore, bear in mind that nature, whether pristine or less than fully pristine, biotic or abiotic, is ontologically independent and autonomous of humankind—natural forms and natural processes are capable of undertaking their own trajectories of existence. We should also remind ourselves that we are the controllers of our science and our technology, and not allow the products of our intellecutal labor to dictate to us what we do to nature itself without pause or reflection. However, it is not the plea of this book that humankind should never transform the natural to become the artefactual, or to deny that artefacticity is not a matter of differing degrees or levels, as such claims would be silly and indefensible. Rather its remit is to argue that in systematically transforming the natural to become the artefactual through our science and our technology, we are at the same time systematically engaged in ontological simplification. Ontological impoverishment in this context is wrong primarily because we have so far failed to recognize that nature embodies its own fundamental ontological value. In other words, it is not true, as modernity alleges, that nature is devoid of all value and that values are simply humanly conferred or are the projections of human emotions or attitudes upon nature. Admittedly, it takes our unique type of human consciousness to recognize that nature possesses ontological value; however, from this it would be fallacious to conclude that human consciousness is at once the source of all values, or even the sole locus of axiologically-grounded intrinsic values. But most important of all, human consciousness does not generate the primary ontological value of independence in nature; nature's forms and processes embodying this value exist whether humankind is around or not.

Note

1. Evidence for this includes at least the following: (a) In June 1992, Drexler was invited to testify before the Senate Committee on Commerce, Science and Transportation, Subcommittee on Science, Technology and Space; (b) *Science*, in 1991, devoted a number, more or less, to the subject; (c) research in it and related fields are being carried out in the USA, Japan and elsewhere.

Appendix 1

Nature as Work of Art

This appendix examines the notion of nature as work of art and attempts to show that it is fundamentally misconceived. While art is paradigmatically an (human) artefact, that is to say, an (human) intentional structure, nature is paradigmatically autonomous and independent of human agency and intention. To assimilate the latter to the former is tantamount to making an ontological category mistake, as the artefactual and the natural are essentially different ontological entities. Yet such an analogy between nature and art is pervasive in both the literature of environmental ethics/philosophy as well as in the management policies of certain major conservation bodies operating today.[1]

Theorists who rely on the analogy are on the whole keen to argue against the need for a radically new, environmental ethic.[2] For instance, Passmore maintains that some acts of destroying nature can be condemned as unwarranted, being imprudent and ignorant, purely by invoking the interests of humans, both long and short-term, of this and future generations, and/or other acts of destruction are just acts of vandalism.[3] Moral condemnation in terms of either category would, between them, exhaustively take care of the attitude of humans toward nature and provide adequate constraint upon human actions in their appropriation of nature.

According to Passmore, nature (whether as wilderness, individual organism or species) is valuable, as a work of art may be said to be valuable. We revere it as we revere art. Passmore says it is "the sort of reverence one feels for a great building, a great work of art."[4] Great works of art are to be admired, cherished and protected. To damage or destroy them amounts to an act of vandalism, like that of scratching one's initials on the facade of Wells Cathedral, or taking a hammer to the Piéta.

Passmore also claims that the concept of vandalism exists in the Western Judaeo-Christian ethical tradition, and that even though it might not have featured too large so far, it is not impossible to lean on it and to develop it further for the purpose of extending the scope of an environmental ethic.[5] Critics of Passmore are right in saying that reliance on the concept of vandalism would not do.[6] For a start, the object upon which vandalism should not be practiced must possess some value independent of that generated by the act of protecting it from destruction. In other words, vandalism as a concept only makes sense if it presupposes a prior value. What could that prior value be in the case of art and in the case of nature according to Passmore's line of argument?

In the case of art, that value may be spelled out as follows: we appreciate art, not in narrow utilitarian or crudely instrumental terms—neither as a form of economic investment nor purely in functional terms (the palazzo is not merely a shelter against the weather). On the contrary, we appreciate it because it may be beautiful, inspiring. Analogously, in the case of nature—we may also find the Grand Canyon, the Great Barrier Reef inspiring or beautiful, etc.

If being awe-inspiring or beautiful underpins nature as art, then such values *par excellence* depend on human consciousness and are its by-products. A world without human consciousness is a world without such values. We have already seen earlier in the book, the last human survivor (in the absence of any equivalent consciousness appearing in the future) would have no qualms pushing the 'destroy' button as far as non-biotic artefacts (of which works of art are a subset) are concerned. Michelangelo's David has no value independently of its human creator and their appreciators. It is indeed fitting that works of art go with their creators and appreciators, in the same way as some cultures consider it appropriate to bury the deceased with their most cherished objects. If nature itself is a work of art or parts of nature are works of art, then, analogously, the Great Canyon, in the absence of human beings capable of finding it awe-inspiring, has no value whatsover and may, therefore, like the contents of the Prado Museum be consigned to destruction. So long as humans live, then and only then, would nature have such values. This is entirely in keeping with the spirit of Humean projectivism—all values, including aesthetic values, are human feelings or attitudes projected upon nature by human consciousness.[7] If nature is a work of art, its value is straightforwardly humanly bestowed, which amounts to strong anthropocentrism, a theme which will be taken up again later.

But if nature is not a work of art but ontologically is distinct from artefacts, this entails that its existence does not depend on humankind, and any value it may possess is not necessarily bestowed by or dependent on the continuing existence of human consciousness.

Ex hypothesi works of art are not naturally-occurring entities. In admiring and cherishing them, we are committed to doing all we can to protect them from destruction, damage and decay, whether these be brought about through natural or anthropogenic causes. Of course, all material things are subject to eventual decay but, all the same, as works of art are consciously designed, produced and maintained by humans for human ends and goals, they are also fit objects for humans to preserve from change and decay to the best of their technological ability.[8] But naturally-occurring entities are not human artefacts, even though it is true that, sometimes, we humans also find them beautiful or inspiring. They have come into existence neither as the result of human design and agency nor to fulfill human ends or purposes, although it is true that humans have found some of them very useful as means to achieve their own goals. It is, therefore, a philosophical error to regard works of art and naturally-occurring objects as belonging to the same ontological category, or to reduce the latter to the category of the artefactual. What exists because of human volition is very different from what exists independent and regardless of human volition—works of art are paradigmatically the products of human volition while natural physical objects which form a landscape are the product of geological, geomorphological and climatological forces which paradigmatically belong to nature, not to culture.

To regard (certain) landscapes as 'nature's works of art' can, therefore, be philosophically very misleading if not handled with care. If the epithet were understood merely figuratively, as a manner of speaking, it is innocuous enough. But one is not justified in sliding from the fact that we, humans, do at times, find certain landscapes beautiful or awe inspiring to implying that like aesthetic artefacts, they, too, should be protected from even nonanthropogenically-induced change so that we may continue to derive aesthetic satisfaction in their presence.

Change in nature is endemic and may be slow or abrupt. Geological changes of either kind throw up structures which alter the landscape. To arrest or deflect geological change where it could lead to unaesthetic or less aesthetic structures amounts to treating geological formations, the products of such processes of change, as mere artefacts in the name of what is beautiful or awe-inspiring. It is, to adapt a phrase, 'to pervert the course of nature' in order to serve our human purposes and ends. It may also be apt to quote a point made by Hargrove:

> We cannot help nature with its plans, for it does not have any plans. When we make plans to help or improve nature, the plans are not nature's but our own, and the result is the stifling of natural creativity and the transformation of the natural objects influenced into human artifacts. No matter how natural they may look, they are no longer original, no longer authentic. Their ontological status has been altered: they have become objects whose essence has preceded their existence.[9]

Such aestheticism, no less than crude resource conservation, are forms of instrumentalism. They are also forms of strong anthropocentrism, namely, that human consciousness is the sole source of all values and the sole locus of intrinsic value—see Callicott, "On the Intrinsic Value of Nonhuman Species."[10] All forms of instrumentalism are instantiations of strong anthropocentrism, for they assume that ultimately it is the presence of humans with their particular consciousness which endows the world with value. As such, nature, independent of humans and their bestowal of values, is itself valueless. It follows that if humans find some parts of the natural world useful (or potentially useful) as factors of production, and other parts capable (or potentially capable) of generating aesthetic satisfaction in them, then and only then is nature valuable.

Passmore (*Man's Responsibility*) distinguishes conservationism from preservationism by saying that the former amounts to "conserving for" and the latter "preserving from." Conservationism is said to take a longer term view of human interests, being concerned with posterity and with intergenerational justice. It, therefore, outlaws the short-term economic interests of one group in society, or of a particular generation of humans. For instance, a typical conservationist response to overfishing would be to ensure that no one group or generation take out so much from the fishery grounds as to cause a collapse in the fish population which would leave nothing for other groups or for other generations to follow—it canvasses for sustainable fishing in this sense. It also follows that a conservationist perspective would generate a management policy, which sees to it that a lake with plenty of fish in it would not be allowed to dry out (assuming that extant technology could be of help), as the replacement of a lake by a reed bed or a meadow could mean the loss of a useful resource, and be of less economic benefit both short and long term. But from the standpoint of preservationism, the argument from resource conservation would not cut any ice—it would tolerate the natural drying out of the lake. Likewise, it would tolerate the natural drying out of the lake even when the instrumental value derived from the lake is not in terms of resource conservation but resource preservation, that is to say, to use the lake as an aesthetic resource, as both resource conservation and resource preservation are aspects of the philosophy of conservationism. Any attempt to prevent natural change in order to maintain the landscape's ability to provide the right aesthetic experience may be said to put in practice the philosophy of conservationism rather than preservationism.

This leads on to a related point which needs further clarification, namely, whether the transformation of nature to become an artistic artefact in the sense discussed above is tantamount to treating nature in a purely instrumental way, as the argument presented claims. The claim is resisted by those who, though sympathetic to the need to rethink the attitude of humans to nature, nevertheless, are suspicious of the view that nature could be an object with moral standing in its own right, independent of that of humans. For instance, Ted Benton in his very favorable review of Eckersley's book, *Environmentalism and Political Theory: Toward an Ecocentric Approach*, nevertheless, thinks that she "needs to rethink the dichotomy, which structures so much of her argument, between anthropocentric and ecocentric approaches" and which also includes "valuing the non-human world for its own sake, rather than as a means to human ends."[11]

Benton is of the opinion that it is not helpful to assign both the resource use of nature on economistic grounds and appreciating nature on aesthetic (as well as spiritual, psychological) grounds to the category of instrumental value.[12] He uses two arguments against this assimilation. First, it is a forced use of the word 'instrumental.' Second, even conceding that aesthetic satisfaction is an instance of anthropocentric instrumental reasoning, it remains true that the instrumental justification is itself predicated upon "pervasive non-instrumental, nonanthropocentric valuations," namely, that it is true humans are indeed spiritually or aesthetically nourished in the way indicated.

Is Benton right? He might not be. To see why, it is necessary to clarify certain key terms, such as means/end rationality or instrumental rationality. These may be ambiguous.[13] First, consider the resource use of nature as an instance of instrumental rationality. In its simplest form, it amounts to no more than this—in order to keep healthy physically, their metabolism intact, adult humans (generally speaking) must take in 2500 calories per day, to be made up in the main of fat, protein and carbohydrates. In other words, we have to eat to remain alive and to function reasonably well as organisms. There is nothing wrong with instrumental rationality and the notion of nature as a resource for humans when these are understood in this way. What Benton as well as ecocentrists like Eckersley object to is the appropriation of natural resources on economistic terms—to regard everything in nature as a potential resource for increasing human consumption, thereby raising the material standard of living, comfort and convenience regardless of the consequences of degrading and destroying nature. And it is this sense of instrumental rationality as understood by Habermas that Benton presumably objects to. It is similar to Heidegger's point, that is, to regard nature as **standing reserve**, appropriating nature through the technological mode of **enframing**. But this sense has nothing to do with the means/end model of rationality. That simply says: for any given end the human agent postulates, the agent seeks a rational means to attain it. It is true that the means/end model in the history of modern Western philosophical thought has acquired a further gloss, namely, that ends are irrationally or non-rationally postulated. As such no rational judgment can be passed either upon the end of maintaining the metabolic integrity of the human agent or that of ever-increasing human consumption and the material standard of living. To achieve the latter goal, obviously, very powerful technologies of controlling nature are required as means, whereas the more modest goal of maintaining metabolic integrity could, by and large, be achieved with much less technologically powerful means.

In other words, there are four separate theses which ought to be distinguished in talking about the related terms 'instrumental rationality,' 'instrumental reasoning,' 'means/end rationality' and 'efficiency':

1. Rational (human) agents in declaring some future state as their end must set about adopting a causally appropriate and efficient means to achieve it. An

agent could not be judged to be rational if having avowed the destruction of Buckingham Palace or The White House to be her/his end, then sets about demolishing it with a feather duster. Either the agent's end is not to be taken as a serious one, or if it were indeed serious, an observer would have to conclude that the agent is rationally defective in some way. If the agent had used an amount of dynamite calculated to do the job of demolishing a building of that size and structure, without causing damage other than to the targeted object, we would all agree that the agent was eminently rational and sane.[14] This sense of means/end rationality may be spelled out in terms of two elements: the conceptual aspect—any agent, to be deemed rational in her/his action, in adopting an end, must also will to adopt an efficient means to achieve it; and the empirical aspect—the means adopted by the rational agent must be deemed to be causally appropriate and adequate for the task in hand. But this sense of rationality and efficiency may be said to be ideologically innocent or innocuous. It is agnostic, and has nothing to say about whether it makes sense to determine the rationality/desirability or otherwise of ends.

2. The Humean thesis about the irrationality/non-rationality of ends is not so ideologically innocent.[15] Many would argue that it lays the philosophical foundation for a worldview which ultimately permits the unlimited and unconstrained exploitation of nature for human ends alone. On this view, it is not irrational, or it is just as rational to prefer a lifestyle of infinitely expanding consumption of material goods at the expense of nature's integrity as it is to prefer one of simple frugality but respecting nature's integrity.
3. Regarding the Humean-derived thesis about means/end rationality, while there is no rationality of ends, there is rationality of means to achieve ends. The rationality here, too, refers to efficient means to accomplish ends. But this sense of efficiency must be distinguished from that mentioned in 1 above, which is about causal adequacy and appropriateness involved in the ideologically innocuous thesis of instrumental reasoning. This other sense, however, is covertly, though not overtly, ideological, as it is understood in economistic terms, spilling over into contexts where such an understanding is inappropriate. Cost benefit analysis when applied to environmental public goods like clean air, peace and quiet illustrates the impropriety of this kind of extrapolation.

   However, the thesis itself may be charged with internal incoherence. If all ends are necessarily irrational or non-rational, and if efficiency itself is an end, then choosing the most efficient means to achieving one's end is itself a rational thing to do only if the ends of efficiency and rationality themselves are presupposed to have been irrationally or non-rationally endorsed. This then reduces the thesis to a kind of absurdity.
4. The thesis of progress as championed by modernity is the most explicitly ideological—every part and aspect of nature is, in principle, a resource for human use and consumption in supporting an ever-increasing material

standard of living, and in practice, would be turned into just such a resource provided the technology is available and the economic costs are less than the estimated benefits. This is the sense of instrumental rationality as understood by Habermas or of standing reserve as understood by Heidegger.

Benton is correct when he says it strains the use of the word 'instrumental' in assimilating aesthetic benefits to resource benefits if he has at the back of his mind the thesis of progress. Nature's spiritual and aesthetic effects on us humans are meant to be opposed to nature's resource use for us on materialistic/economistic grounds, and should also not be considered within the framework of cost/benefit analysis. Moreover, the two demands on nature—sustaining material progress on the one hand and catering for spiritual/aesthetic needs on the other—normally cannot be satisfied simultaneously given the world as we know it to be. The more we use up nature as a resource to further our material well-being, the less there may be of nature left for us to derive aesthetic benefits from it to sustain our mental, psychological and spiritual well-being.[16] Those who believe in the thesis of progress as purveyed by modernity, admittedly, do not care much, if at all, for our spiritual flourishing, as they assume that material affluence exhausts the notion of human happiness. This much Benton and Eckersley certainly have in common.

But is Benton correct in implying that the pursuit of non-material well-being falls outside the model of means/end rationality when understood in the 'ideologically innocuous' sense? In this sense, it merely says that for any agent with any avowed end whatsoever, the agent must—a hypothetical imperative—adopt causally appropriate/efficient means to attain it. Just as human agents must take in enough food to sustain their metabolic integrity, in the same way, in order to maintain their spiritual, psychological, mental well-being, they must have access to nature which can bestow such benefits on them. Therefore, they ought to preserve nature for just such a purpose. The means/end model of instrumental reasoning covers it neatly. So, contrary to what Benton maintains, it is paradigmatically instrumental valuation. Also contrary to his other claim, it is paradigmatically anthropocentric valuation. It is concerned with promoting the interests or welfare of humans. It is about using nature as a means to secure a human end, namely, human flourishing. The *raison d'être* for preserving nature is precisely not for its own sake but to maintain and sustain a human good, that is, for the sake of us, humans. The difference between Benton and the instrumentalist who regards nature simply as a storehouse of raw resources for production and consumption is that the latter's conception of human flourishing is a much narrower one confined to promoting material well-being only, whereas Benton's is a much richer one encompassing the spiritual and aesthetic sides of human nature. However, under both conceptions, humankind appropriates nature to promote its own ends and well-being. Admittedly, cutting down a rainforest to create pastureland for raising cattle to supply the demand for hamburgers involves the destruction of the rainforest, whereas deriving psychical satisfaction from the rainforest requires that it remain intact. But all the same, satisfaction from eating hamburgers or from watching wildlife in a rainforest is satisfaction for humans generated by regarding nature either as resource conservation or resource preservation.[17]

Benton has also written:

> Would we count arguments for the preservation of human artifacts such as cathedrals, sculptures, paintings, or regional dialects on the grounds that they were valuable as symbols, or for spiritual well-being as examples of instrumental reasoning? Surely it is important to differentiate among human ends, and if we do, we might find that the satisfaction of many of them does not conform to the means/ends schema in the way required by the paradigm of instrumental rationality.[18]

He is here talking about cultural artefacts, and the examples he has cited are mainly non-biotic artefacts. But as we have seen these are total human creations with no *telos* (not even a residual one) of their own, being entirely dependent upon humans in their conception, execution and maintenance. They have the significance, whatever it may be, humans choose to endow them with. Their final cause is designed and built into them. As Hargrove says, their essence precedes their existence. Once brought into existence, their original significance may be lost but if they are to remain as cultural objects, they must be assigned another significance. A cathedral was originally constructed as an expression of the Christian attempt to glorify the Almighty. To secularists of the modern era, it is regarded differently, perhaps as an expression of the human spirit to reach something beyond and greater than themselves, like the beautiful and the sublime.[19]

These valuations are unquestionably anthropocentric. But are they also non-instrumental judged against the ideologically innocuous sense of instrumental reasoning? All artefacts are designed and executed with an end or purpose in mind. In the case of a cathedral, the end is to glorify (Christian) God. As building a cathedral is an immense undertaking which often takes several centuries to complete, the purpose must be very firmly carried, from generation to generation, in the minds of those engaged in the enterprise. Of course, this is not to say that other less pure motives might not also have operated along the way to assist its completion. At least, the prime and fundamental motive cannot be venial ones.

Building a cathedral in order to glorify God is an eminently rational thing to do, in one sense of the term 'rational,' as it is an activity which conforms to the means/end model (in the ideologically innocuous sense). So is launching a competition for an anthem in order to encourage the growth of a national identity. In other words, cultural constructs are both a means to an end as well as an expression of the end itself. A song (like *Advance Australia Fair*) once adopted as the national anthem is both a means of constructing and/or reinforcing a sense of national identity as well as an expression of that identity itself. An artist in drawing a portrait may merely be earning her/his bread and butter; the patron of the artist in commissioning the portrait may be commemorating his dog or his mistress—but the respective ends of the artist and the patron do not, however, necessarily preclude the artist from using the opportunity to create a work of aesthetic worth. To achieve the latter end, s/he might have to lavish more time, effort and energy on the commission than if merely fulfilling the contract were the only goal.

More mundane artefacts clearly are covered by the means/end mode. An ax is an implement constructed with the aim of cutting down a tree or chopping wood. Of course, once an ax or a sword has been made, its original function might become obsolete and it may only be admired as an object of beauty—*samurai* swords in feudal Japan were designed presumably as functional, ritualistic as well as aesthetic objects but today they are appreciated only as the latter.[20] So could a house. Indeed, one might want to build not any old house but a beautiful one. In that case, the artefact also conforms to the means/end model—here the end is not merely a house which fulfills its functional purpose but one which also satisfies that of being aesthetically pleasing.

To conclude: artefacts, whether mundane or invested with cultural significance, embody instrumental reasoning or means/end rationality understood in an ideologically innocuous way. Such a mode also covers the context in which nature is regarded as a source of non-material benefits. *Contra* Benton, the preservation of non-biotic artefacts as well as of nature for the reason just indicated is deeply anthropocentric. In the case of the former, such artefacts have no *telos* of their own (unlike biotic artefacts which have a residual *telos*); it makes no sense to talk of preserving them for their own sakes. And in the case of natural entities—both biotic and abiotic—their preservation is definitely not for their own sake, but for the sake of humans to enrich themselves aesthetically, spiritually, psychologically.

But as nature is not an artistic artefact, although it happens that we, humans, at times find some parts of it aesthetically pleasing, its 'deepest' value, if such exists, can neither be anthropogenically nor anthropocentrically grounded.[21] In principle, it is totally independent and autonomous of humans, which puts it in a different ontological category from the artefactual. But it is one of the main contentions of this book that its ontological status can be said to generate a 'deep' value, namely, the value of independence.

Notes

1. For a critique of the implications of such a thesis for environmental management, see Keekok Lee, "Beauty For Ever?" 213-25.

2. There is another reference in environmental theorizing which also relies on the analogy between nature and art. Robert Elliot (in "Faking Nature," *Inquiry* 25 1982: 81-93) first raises it by arguing against the restoration of damaged natural ecosystems and landscapes on the grounds that the restoration, no matter how perfect, would have less va-

lue than the original—the restored area would be analogous to an art forgery. (Elliot has since elaborated on his 1982 article—see Chapter 3 in *Faking Nature*. Robert Goodin (*Green Political Theory* Oxford: Blackwell, Polity Press, 1992) has also added his voice to this line of thought, emphasizing that the different histories of the original and the fake give rise to differential, non-substitutable values. This account agrees with Elliot and Goodin on this limited point they are making, but goes beyond it to argue more fundamentally that the very analogy between nature and art is misconceived in principle. In this sense, such an attempt is more in line with Eric Katz's thinking ("The Big Lie: Human Restoration of Nature") than with Elliot's and Goodin's.

3. John Passmore, *Man's Responsibility for Nature* (London: Duckworth, 1980).

4. Passmore, *Man's Responsibility*, 124. For another voice similar to Passmore's, see Timothy Sprigge, "Respect for the Non-Human," in *The Philosophy of the Environment*, T. D. J. Chappell, ed. (Edinburgh: Edinburgh University Press, 1997), 127.

5. Other arguments against Passmore will not be pursued, such as that not all parts of nature could be construed as works of art, never mind as great works of art, as humans would not remotely find them beautiful or inspiring.

6. See Val Routley, "Critical Notice of John Passmore's *Man's Responsibility for Nature*," *Australasian Journal of Philosophy* 53 (1975): 177; Peter Singer, "Not For Humans Only: The Place of Nonhumans in Environmental Issues," in *Ethics and Problems of the 21st Century*, edited by K. E. Goodpaster and K. M. Sayre, 203-04; Tom Regan, "The Nature and Possibility of an Environmental Ethic," *Environmetal Ethics* 3 (1981), 16-21; Robin Attfield, *The Ethics of Environmental Concern*, 152-53.

7. See Chapter 5 section entitled **Resisting Humean Projectivism** for a more detailed discussion.

8. Humans also may decide in certain cases to allow works of art to change and even to decay. An instance of the former would involve a decision not to restore a painting (even if it were technologically feasible to do so) to its original colors—the recent restoration of Michelangelo's paintings on the ceilings of the Sistine Chapel has proved controversial. An instance of allowing an artefact to decay would be to let a monument transform itself into a ruin—as a ruin, it could, after all, be 'perfect.'

9. Eugene C. Hargrove, *Foundations of Environmental Ethics* (Englewood Cliffs, New Jersey: Prentice Hall, 1989), 195-96.

10. See J. Baird Callicott, "On the Intrinsic Value of Nonhuman Species".

Resisting Humean Projectivism in Chapter 5 looks critically at the twin theses which constitute this kind of strong anthropocentrism. Callicott rejects the second but upholds the first; Holmes Rolston (*Environmental Ethics: Duties to and Values in the Natural World*, 112-17) upbraids him for conceding too much to anthropocentrism by his truncated account—for a way of resolving the disagreement between Callicott and Rolston whilst going beyond what they each have to say, see Keekok Lee, "The Source and Locus of Intrinsic Value," 297-309.

11. Ted Benton, "Review of Robyn Eckersley's *Environmentalism and Political Theory: Toward an Ecocentric Approach*," *Environmental Values* 2 (1993): 279.

12. On this, Benton appears to agree with Hargrove. But see Lee, "Instrumentalism and the Last Person Argument" (*Environmental Ethics* 15 1993, 333-44), where three types of instrumentalism are distinguished, namely, strong instrumentalism or instrumentalism (I), weak instrumentalism (II) which itself is divided into instrumentalism IIa and IIb. I refers to the resource use of nature on economistic grounds to promote progress as material affluence; IIa to the contribution of nature to human nonmaterial well-being and IIb to the

'public service' function of nature in acting as a sink to absorb waste, etc.

13. Their relationship in turn to the notion of instrumental value itself remains unclear.

14. For the purpose of this discussion, there is no need to be embroiled with the more complicated issue about the rationality of belief-systems. One is here merely focusing on the rationality of a single action informed by a single belief.

15. This Humean thesis about the irrationality/ non-rationality of ends (call this II) is to be distinguished from the earlier mentioned Humean thesis of projectivism, that is, that all values are dependent on human consciousness and are humanly bestowed upon nature (call this I). However, II is an aspect of and follows on from I.

16. This argument certainly holds true given the extant technology of today.

17. There is one further comment worth making. As far as aesthetic pleasure from nature is concerned, if nature is regarded as a mere trigger of the aesthetic experience, then technological improvements could soon deliver us a simulated experience, no different, it is said, from the real thing. For instance, a virtual reality presentation of the Great Barrier Reef could act as an equally effective trigger of an aesthetic response. If this were really so, then we could destroy the Great Barrier Reef itself (to serve some other end of ours) without undermining our aesthetic experience of it.

18. Benton, "Review of Robyn Eckersley's *Environmentalism and Political Theory*," 280.

19. Naturally-occurring objects too may be endowed with cultural significance, although they are not artefacts. Lake Baikal is a cultural symbol for Russians just as Ayrs Rock is a religious symbol for the aboriginal peoples of Australia. But for the purpose of the discussion, only cultural artefacts will be considered.

20. Indeed, in the artisanal tradition in general, there is neither antagonism nor divorce between functional utility and beauty. The two were intertwined. But in the age of modernity, especially under industrialization involving methods of mass production, the two were sundered. Social theorists, like William Morris, argue that even ordinary household objects should be designed and executed to embody both functional and aesthetic desiderata, and that the severance of the one from the other is itself a symptom of what is profoundly wrong with industrial civilization embedded in a capitalist economy in which everything, including art, is commodified. (For further discussion on Morris, see Keekok Lee, *Social Philosophy*, 270-78.)

21. Such a response is largely culturally determined—see Lee, "Beauty For Ever?" for further discussion.

Appendix 2

Is Nature a Mere Social Construct?

Chapter 3 has isolated several meanings of the word 'nature,' and has argued that one of them (namely, $nature_c$) is too wide to be of any help in shedding light on the problem of how humankind ought to relate to nature, and that another (namely, $nature_{fa}$) is more fundamental than the others delineated. However, this kind of clarification appears to have been called into question by a strand of postmodernist thinking which maintains that nature is a heavily contested site, that all accounts of nature are necessarily socially constructed implying that deconstruction of such socially constructed accounts is the only legitimate activity or that it is the prelude to undermining one while championing another.[1] However, either interpretation in turn implies which ever account becomes the dominant one depends on politics both in the narrower and larger senses. Sociology of knowledge is all; constructing a systematic and critical epistemology and metaphysics is neither possible nor viable.[2] But here at best only an outline of a response to such a challenge can be given.

What needs pointing out immediately is that it is not clear what the thesis that nature necessarily is a social construction really amounts to. Two things in particular are unclear: what constititues the thesis and what could be its target, witting or unwitting? Take the latter question first. It appears to undermine any attempt on the part of environmental thinkers to defend in any way the integrity of nature. If all accounts of nature are but mere human representations, then in one important sense, there is nothing to defend apart from the human interests behind these various representations. It thereby rules out any attempt to argue for non-anthropocentrism, namely, that nature may be morally considerable irrespective of humans and their interests. It will be argued that the thesis does not succeed if that were its (unwitting) objective.

But what actually constitutes the thesis? As a general thesis, it seems itself to be a derivation from another which is meant to be universal in scope, namely, that all definitions and meanings of terms, all categories are social constructions which in turn appears to be an implication of the theory of deconstructionism.[3] The remarks which follow are concerned not so much with the generic thesis but when it is applied in the context of attempting to elucidate the meanings of 'nature.' This more limited derivative is susceptible to at least four different interpretations which, unfortunately, upholders of the thesis often fail to distinguish:

1. Any account is necessarily anthropogenic as only humans can engage in delineating terms and defining them in certain ways and, therefore, humans are the source of all accounts.[4] However, defining terms is a linguistic activity; furthermore, human consciousness itself is necessarily mediated *via* language. But language is public and social, not private and individual.
2. Any account is necessarily anthropocentric. As humans are beings which engage in delineating and defining terms in certain ways, these delineations and definitions are bound, therefore, to be filtered through the human perspective, thereby reflecting and embodying their interests, preferences or biases.
3. Any account necessarily reflects, embodies and, therefore, implicitly advances the interests, preferences or biases of particular groups in society. Socially and politically, humans do not constitute a single group. On the contrary, they are fragmented into different groups which are delineated in terms of color, gender, race, class, money, status, historical ages and periods, etc. An individual necessarily falls under more than one of the various groups thus constructed. Each group may give its own account of nature, but the dominant account is that propagated by the most powerful group in social and political terms. In the global context, that group today consists of those in charge of transnational corporations, the World Bank (and similar organizations), national governments including key bodies like their central banks who are, in the main, males and white.
4. Any account of nature, whether anthropocentric or nonanthropocentric in character, is necessarily advanced by a particular individual or group whose behavior or conduct, in turn, is necessarily socially and normatively constituted.

Under the first interpretation, the thesis that nature is socially constructed is true. However, it is but a truism and, therefore, innocuous. It cannot as such undermine the possibility of articulating a critical epistemology and metaphysics concerning nature of which the delineation and definition of terms is a part. Language is the collective product of human consciousness which in turn shapes and informs that consciousness. But as thought cannot be articulated without language, language in that sense lays down the limits of what can be said at any particular time. But so long as this is not taken to imply that what can be said exhausts what is (and what can be thought), it remains innocuous—language can be stretched *via* metaphors, the introduction of new terms minted from old or dreamed up afresh, etc., to enable new thoughts and insights to be articulated, discussed and critically assessed.

Under the second interpretation, the thesis appears to have failed to distinguish between anthropocentric and anthropogenic. 'Anthropogenic' means 'caused, produced by humans' or that humans are the source of something. 'Anthropocentric' means 'putting the interests, preferences of humans at the center of things'—an anthropocentric worldview is one which celebrates only human achievements, promotes only human interests, preferences and values, which always prioritizes human over nonhuman interests, or which claims humans alone to have intrinsic value while nonhumans are only of instrumental value to humans.[5]

Failure to distinguish between the two notions makes it possible for the thesis to pass off as a true one. A nonanthropocentric viewpoint can only be formulated and articulated by humans who possess language. Other nonhuman naturally-occurring beings, without our unique type of consciousness, *ex hypothesi*, cannot do so. If these do have interests or embody certain values, then only humans can say so on their behalf.[6] In this sense, any nonanthropocentric perspective or viewpoint is necessarily anthropogenic, just as any anthropocentric perspective or viewpoint is also necessarily anthropogenic. 'Anthropocentric' is not the antonym of 'anthropogenic'; the antonym of the latter is 'nonanthropogenic,' which means 'not caused, produced or generated by humans.'

The failure to grasp the distinction could also lead to the following reasoning: as humans alone are capable of articulating interests or preferences, the only interests or preferences they can articulate are those which directly concern themselves only. From an anthropogenic premise, an anthropocentric conclusion is derived. But the reasoning as it stands is unsound. The conclusion does not necessarily follow unless it already presupposes that only beings which are capable of articulating their interests or preferences *via* linguistic categories are morally considerable beings. This additional premise would then rule out nonhuman beings from the domain of moral considerability, as they necessarily cannot articulate their interests or preferences.[7] It also follows that, *ceteris paribus*, morally considerable beings alone would enjoy rights protecting their interests or preferences, or are owed duties by fellow morally considerable beings not to have those interests trampled upon. But beings which are not morally considerable neither have rights against morally considerable beings nor do morally considerable beings owe them any direct duties to safeguard their unarticulated interests or preferences. At best morally considerable beings owe them indirect duties.[8] But the additional premise and its implications are precisely those which nonanthropocentrists, of one sort or another, strenuously challenge.

Furthermore, once the clarifications made above have been grasped, it is clear that not all accounts of nature are socially constructed in the sense of necessarily advancing anthropocentric interests and values. Some accounts are distinctly nonanthropocentric, including the one advanced in this book. These deny, each in its own ways, that humans alone are morally considerable, that human interests necessarily trump nonhuman ones in all conflicts between the two. Each affirms, in its own ways, the moral standing of nonhuman nature and pleads that nature or their targeted bit of nature be treated by fellow humans with respect because of their moral considerability. As such, the thesis under scrutiny is simply false.

To conclude, the second interpretation is, therefore, either an innocous truism—all accounts of nature are anthropogenic—or simply false, as not all accounts of nature are socially constructed in the sense of necessarily advancing some version of anthropocentrism or another. This then means there is room to argue for a nonanthropocentric perspective which can serve to challenge the still dominant anthropocentric paradigm.

Under the second interpretation, humankind is assumed to form a single cohesive group whose interests are clearly identifiable under the term 'anthropocentric.' In contrast, under the third interpretation, humankind is splintered along certain fault lines. According to Marxist thought, for instance, the most important of these fault lines is class—the capitalist class will therefore have a view about nature which is different from that of those who do not own capital. According to ecofeminine thought, the male perspective of nature is different from that of the female—to the former, nature is female, to be subdued, raped and controlled, while to the latter, nature is the nurturing mother, the reproductive womb.[9] According to the affluent, nature is a place for recreation, solitude, for recharging one's spiritual or aesthetic batteries; according to the poor, nature is a place in which one struggles to ensure bare survival, where there is no time to look up from back-breaking chores to admire the beautiful or the sublime scenery in the distance. According to white people who rule the land (as in North America[10]), nature (as wilderness) is where people do not live, only visit; according to the dispossessed native Americans, nature as wilderness means physical and spiritual exile.[11] Nature for some means the recovery of the lost Garden of Eden—on this social construction, 'tame nature' with its meadows and meandering rivers or burbling streams is what is 'natural'; but so-called 'wild nature' is 'unnatural' and therefore frightening, to be shut out of sight as exemplified by the seventeenth-century coach drivers who pulled down the blinds while driving through the Peak District in England for fear the passengers might be scared by what they saw out of their windows. Pre-Romantic (English) thought favored tame nature while Romanticism put its premium on wild nature, thereby bringing about a revolution in nature aesthetics.

As a matter of fact, many of the case-studies and arguments for the thesis of nature as a social construction may be traced to this third interpretation—for instance, for a feminine deconstruction of the dominant account of nature, see Carolyn Merchant,[12] for a Marxist version, see William Leiss,[13] for a deconstruction of the wilderness account, see William Cronon.[14] This approach does seem to yield some very rich and interesting rewards.[15]

Furthermore, these studies also attempt to raise the issue of social justice between groups. Developing nations are not so keen to refrain from development using technologies which could damage nature, arguing that it is more important to raise the material standard of living of their citizens, and pointing out that such ecologically insensitive technologies are all they could afford. In the presence of extreme inequalities in the world and with no serious transfer of wealth or improved methods to cause less environmental damage, poorer nations are simply being asked to carry the burden of environmental global health. It is all very well saying that tigers in India should be saved but ought they to be saved at the expense of those people whose livelihoods would be destroyed if more space were to be given to the tigers? An analogous debate arises within developed economies; for instance, those who work in the timber industry which cut old-growth forests in North America are being asked to shoulder the cost of preserving such forests for the sake of those who want to enjoy them as they are. On this understanding, it looks as if all environmental disputes turn out to be social/political disputes about unequal power and economic relations between human groups.

On this view, the moral plea to defend nature (as individual mammals or organisms, species or ecosystems) is not to be understood as a direct moral defense but as a plea on the part of a certain group like the white affluent professional middle classes in the North to protect their own interests and preoccupations. These do not earn their living mining the Earth or cutting down trees but their lifestyle recreational activities like hiking, mountaineering, canoeing or whatever, could be undermined by those who earn their living in ways which involve destroying or damaging the nature they (the well-heeled brain-workers) are pleading to save. Environmental justice is about justice between different human groups but not between human communities on the one hand and nonhuman groups or individuals on the other, as the latter falls outside the pale of morality. Environmental justice is distinctly anthropocentric in orientation. It is no different from justice in general when that is understood to involve the problem of distributing goods which are scarce between individuals or groups in the presence of competing claims for them.

Such issues, obviously, are legitimate and should be raised and debated in one form or another. However, their existence and the urgency of addressing them do not mean that every account of nature is necessarily socially constructed in the sense that it is the result of differential economic and power relations between groups. For instance, a nonanthropocentric account of nature in favor of saving biodiversity does not *per se* advance any sectional human interests, as its avowed aim is to protect certain values in nature which are embodied in species. But any cause could be hijacked by some other to promote its own sectional interests —religion, it is said, exists for humans to celebrate their god(s) but it is true that in religion's long history, more often than not, their God or gods have been commandeered to serve the interests of a certain class, race, sex, or nation. In spite of this unfortunate liaison, religion cannot be reduced to mammon, imperialism or patriarchy. No more can all accounts of nature be reduced to the conflict of sectional human interests. To try to do so, once again, runs into the danger of confusing anthropogenic with anthropocentric.

However, some theorists who may be said to fall under the third interpretation wish to dissociate themselves from certain epistemic implications which appear to follow from the more general thesis of nature as social construct(s). For instance, in Cronon's recent *Uncommon Ground*:

1. At least two of its contributors, James Proctor and Richard White, have explicitly distanced themselves from philosophical or normative relativism—see pages 449 and 45.[16]
2. Several like William Cronon himself, Michael Barbour and Richard White acknowledge that "nature is real,"[17] that "there is more to the world than just words,"[18] although we need words to talk about the world, or that the natural world is more than just "our representations of it."[19]
3. Merchant acknowledges that "nonhuman nature ... is a free and autonomous actor, just as humans are free autonomous actors. But nature limits human freedom to totally dominate and control it, just as human power limits nature's and other humans' freedom."[20] Cronon says that "the nonhuman world is real and autonomous, a place always worthy of our respect and care."[21]

These acknowledgments that nature (or what they prefer to call the "nonhuman world") does exist out there, is real, autonomous and free, and deserving our respect would appear to make them sympathetic to the fundamental meaning of nature given in this book. Cronon, furthermore, recognizes nature as 'the Other': "In the broadest sense, wilderness teaches us to ask whether the Other must always bend to our will, and, if not, under what circumstances it should be allowed to flourish without our intervention."[22] This book, too, argues for nature as 'the Other' or to be more precise as 'the ontological other.' Given these important areas of agreement and overlap, wherein lies the difference between those who see themselves as social constructivists of one description or another and those who do not?

One outstanding difference focuses on the human/nonhuman or culture/nature distinction. Chapter 5 deals with this matter; so only a few brief words need be said here. The distinction is normally understood as a dualism. Postmodernist thought regards this and other dualisms to be the cornerstone of the modern (Western-inspired) worldview which is implicated in the ills of the modern age, including environmental damage and destruction.[23] The dualism must, therefore, be challenged and rejected if the world is to be put right. This book, too, rejects the human/nonhuman distinction as a dualism; however, it argues for the retention of the distinction as ontological dyadism. In dualism, one side is privileged and the other inferiorized; historically, the privileged is the human, but social constructivists hold that those who challenge the privileged master class by simply substituting the nonhuman for it are still adhering to dualist thinking. The old power relationship has simply been inverted—nature as wilderness is now privileged, but humans (or at least the politically weak humans) are inferiorized and, therefore, driven out of such places except as visitors. According to social constructivists, dualism is philosophically untenable because no ready distinction can be made between culture and nature—humans are a part of nature (not apart from nature), as nature (at least on Earth) exists only as humanized nature. On the other hand, ontological dyadism, while rejecting the power relationship of domination and oppression inherent in dualism, argues for the retention of the distinction as one based on an ontological difference between what is human and what is nonhuman—humans, given their unique consciousness, their ability to think abstractly, and their increasingly more and more powerful technologies which enable them systematically to transform the natural to become the artefactual, have set themselves apart from nature. Our human consciousness is capable of realizing that nature, its processes and its products belong to a different ontological category from that of humankind itself, its intentional activities and the artefacts generated from such activities. Such realization in turn leads to the further recognition that there may be something morally amiss about the totalizing project of humanizing nature entailing the ultimate elimination of nature as the ontological other which humankind in its history appears to intensify as it acquires increasingly more powerful technologies.

The fourth interpretation may be attributed to Steven Vogel in his recent attempt (*Against Nature*) to formulate an account of nature based on Habermas's discourse ethics as he has argued and developed it.[24] Nature, *ex hypothesi*, cannot speak and, therefore, cannot be participants in **communication**. It follows they are not moral subjects.[25] Only humans, as individuals or as groups, can put forward accounts of nature. Vogel has no difficulty in recognizing that this in itself does not entail that only humans have value. In that sense, any account of nature advanced is necessarily anthropogenic (in the terminology of this book but not in Vogel's), though not necessarily anthropocentric, as it can be nonanthropocentric. If this is so, where then lies the charge of anthropocentrism in Vogel's position? The charge appears to lie in the fact that humans, in speaking on behalf of nature as it were, nevertheless, are socially and normatively constituted. Some speak as scientists, others like Vogel himself, as philosophers or theorists of one description or other, yet others, as activists or advocates like animal welfarists. But all activities, whether theoretical or practical, are conducted within a social, normative framework. Science itself is certainly no exception and so neither are scientists posing as 'experts.' Their expertise is not entirely value-free in the way modern science likes to propagate. Science is a social institution; its practitioners have been successfully initiated, consciously and otherwise, into its methodological, epistemic, metaphysical and value (including even political) requirements. In other words, those who claim to tell us what nature is, are themselves socially constructed.[26] But to make Vogel's position clear, it appears one needs to distinguish not merely anthropocentric from anthropogenic but also between two senses of anthropocentrism itself.

The first may be called 'axiological anthropocentricism,' the view that only humans are (intrinsically) valuable. The second may be called 'existential anthropocentrism,' the view that humans who alone can be participants in **communication**—as moral subjects—are necessarily socially and normatively constituted. The former is incompatible with a nonanthropocentric account of nature; the latter is perfectly compatible with such an account.[27] Existential anthropocentrism is, undoubtedly, correct; all accounts of nature are necessarily socially constructed in the sense that humans who articulate them are themselves socially and normatively constituted beings. Like the thesis closely related to it, namely, that all accounts of nature are anthropogenic as they emanate from humans, it is true but innocuous, provided it is not understood to mean that all knowledge is reducible to the sociology of knowledge, to embrace philosophical or normative relativism, or that reality or nature does not exist outside the activities of such socially and normatively constituted beings and their representations of reality or nature. In other words, all accounts of nature are necessarily anthropogenic as well as anthropocentric in the existential sense, but not necessarily anthropocentric in the axiological sense just identified. But Vogel does not endorse axiological anthropocentrism, only existential anthropocentrism.

To conclude: the thesis that nature is socially constructed turns out under critical scrutiny not to be the potent knock-down argument against the possibility of articulating a nonanthropocentric account of nature to challenge the current anthropocentric paradigm. The terms 'anthropogenic' and 'anthropocentric' need to be carefully distinguished. The apparent potency of the thesis lies primarily in confusing these notions. Another distinction implicit in the writings of some social constructivists must also be distinguished, namely, between axiological and existential anthropocentrism. While the former is incompatible with any nonanthropocentric account of nature, the latter is not. Furthermore, it transpires that not all who might call themselves social constructivists of one description or another embrace philosophical or normative relativism, or elide the distinction between representation and that which is represented, although they would hold that there is always an epistemological gap that cannot be totally bridged. However, they seem to be committed to rejecting dualistic thought. But Chapter 5 argues in greater detail that one should not confuse rejecting the human/nonhuman distinction as dualism with rejecting it as ontological dyadism. Rejecting the former is right but rejecting the latter would end with the ontological (as well as the empirical) elimination of nature as 'the Other.'

In the light of the above clarification, interpretations one, two and four lose the challenging bite they seem to have at first sight against any attempt to argue for nonanthropocentrism. Interpretation three does seem to pick up on a theme which has nothing to do with potential philosophical challenge but addresses itself to an important problem of political justice. This interpretation has merits provided it is not understood as holding the indefensible position that any championing of nonanthropocentric concerns is merely and necessarily a disguise for championing anthropocentric ones whether sectionally construed or not.

Notes

1. For another critique, somewhat different from this one, see Holmes Rolston, "Nature For Real: Is Nature a Social Construct?" in *Philosophy of the Environment*, ed. T. D. J. Chappell. See also Searle, *The Construction of Social Reality*.

2. See William Cronon, ed. *Uncommon Ground: Toward Reinventing Nature* (New York and London: W. W. Norton & Company, 1995); Soper, *What Is Nature?: Culture, Politics and the Non-Human*; Soulé and Lease, *Reinventing Nature?*; Steven Vogel, *Against Nature: The Concept of Nature in Critical Theory* (Albany, New York: State University of New York Press, 1996); Arran Gare, *Postmodernism and the Environmental Crisis*.

3. Deconstructionism is current in literary criticism, history and philosophy. It holds that no text (word) has any fixed meaning; each reading may bring forth a new one. However, none of the multiplicity of meanings is definitive.

4. The philosophical significance of the distinction between anthropogenic and nonanthropogenic is discussed in Chapter 5 section entitled **Anthropogenic and Non-anthropogenic**.

5. At first sight, Steven Vogel might be said to have failed to distinguish between these two notions. But on closer reading, it could simply be that he has not so much failed to distinguish the two, as to have used the word 'anthropocentric' to stand for 'anthropogenic' as well. He says:

> These conclusions, which one might well want to call 'anthropocentrist,' seem to me inescapable once one accepts the arguments that ground discourse ethics. But it is important here to distinguish two senses that the word *anthropocentrism* might have. As already suggested, to assert that *value can be determined only by humans* is not to assert that *only humans have value*. Nor is it to assert that all value is determined only relative to human 'interests,' narrowly construed. (*Against Nature*, 164)

The fault, then, with Vogel's account from the standpoint of this book, lies in its failure to invoke the term 'anthropogenic' to refer to his first sense of anthropocentrism, namely, that value can be determined only by humans. However, a more serious disagreement between Vogel and this author lies in that very sense identified—Chapter 5 in the section entitled **Resisting Humean Projectivism** argues that humans are the source only of recognized-articulated values but are neither the source nor loci of mutely-enacted values in nature.

Vogel's account will be examined again under the fourth interpretation, where two senses of the term 'anthropocentrism' will be distinguished which do not coincide with Vogel's own two-fold meanings of the term.

6. For detailed arguments on this and related points, see section entitled **Resisting Humean Projectivism** in Chapter 5.

7. This critique is neutral as to whether all or only some nonhuman beings have (unarticulated) interests and/or preferences. The cut-off point varies—some theorists (like Regan) draw it at the level of mammals or even birds, others (like Singer) possibly at the level of crustaceans. Yet others (like Taylor and Attfield) claim that both animals and plants have interests and/or preferences. Some (like Plumwood) even hold that abiotic entities such as rivers also qualify.

8. The distinction between direct and indirect duties may be illustrated by a story about St. Francis as retold by Passmore (*Man's Responsibility*, 112). Francis had a disciple called Jonathan. One of the brethen who fell ill and was convalescing told Jonathan he fancied pig trotters to whet his appetite. Jonathan, eager to see him regain his health, went out and cut off the trotters from a living pig to cook and serve them up to the invalid as a delicious meal. When Francis came to hear of the episode he reprimanded Jonathan, surprisingly, not for having inflicted cruelty on the pig but for having violated the interests of the pig owner. In other words, even for St. Francis (at least in the light of this particular tale recounted), we have no direct duties to avoid cruelty to pigs; we ought not to be cruel to them because cruelty could lead to their death or injury which amount to inflicting economic loss on their owners. This means that the duty not to harm them is only an indirect duty; the direct duty is not to harm or undermine the interests of the humans who own or value them in some instrumental way.

9. This view is said to be 'feminine' rather than 'feminist'—see Victoria Davion, "Is Ecofeminism Feminist?" in *Ecological Feminism*, Karen J. Warren, ed. (London and New York: Routledge, 1994), where she argues that the ecofeminine perspective uncritically glorifies certain feminine qualities as the way to solve the ecological crisis. In contrast, the ecofeminist perspective (like Plumwood's) argues that there are important philosophical links between the domination of women and the domination of nature, upon which it alone can throw adequate light.

10. Another example is apartheid South Africa. Under that regime, large areas of land were put aside as nature reserves but by expelling tribal peoples from their traditional habitats.

11. For a recent expression of this view, see Ramachandra Guha's and Juan Martinez-Alier's *Varieties of Environmentalism: Essays North and South* (London: Earthscan Publications Ltd., 1997) in which the authors maintain that the notion of wilderness exported to poor economies of the South is but a late twentieth century form of American (Western) cultural imperialism.

12. Carolyn Merchant, *The Death of Nature: Women, Ecology, and the Scientific Revolution* (San Francisco: Harper and Row, 1980).

13. William Leiss, *The Domination of Nature* (Boston: Beacon Press, 1974).

14. William Cronon, "The Trouble with Wilderness; or Getting Back to the Wrong Nature," in *Uncommon Ground*, ed. William Cronon.

15. Cronon's edited volume, *Uncommon Ground*, is a recent instance.

16. Christine Hayles holds a similar view—see "Searching for Common Ground," in *Reinventing Nature?* Soulé and Lease, eds. She calls her position "constrained constructivism."

Jacques Derrida, a key founder of deconstructionism, has denied in an interview that there is no world beyond the text—see Richard Kearney, *Dialogues with Contemporary Continental Thinkers* (Manchester: Manchester University Press, 1984), 123. However, Bruno Latour (another leading light in this postmodernist tendency) thinks it naive to believe that there can be reality beyond the linguistic. However, Elizabeth Ann R. Bird ("The Social Construction of Nature: Theoretical Approaches to the History of Environmental Problems," *Environmental Review* 11 1987: 255-64) holds that Latour's later work (on Louis Pasteur) argues instead that "nature—in the form of Pasteur's 'natural-technical object'—becomes an actor negotiating a *new reality*. In the terms of that context, the microbes became actors in shaping a new environment."

17. Cronon, *Uncommon Ground*, 457.

18. Cronon, *Uncommon Ground*, 458.

19. Cronon, *Uncommon Ground*, 457.

For a clarification of the distinction between matters of epistemology and matters of ontology as well as the epistemic sense of the objective-subjective distinction and the related ontological sense of that distinction, see Searle, *The Construction of Social Reality*, 5-9.

20. Cronon, *Uncommon Ground*, 453. However, in some of her other writings, Merchant does focus on the subjectivity of scientific knowledge, lending substance to the view that nature is essentially unknowable—see her *Radical Ecology: The Search for a Livable World* (London and New York: Routledge, 1992).

21. Cronon, *Uncommon Ground*, 458.

22. Cronon, *Uncommon Ground*, 88.

23. For a clarification of what postmodern thought might mean, see J. Baird Callicott and Fernando J. R. da Rocha, *Earth Summit Ethics*, ix-xi. There they distinguish between two types—eliminative or deconstructive postmodernism (or ultramodernism) and constructive or revisionary postmodernism. The former issues in relativism, even nihilism, by overcoming the modern worldview through an anti-worldview. They, however, advocate the latter, which seeks to surplant the modern worldview with a substantively different new postmodern worldview which is to be contrasted with both the modern and the premodern worldviews. But even this brand of postmodernism, in rejecting the nature/culture dualism of modernity, rejects the very nature/culture distinction itself.

24. While one clear source of social constructivism comes from deconstructionism, another source may be traced to Critical Theory. The former is largely Gallic in inspiration and origin (see Gare, *Postmodernism*); the latter, Germanic/Central European and is largely an off-shoot of Marxism.

25. Nor are they moral agents (who can be held responsible for their behavior). Three different categories of moral beings need to be distinguished:

1. Moral agents are those capable of grasping the difference between good and evil and, therefore, can be held responsible for their conduct. Only humans qualify to be moral agents, although not all humans may be said to be moral agents.
2. Moral subjects are those who can participate in discourse, using language to express a point of view, declare an interest, etc. Only humans can be moral subjects, although again not all humans may be said to be moral subjects. But even those who are not full moral agents may be moral subjects—children are certainly moral subjects. (If chimpanzees could be taught a version of human language, then they, too, would count as moral subjects.)

3. Moral patients are the widest category, as it includes any being which can be said to be morally considerable. Some moral patients are ex moral subjects or ex moral agents—for instance, those humans who are severely brain damaged or more or less permanently comatose. But moral patients include nonhuman beings. According to the argument of this book, these may be either biotic or abiotic entities. As these nonhuman others are morally considerable beings, moral agents *prima facie* have moral duties and responsibilities to them to ensure that their very existence and their mode of existence are not (systematically) subverted or damaged by human conduct.

For a further discussion of the philosophical significance of the distinction between moral agent and moral patient, see Chapter 5 section entitled **Anthropogenic and Non-anthropogenic**.

26. For a detailed, intricate account of these issues, see David Takacs's *The Idea of Biodiversity: Philosophies of Paradise* (Baltimore and London: The Johns Hopkins University Press, 1996). Takacs shows the difficult terrain which conservation biologists are trying to navigate, caught between the methodological demand that their science be objective and value-free on the one hand, and their missionary goal of using that very cognitive authority they as scientists possess, derived from their so-called value-free activity, to advocate the saving of biodiversity on the other. In so doing, he argues that these scientists are also trying to renegotiate the boundaries of what constitutes science so that science as facts and science as advocacy would no longer be as sharply compartmentalized as they are expected to be today.

27. Throughout this book, unless stated otherwise, the term 'anthropocentrism' is used in the sense of 'axiological anthropocentrism' and not 'existential anthropocentrism.'

Bibliography

Agazzi, Evandro. "Nature and the Natural: Some Philosophical Reflections." In *The Concept of Nature in Science and Theology (Part 1)*, edited by Niels H. Gregersen *et al.* Geneva: *Labor Et Fides*, 1995.

Alford, C. Fred. *Science and the Revenge of Nature: Marcuse and Habermas*. Florida: University Presses of Florida, 1985.

Allaby, Michael, and James Lovelock. *The Greening of Mars*. London: Andre Deutsch, 1984.

Allen, Garland E. *Life Science in the Twentieth Century*. Cambridge: Cambridge University Press, 1979.

Amherst, Alicia. *A History of Gardening in England*. London: B. Quaritch, 1896.

Anderson, Ian. "Metal Guru Predicts a Future of Mini Miners." *New Scientist* (8 October 1994): 10.

Aquinas, Thomas. *Summa Contra Gentiles*. Translated by Vernon J. Bourke. Notre Dame, Indiana: Notre Dame University, 1975.

Arendt, Hannah. *The Human Condition*. New York: Doubleday Anchor Books, 1958.

Aristotle. *The Basic Works of Aristotle*. Edited by Richard McKeon. New York: Random House, 1941.

Armbruster, Peter, and Fritz Peter Hessberger. "Making New Elements." *Scientific American* (September 1998): 50-55.

Arnhart, Larry. "Aristotle, Chimpanzees and Other Political Animals." *Biologie et Vie Sociale* 29 (1990): 477-557.

Ashoori, R. C. "Single-Electron Capacitance Spectroscopy of a New Electron Box." *Physica B* 189 (1993): 117-24.

Ashoori, R. C., H. L. Stormer, J. S. Weiner, L. N. Pfeiffer, S. J. Pearton, K. W. Baldwin, and K. W. West. "Single-Electron Capacitance Spectroscopy of Discrete Quantum Levels." *Physical Review Letters* 68 (18 May 1992): 3088-

91.

Ashoori, R. C., H. L. Stormer, J. S. Weiner, L. N. Pfeiffer, K. W. Baldwin, and K. W. West. "Energy Levels of an Artificial Atom Probed with Single-Electron Capacitance Spectroscopy." *Surface Science* 305 (1994): 558-65.

Atkins, F. W. *The Second Law*. New York: Scientific American Books Inc., 1984.

Attfield, Robin. "The Good of Trees." *The Journal of Value Inquiry* 15 (1981): 35-54.

—. *The Ethics of Environmental Concern*. 2d ed. Athens and London: The University of Georgia Press, 1991.

—. "Genetic Engineering: Can Unnatural Kinds Be Wronged?" In *Animal Genetic Engineering: Of Pigs, Oncomice and Men*, edited by Peter Wheale and Ruth McNally. London: Pluto Press, 1995.

Attfield, Robin, and Andrew Belsey. *Philosophy and the Natural Environment*. Cambridge: Cambridge University Press, 1994.

Ayala, Francisco J. "The Difference of Being Human." In *Biology, Ethics, and the Origins of Life*, edited by Holmes Rolston, III. Boston and London: Jones and Bartlett Publishers, 1995.

Ball, Philip. *Made to Measure: New Materials for the 21st Century*. Princeton: Princeton University Press, 1998.

Barnes, Jonathan. "Life and Work." In *The Cambridge Companion to Aristotle*, edited by Jonathan Barnes. Cambridge: Cambridge University Press, 1995.

—. "Metaphysics." In *The Cambridge Companion to Aristotle*, edited by Jonathan Barnes. Cambridge: Cambridge University Press, 1995.

—, ed. *The Cambridge Companion to Aristotle*. Cambridge: Cambridge University Press, 1995.

Barnet, J., and C. Morse. *Scarcity and Growth*. Baltimore: Johns Hopkins University Press, 1963.

Baron, Marcia W., Philip Pettit and Michael Slote. *Three Methods of Ethics: A Debate*. Oxford and Malden, Massachusetts: Blackwell, 1997.

Benton, Ted. "Marxism and Natural Limits: An Ecological Critique and Reconstruction." *New Left Review* 178 (November/December 1989): 51-86.

—. "Ecology, Socialism and the Mastery of Nature: A Reply to Reiner Grundmann." *New Left Review* 194 (July/August 1992): 55-74.

—. "Review of Robyn Eckersley's *Environmentalism and Political Theory: Toward an Ecocentric Approach*." *Environmental Values* 2 (1993): 277-80.

Bergson, Henri. *Creative Evolution*. Authorized translation by Arthur Mitchell. London: Macmillan and Co. Limited, 1911.

Birch, Paul. "Terraforming Venus Quickly." *Journal of the British Interplanetary Society* 44 (1991): 157-67.

—. "Terraforming Mars Quickly." *Journal of the British Interplanetary Society* 45 (1992): 331-40.

—. "The Customised Cosmos." *Guardian* (20 May 1993): 12.

Birch, Thomas H. "Moral Considerability and Moral Universal Consideration."

Environmental Ethics 15 (1993): 313-32.

Bird, Elizabeth Ann R. "The Social Construction of Nature: Theoretical Approaches to the History of Environmental Problems." *Environmental Review* 11 (1987): 255-64.

Borgmann, Albert. "The Nature of Reality and the Reality of Nature." In *Reinventing Nature? Responses to Postmodern Deconstruction*, edited by Michael E. Soulé and Gary Lease. Washington, D.C.: Island Press, 1995.

Bowler, Peter J. *The Mendelian Revolution: The Emergence of Hereditarian Concepts in Modern Science and Society*. London: The Athlone Press, 1989.

Brennan, Andrew. "The Moral Standing of Natural Objects." *Environmental Ethics* 6 (1984): 35-56.

Brooks, Michael. "Take One Quantum Dot." *New Scientist* (29 August 1998): 22-25.

Bud, Robert. *The Uses of Life: A History of Biotechnology*. Cambridge: Cambridge University Press, 1993.

Bunge, Mario. "Philosophical Inputs and Outputs of Technology." In *The History and Philosophy of Technology*, edited by George Bugliarello and Dean B. Doner. Urbana: University of Illinois Press, 1979.

—. "Toward a Philosophy of Technology." In *Philosophy and Technology: Readings in the Philosophical Problems of Technology*, edited by Carl Mitcham and Robert Mackey. New York and London: The Free Press, 1983.

Callicott, J. Baird. "Animal Liberation: A Triangular Affair." *Environmental Ethics* 2 (1980): 311-28.

—. "On the Intrinsic Value of Nonhuman Species." In *The Preservation of Species*, edited by Bryan G. Norton. Princeton: Princeton University Press, 1986.

—. "The Wilderness Idea Revisited: The Sustainable Development Alternative." *Environmental Professional* 13 (1991): 235-47.

—. "La Nature Est Morte, Vive La Nature!" *Hastings Center Report* 22 (1992): 16-23.

—. "Rolston on Intrinsic Value: A Deconstruction." *Environmental Ethics* 14 (1992): 129-43.

—. "The Role of Technology in the Evolving Concept of Nature." In *Ethics and Environmental Policy: Theory Meets Practice*, edited by Frederick Ferré and Peter Hartel. Athens: The University of Georgia Press, 1994.

—. *Earth's Insights: A Survey of Ecological Ethics from the Mediterranean Basin to the Australian Outback*. Berkeley and London: University of California Press, 1994.

—. "The Value of Ecosystem Health." *Environmental Values* 4 (1995): 345-61.

—. "Benevolent Symbiosis: The Philosophy of Conservation Reconsidered." In *Earth Summit Ethics*, edited by J. Baird Callicott and Fernando J. R. da Rocha. Albany: State University of New York Press, 1996.

Callicott, J. Baird, and Karen Mumford. "Ecological Sustainability as a

Conservation Concept." *Conservation Biology* 11 (February 1997): 32-40.

Callicott, J. Baird, and Fernando J. R. da Rocha. *Earth Summit Ethics: Toward a Reconstructive Postmodern Philosophy of Environmental Education*. Albany: State University of New York Press, 1996.

Capra, Fritjof. *The Turning Point*. New York: Simon and Schuster, 1982.

Carruthers, Peter. *The Animals Issue: Moral Theory in Practice*. Cambridge: Cambridge University Press, 1992.

Caspar, Max. *Kepler*. Edited and translated by C. Doris Hellman. London and New York: Aberlard-Schuman, 1959.

Chappell, T. D. J., ed. *The Philosophy of the Environment*. Edinburgh: Edinburgh University, 1997.

Cherfas, Jeremy. *Man Made Life: A Genetic Engineering Primer*. Oxford: Basil Blackwell, 1982.

Cloning Issues in Reproduction, Science and Medicine. London: Human Genetics Advisory Commission and Human Fertilisation & Embryology Authority, January 1998.

Coghlan, Andy. "Clotted Milk." *New Scientist* (27 September 1997): 10.

—. "Sensitive Flowers." *New Scientist* (26 September 1998): 24-28.

Collingwood, R. G. *The Idea of Nature*. Oxford: Clarendon Press, 1945.

Commoner, Barry. *The Closing Circle: Nature, Man & Technology*. New York: Alfred A. Knopf, Inc., 1971.

Comte, Auguste. *Cours de Philosophie Positive*, 6 vols. Paris, 1830-1842.

—. *The Essential Comte*, edited by S. Andreski. London: Croom Helm, 1974.

Concar, David. "The Organ Factory of the Future?" *New Scientist* (18 June 1994): 24-29.

Costanza, Robert, Bryan Norton and Ben Hassell. *Ecosystem Health: New Goals for Environmental Management*. Washington, D.C.: Island Press, 1992.

Crick, Francis H. C. *Of Molecules and Men*. Seattle and London: University of Washington Press, 1966.

Cronon, William. "The Trouble with Wilderness; or, Getting Back to the Wrong Nature." In *Uncommon Ground: Toward Reinventing Nature*, edited by William Cronon. New York and London: W. W. Norton & Company, 1995.

Cullen, James. *A Systems Approach to Environmental Values: Systems Process and the Bifurcation of Nature*. MA diss., University of Lancaster, 1994.

Davies, Hunter. *George Stephenson*. Middlesex: Hamlyn Paperbacks, 1980.

Davion, Victoria. "Is Ecofeminism Feminist?" In *Ecological Feminism*, edited by Karen J. Warren. London and New York: Routledge, 1994.

Dawkins, Richard. *The Selfish Gene*. Oxford: Oxford University Press, 1976.

—. *The Extended Phenotype*. San Francisco: Freeman, 1982.

DeGrazia, David. *Taking Animals Seriously: Mental Life and Moral Status*. Cambridge and New York: Cambridge University Press, 1996.

Dennett, Daniel. "Conditions of Personhood." In *Identities of Persons*, edited by A. Rorty. Berkeley: University of California Press, 1976.

—. *The Intentional Stance*. Cambridge, Massachusetts: MIT Press, 1987.

Derrida, Jacques. *Dissemination*. London: Athlone Press, 1981.

Descartes, René. "Discourse on the Method." In *The Philosophical Writings of Descartes*, translated by John Cottingham, Robert Stoothoff and Dugald Murdoch, vol. 1. Cambridge: Cambridge University Press, 1992.

Devall, Bill, and George Sessions. *Deep Ecology: Living as if Nature Mattered*. Salt Lake City, Utah: Peregrine Smith Books, 1985.

Dickens, Peter. *Reconstructing Nature: Alienation, Emancipation and the Division of Labour*. London and New York: Routledge, 1996.

Dijksterhuis, E. J. *The Mechanization of the World Picture*. Oxford: The Clarendon Press, 1961.

Doeleman, J. A. "Environment and Technology: Speculating on the Long Run." In *Technology and the Environment* (*Research in Philosophy and Technology Series, Volume 12*), edited by Frederick Ferré. Connecticut and London: JAI Press Inc., 1992.

Donnelly, Strachan, Charles R. McCarthy, and Rivers Singleton Jr. "The Brave New World of Animal Biotechnology." *Hastings Centre Report* 24 (February-January 1994): S1-S31.

Drake, Stillman. *Discoveries and Opinions of Galileo*. New York: Doubleday Anchor Books, 1957.

—. *Galileo*. Oxford: Oxford University Press, 1980.

Drake, Stillman, and C. D. O'Malley, trans. *The Controversy on the Comets of 1618*. Philadelphia: University of Pennsylvania Press, 1960.

Drexler, K. Eric. "Exploring Future Technologies." In *Doing Science: The Reality Club*, edited by John Brockman. New York: Prentice-Hall, 1991.

—. *Foresight Survey*. California: The Foresight Institute, 1991.

—. *Engines of Creation: The Coming Era of Nanotechnology*. New York: Doubleday/Oxford: Oxford University Press, 1992.

—. *Nanosystems: Molecular Machinery, Manufacturing and Computation*. New York and Chichester: John Wiley and Sons, 1992.

Drexler, K. Eric, Chris Petersen and Gayle Pergamit. *Unbounding the Future*. London: Simon and Schuster, 1992.

Dreyfus, Hubert L. "Heidegger on Gaining a Free Relation to Technology." In *Technology and Values*, edited by Kristin Shrader-Frechette and Laura Westra. Lanham, Maryland, and Oxford: Rowman & Littlefield Publishers, Inc., 1997.

Dryzek, John. *Rational Ecology*. Oxford: Blackwell, 1987.

D'Silva, Joyce. "A Critical View of the Genetic Engineering of Farm Animals." In *Animal Genetic Engineering: Of Pigs, Oncomice and Men*, edited by Peter Wheale and Ruth McNally. London: Pluto Press, 1995.

Easterbrook, Gregg. *A Moment on the Earth: The Coming Age of Environmental Optimism*. London: Penguin Books Ltd., 1996.

Eckersley, Robyn. *Environmentalism and Political Theory: Toward an Ecocentric*

Approach. London: University College Press, 1992.
The Economist 313 (9 December 1989): 113-14. "The Invisible Factory."
Elliot, Robert. "Faking Nature." *Inquiry* 25 (1982): 81-93.
—. "The Value of Wild Nature." *Inquiry* 26 (1983): 359-61.
—. "Environmental Degradation, Vandalism and the Aesthetic Object Argument." *Australasian Journal of Philosophy* 67 (1989): 191-204.
—. "Intrinsic Value, Environmental Obligation and Naturalness." *Monist* 75 (1992): 138-59.
—. "Ecology and the Ethics of Environmental Restoration." In *Philosophy and the Natural Environment*, edited by Robin Attfield and Andrew Belsey. Cambridge: Cambridge University Press, 1994.
—. "Extinction, Restoration, Naturalness." *Environmental Ethics* 16 (1994): 135-44.
—. *Faking Nature: The Ethics of Environmental Restoration*. London: Routledge, 1997.
Evans, Edward P. *The Criminal Prosecution and Capital Punishment of Animals: The Lost History of Europe's Animal Trials*. London: William Heinemann, 1906.
Evernden, Neil. *The Social Creation of Nature*. Baltimore and London: The Johns Hopkins University Press, 1992.
Feenberg, Andrew. *Critical Theory of Technology*. Oxford and New York: Oxford University Press, 1991.
Feinberg, Joel. "The Rights of Animals and Unborn Generations." In *Philosophy and Environmental Crisis*, edited by William T. Blackstone. Athens: University of Georgia Press, 1974.
Ferkiss, Victor. *Nature, Technology and Society: Cultural Roots of the Current Environmental Crisis*. New York and London: New York University Press, 1993.
Ferré, Frederick. *Philosophy of Technology*. Englewood Cliffs, New Jersey: Prentice Hall, 1988.
—, ed. *Technology and the Environment* (*Research in Philosophy and Technology Series, Volume 12*) Connecticut and London: JAI Press Inc., 1992.
—. *Being and Value: Toward a Constructive Postmodern Metaphysics*. Albany: State University of New York Press, 1996.
Ferry, Luc. *The New Ecological Order*. Chicago and London: Chicago University Press, 1995.
Feynman, Richard. "There's Plenty of Room at the Bottom" was published in a shorter version as "The Wonders That Await a Micro-micro-scope." *Saturday Review* 43 (2 April 1960): 45-47; reproduced in a longer version under the original title in *Miniaturization*, edited by H. D. Gilbert. New York: Reinhold, 1961.
Fichte, Johann Gottlieb. *The Vocation of Man*. London: John Chapman, 1848.
Fisher, Anthony C., and Frederick M. Peterson. "The Environment in Economics:

A Survey." *Journal of Economic Literature* 14, (March 1976): 1-33.

Fox, Michael. "Transgenic Animals: Ethical and Animal Welfare Concerns." In The Bio-Revolution: Cornucopia or Pandora's Box? edited by Peter Wheale and Ruth McNally. London: Pluto Press, 1990.

Fox, Warwick. *Toward a Transpersonal Ecology: Developing New Foundations for Environmentalism*. Boston: Shambhala, 1990.

Frankel, Charles. "The Rights of Nature." In *When Values Conflict: Essays on Environmental Analysis, Discourse, and Decision*, edited by Laurence H. Tribe, Corinne S. Schelling and John Voss. Cambridge, Massachusetts: Ballinger Publishing Company, 1976.

Frey, Raymond G. "Rights, Interests, Desires and Beliefs." *American Philosophical Quarterly* 16 (1979): 233-39.

—. *Interests and Rights: The Case Against Animals*. Oxford: The Clarendon Press, 1980.

—. "The Significance of Agency and Marginal Cases." *Philosophica* 39 (1987): 39-46.

Galileo, Galilei. "The Assayer." In *The Controversy on the Comets of 1618*, edited by Stillman Drake and C. D. O'Malley. Philadelphia: University of Philadelphia Press, 1960.

Gare, Arran E. *Postmodernism and the Environmental Crisis*. London and New York: Routledge, 1995.

Geras, Norman. *Marx and Human Nature: Refutation of a Legend*. London: Verso Books, 1983.

Gibbs, W. Wyat, and Corey S. Powell. "Bugs in the Data? The Controversy over Martian Life Is Just Beginning." *Scientific American* 275 (October 1996): 12-13.

Goodin, Robert. *Green Political Theory*. Oxford: Blackwell, Polity Press, 1992.

Goodpaster, K. E., and K. M. Sayre, eds. *Ethics and Problems of the 21st Century*. Notre Dame and London: University of Notre Dame Press, 1979.

Gotthelf, Allan. "Aristotle's Conception of Final Causality" In *Philosophical Issues in Aristotle's Biology*, edited by Allan Gotthelf and James G. Lennox. Cambridge and New York: Cambridge University Press, 1987.

Gotthelf, Allan, and James G. Lennox, eds. *Philosophical Issues in Aristotle's Biology*. Cambridge and New York: Cambridge University Press, 1987.

Gould, Stephen Jay. "The Evolution of Life on the Earth." *Scientific American* 271 (1994): 62-69.

Grant, Edward. *The Foundations of Modern Science*. Cambridge and New York: Cambridge University Press, 1997.

Guha, Ramachandra, and Juan Martinez-Alier. *Varieties of Environmentalism: Essays North and South*. London: Earthscan Publications Ltd., 1997.

Halfpenny, Peter. *Positivism and Sociology: Explaining Social Life*. London: Allen and Unwin, 1982.

Haldane, John. "Admiring the High Mountains: The Aesthetics of Environment."

Environmental Values 3 (1994): 97-106.
Hall, Nina. "Heavy Metal Mystery." *Observer* (12 February 1995), Life Section: 67.
Hankinson, R. J. "Philosophy of Science." In *The Cambridge Companion to Aristotle*, edited by Jonathan Barnes. Cambridge: Cambridge University Press, 1995.
—. "Science." In *The Cambridge Companion to Aristotle*, edited by Jonathan Barnes. Cambridge: Cambridge University Press, 1995.
Hannah, Lee, David Lohse, Charles Hutchinson, John L. Carr, and Ali Lankerani. "A Preliminary Inventory of Human Disturbance of World Ecosystems." *Ambio* 23 (1993): 246-50.
Hare, R. M. *Freedom and Reason*. Oxford: The Clarendon Press, 1963.
Hargrove, Eugene C., ed. *Beyond Spaceship Earth: Environmental Ethics and the Solar System*. San Francisco: Sierra Club Books, 1986.
—. *Foundations of Environmental Ethics*. Englewood Cliffs, New Jersey: Prentice Hall, 1989.
—. "Weak Anthropocentric Intrinsic Value." *Monist* 75 (1992): 183-207.
Harlan, Jack R. *Crops and Man*. Madison, Wisconsin: American Society of Agronomy, 1975.
Harvey, H. J. "The National Trust and Nature Conservation: Prospects for the Future." [Personal Communication in August 1994.]
Harvey, John. *Medieval Gardens*. London: B. T. Batsford, Ltd., 1981.
Hawkes, Nigel. "Greening the Red Planet." *Geographical Magazine* (February 1993): 20-23.
Hayles, N. Katherine. "Searching for Common Ground." In *Reinventing Nature? Responses to Postmodern Deconstruction*, edited by Michael E. Soulé and Gary Lease. Washington, D.C.: Island Press, 1995.
Heath, M. J. "Terraforming: A Down to Earth Perspective." *Terra Nova* 3 (1991): 655-58.
Hegel, G. W. F. *The Philosophy of Right*. London: Oxford University Press, 1942.
Heidegger, Martin. "The Question Concerning Technology." In *The Question Concerning Technology and Other Essays*. New York: Harper and Row Ltd., 1982.
Hilpinen, Rosto. "Belief Systems as Artifacts." *Monist* 28 (1995): 136-55.
Hobbes, Thomas. *Leviathan*. New York: Collier Books, 1962.
Ihde, Don. *Instrumental Realism: The Interface between Philosophy of Science and Philosophy of Technology*. Bloomington: Indiana University Press, 1991.
—. *Philosophy of Technology: An Introduction*. New York: Paragon House, 1993.
Jamieson, Dale. "Ecosystem Health: Some Preventive Medicine." In *Environmental Values* 4 (1995): 333-44.
Jantsch, Erich. *The Self-Organizing Universe: Scientific and Human Implications of the Emerging Paradigm of Evolution*. Oxford: Pergamon Press, 1980.
Johnson, Lawrence E. *A Morally Deep World*. Cambridge: The University of

Cambridge Press, 1991.

Jonas, Hans. *The Phenomenon of Life: Toward a Philosophical Biology*. New York: Harper and Row, 1966.

Journal of the British Interplanetary Society 43. "Human Exploration of Mars." (1990).

—. "Terraforming." 44 (1991).

—. "Terraforming." 45 (1992).

Judson, H. F. *The Eighth Day of Creation: Makers of the Revolution in Biology*. New York and London: Simon and Schuster Inc., 1979.

Kant, Immanuel. *Lectures on Ethics*. New York and Evanston: Harper and Row, Publishers, 1963.

Kargel, Jeffrey S., and Robert G. Strom. "Global Climatic Change in Mars." *Scientific American* 275 (November 1996): 60-68.

Katz, Eric. "The Big Lie: Human Restoration of Nature." In *Technology and the Environment* (*Research in Philosophy and Technology* Series,Vol. 12), edited by Frederick Ferré. Connecticut and London: JAI Press Inc., 1992.

—. "The Call of the Wild: The Struggle Against Domination and the Technological Fix of Nature." *Environmental Ethics* 14 (1992): 265-73.

—. "Artefacts and Functions: A Note on the Value of Nature." *Environmental Values* 2 (1993): 223-32.

—. *Nature as Subject: Human Obligation and Natural Community*. Lanham, Maryland: Rowman and Littlefield Publishers, Inc., 1997.

Kearney, Richard. *Dialogues with Contemporary Continental Thinkers*. Manchester: Manchester University Press, 1984.

Kevles, Daniel J., and Leroy Hood, eds. *The Code of Codes: Scientific and Social Issues in the Human Genome Project*. Cambridge, Massachusetts: Harvard University Press, 1992.

Kirk, G. S., and J. E. Raven. *The PreSocratic Philosophers*. Cambridge: Cambridge University Press, 1977.

Kloppenburg, Jack Ralph, Jr. *First the Seed: The Political Economy of Plant Technology, 1492-2000*. Cambridge and New York: Cambridge University Press, 1990.

Köhler, W. *The Mentality of Apes*. London: Routledge and Kegan Paul, 1925.

Kolakowski, Lezek. *Positivist Philosophy*. UK: Penguin Books, 1972.

Kolata, Gina. *Clone: The Road to Dolly and the Path Ahead*. New York and London: Penguin Books Ltd., 1997.

Krimsky, Sheldon. *Genetic Alchemy: The Social History of the Recombinant DNA Controversy*. Cambridge, Massachusetts and London: The MIT Press, 1982.

—. *Biotechnics and Society: The Rise of Industrial Genetics*. New York and London: Praeger, 1991.

Kuhn, Thomas S. *The Structure of Scientific Revolutions*. Chicago: Chicago University Press, 1962.

Kullman, Wolfgang. "Different Concepts of the Final Cause in Aristotle." In

Aristotle on Nature and Living Things, edited by Allan Gotthelf. Pennsylvania and Bristol: Mathesis Publications Inc. and Bristol Classical Press, 1985.

Latour, Bruno. *We Have Never Been Modern*. Cambridge: Harvard University Press, 1993.

Leclercq, Jean. "Vers La Société Basée Sur Le Travail." *Revue du Travail* 51 (1950): 3.

Lee, Keekok. *A New Basis for Moral Philosophy*. London: Routledge, 1985.

—. *Social Philosophy and Ecological Scarcity*. London: Routledge, 1989.

—. *The Positivist Science of Law*. Hants: Gower Publishing Company Limited, 1989.

—. *The Legal-Rational State: A Comparison of Hobbes, Bentham and Kelsen*. Hants: Gower Publishing Company Limited, 1990.

—. "Instrumentalism and the Last Person Argument." *Environmental Ethics* 15 (1993): 333-44.

—. "Awe and Humility: Intrinsic Value in Nature. Beyond an Earthbound Environmental Ethics." In *Philosophy and the Natural Environment*, edited by Robin Attfield and Andrew Belsey. Cambridge: Cambridge University Press, 1994. (Royal Institute of Philosophy Supplement: 36.)

—. "Beauty For Ever?" *Environmental Values* 4 (1995): 213-25.

—. "The Source and Locus of Intrinsic Values—A Re-examination." *Environmental Ethics* 18 (1996): 297-309.

—. "Designer Mountains: The Ethics of Nanotechnology." *Terra Nova* 2 (1997): 127-34.

—. "Biodiversity." In *Encyclopedia of Applied Ethics, Volume 1*, edited by Ruth Chadwick. San Diego: Academic Press, 1998.

Leggett, Jeremy K., ed. *Global Warming: The Greenpeace Report*. Oxford: Oxford University Press, 1990.

Leiss, William. *Under Technology's Thumb*. Montreal and London: McGill-Queen's University Press, 1990.

—. *The Domination of Nature*. Boston: Beacon Press, 1974.

Lewis, C. S. *Studies in Words*. 2d ed. Cambridge: Cambridge University Press, 1967.

Lewontin, R. "Agricultural Research and the Penetration of Capital." *Science for the People* 14 (1982): 12-17.

Lloyd, G. E. R. "Empirical Research in Aristotle's Biology." In *Philosophical Issues in Aristotle's Biology*, edited by Allan Gotthelf and James G. Lennox. Cambridge and New York: Cambridge University Press, 1987.

—. "Greek Antiquity: The Invention of Nature." In *The Concept of Nature: The Herbert Spencer Lectures*, edited by John Torrance. Oxford: Clarendon Press, 1992.

Locke, John. *Two Treatises of Government*. Edited by Peter Laslett. Cambridge: Cambridge University Press, 1988.

Lovelock, James. *Gaia: A New Look at Life on Earth*. Oxford and New York:

Oxford University Press, 1982.

—. *The Ages of Gaia*. Oxford: Oxford University Press, 1991.

Mackie, John L. *Ethics: Inventing Right and Wrong*. Harmondsworth, Middlesex: Penguin Books Ltd., 1977.

Maclachlan, James. "Drake v. the Philosophers." In *Nature, Experiment, and the Sciences: Essays on Galileo and the History of Science in Honour of Stillman Drake*, edited by Trevor H. Levere and William R. Shea. (*Boston Studies in the Philosophy of Science*, Vol. 120.) Dordrecht/Boston/London: Kluwer Academic Publishers, 1990.

McKay, Christopher P., Owen B Toon, and James F. Kasting. "Making Mars Habitable." *Nature* 352 (8 August 1991): 489-96.

McKibben, Bill. *The End of Nature*. New York: Random House, 1989.

McKie, Robert. "Russia and America Plan Joint 2001 Mission to Drill for Life-forms on Mars: Roving Robot May Solve Mystery of Thirsty Planet." *Observer* (24 February 1994): 3.

—. "Medicine Man at the $100m Biotech Temple." *Observer* (9 October 1994), Business Section: 6.

McLaughlin, Andrew. *Regarding Nature: Industrialism and Deep Ecology*. New York: State University of New York Press, 1993.

McNally, Ruth. "Mad Dogs or Jackasses: The European Eradication Programme." In *Animal Genetic Engineering: Of Pigs, Oncomice and Men*, edited by Peter Wheale and Ruth McNally. London: Pluto Press, 1995.

McNeilly, F. S. *The Anatomy of Leviathan*. London: Macmillan, 1968.

Magner, Lois N. *A History of the Life Sciences*. New York and Basel: Marcel Dekker, Inc., 1979.

Manzini, Ezio. *The Material of Invention*. Cambridge, Massachusetts: The MIT Press, 1989.

Marcuse, Herbert. *One Dimensional Man: Studies in the Ideology of Advanced Industrial Society*. Boston: Beacon Press, 1964.

Marietta, Don E. *For People and The Planet: Holism and Humanism in Environmental Ethics*. Philadelphia: Temple University Press, 1994.

Mars Alive. Text adapted from the programme, *Mars Alive*, transmitted by BBC Horizon on 8 February 1993 on BBC2.

Martin, C. B., and Karl Pfeifer. "Intentionality and the Non-Psychological." *Philosophy and Phenomenological Research* 46 (1986): 536-56.

Marx, Karl, and Friedrich Engels. *Collected Works*, Vol. 5. London and Moscow: Lawrence and Wishart and Progress Publishers, 1976.

Mathews, Freya. *The Ecological Self*. London: Routledge, 1991.

Maturana, Humberto R., and Francisco J. Varela. *Autopoiesis and Cognition: The Organization of the Living*. Dordrecht/Boston: D. Reidel Publishing Company, 1980.

—. *The Tree of Knowledge: The Biological Roots of Human Understanding*. Boston: Shambhala, 1988.

Mayer, Sue. "Environmental Threats of Transgenic Technology." In *Animal Genetic Engineering: Of Pigs, Oncomice and Men*, edited by Peter Wheale and Ruth McNally. London: Pluto Press, 1995.

Mayr, Ernst. *Animal Species and Evolution*. Cambridge, Massachusetts: Belknap Press of Harvard University Press, 1963.

—. *The Growth of Biological Thought*. Cambridge, Massachusetts: Belknap Press of Harvard University Press, 1982.

—. *Toward a New Philosophy of Biology*. Cambridge, Massachusetts, and London: Belknap Press of Harvard University Press, 1988.

Merchant, Carolyn. *The Death of Nature: Women, Ecology, and the Scientific Revolution*. San Francisco: Harper and Row, 1980.

—. *Radical Ecology: The Search for a Livable World*. London and New York: Routledge, 1992.

Meyer-Abich, K. M. *Revolution for Nature: From the Environment to the Connatural World*. Cambridge, UK, and Denton, USA: The White Horse Press, 1993.

Milbrath, Lester. "Fears and Hopes of an Environmentalist for Nanotechnology." In *Nanotechnology: Research and Perspective*, edited by B. C. Crandall and James Lewis. Cambridge, Massachusetts: MIT Press, 1992.

Mill, John Stuart. "Nature." In *John Stuart Mill: Autobiography and Other Writings*. Edited by Jack Stillinger. Boston: Houghton Mifflin Company, 1969.

Mitcham, Carl. "Philosophy and the History of Technology." In *The History and Philosophy of Technology*, edited by George Bugliarello and Dean B. Doner. Urbana and London: University of Illinois Press, 1979.

—. "Three Ways of Being With Technology." In *From Artifact to Habitat: Studies in the Critical Engagement of Technology* (*Research in Technology Series, Volume 3*), edited by Gayle L. Ormiston. London and Toronto: Associated University Press, 1990.

—. *Thinking Through Technology: The Path between Engineering and Philosophy*. Chicago: Chicago University Press, 1994.

—, ed. *Social and Philosophical Constructions of Technology* (*Research in Philosophy and Technology Series, Volume 15*). Greenwich, Connecticut and London: JAI Press Inc., 1995.

Mitcham, Carl, and Robert Mackey, eds. *Philosophy and Technology: Readings in the Philosophical Problems of Technology*. New York and London: The Free Press, 1983.

Mumford, Lewis. *Technics and Civilization*. London: George Routledge and Sons, Ltd., 1946.

—. *The Myth of the Machine: Technics and Human Development*. London: Secker and Warburg, 1967.

Munro, D. H. *Empiricism and Ethics*. Cambridge: Cambridge University Press, 1970.

Naess, Arne. "The Shallow and the Deep, Long-Range Ecology Movement: A Summary." *Inquiry* 16 (1973): 95-100.

—. "Identification as a Source of Deep Ecological Attitudes." In *Deep Ecology*, edited by M. Tobias. San Diego: Avant Books, 1985.

—. *Ecology, Community and Lifestyle: Outline of an Ecosophy*. Translated by David Rothenburg. Cambridge: Cambridge University Press, 1988.

—. "Self-realization: An Ecological Approach to Being in the World." In *Thinking Like a Mountain*, edited by John Seed, Joanna Macy, Pat Fleming and Arne Naess. Philadelphia: New Society Publishers, 1988.

—. "Man Apart and Deep Ecology: A Reply to Reed." *Environmental Ethics* 12 (1990): 185-92.

Nature. "No 'Nanofossils' in Martian Meteorite." (4 December 1997): 454.

New Scientist. "Martians Have Landed Again." (9 November 1996): 3.

The 1992 Environmental Almanac. Washington: World Resources Institute, 1992.

Norton, Bryan. "Environmental Ethics and Weak Anthropocentrism." *Environmental Ethics* 6 (1984): 131-48.

—. *Why Preserve Natural Variety?* Princeton: Princeton University Press, 1987.

—. *Toward Unity Among Environmentalists*. New York and Oxford: Oxford University Press, 1992.

—. "Sustainability, Human Welfare and Ecosystem Health." *Environmental Values* 1 (1992): 97-111.

—. "Epistemology and Environmental Values." *Monist* 75 (1992): 208-26.

—. "Where Do We Go from Here?" In *Ethics and Environmental Policy: Theory Meets Practice*, edited by Frederick Ferré and Peter Hartel. Athens: The University of Georgia Press, 1994.

Nussbaum, M. *The Fragility of Goodness*. Cambridge: Cambridge University Press, 1986.

Olby, Robert C. *Origins of Mendelism*. London: Constable, 1966.

—. *The Path to the Double Helix*. London: Macmillan, 1974.

—. "The Revolution in Biology." In *Companion to the History of Modern Science*, edited by Robert C. Olby, Geoffrey Cantor, John Christie and Jonanthon Hodge. London and New York: Routledge, 1990.

—. "The Emergence of Genetics." In *Companion to the History of Modern Science*, edited by Robert C. Olby, Geoffrey Cantor, John Christie and Jonathon Hodge. London and New York: Routledge, 1990.

Olby, Robert C., Geoffrey Cantor, John Christie and Jonathon Hodge, eds. *Companion to the History of Modern Science*. London and New York: Routledge, 1990.

Ormiston, Gayle L., ed. *Studies in the Critical Engagement of Technology* (*Research in Technology Series*, Vol. 3). Bethlehem: Lehigh University Press, 1990; London and Toronto: Associated University Press, 1990.

Pacey, Arnold. *Technology in World Civilization: A Thousand Year History*. Cambridge, Massachusetts: The MIT Press, 1990.

Pain, Stephanie. "The Intraterrestrials." *New Scientist* (7 March 1998): 28-32.

Passmore, John. *Man's Responsibility for Nature*. 2d. ed. London: Duckworth, 1980.

Pilkington, Edward. "Fur Starts to Fly as Police Arrest Bull Terrier Which Killed a Cat." *Guardian* (21 August 1995): 1.

Plato. *The Republic*, Vol. 1, Books I-V. Translated by Paul Shorey. Cambridge, Massachusetts, and London: Harvard University Press and William Heinemann, 1982.

Plumwood, Val. *Feminism and the Mastery of Nature*. London: Routledge, 1993.

Popper, Karl. *Conjectures and Refutations*. London: Routledge and Kegan Paul, 1969.

Preus, Anthony. *Science and Philosophy in Aristotle's Biological Works*. New York: Georg Olms Verlag Hildesheim, 1975.

Prigogine, Ilya, and Isabelle Stenger. *Order Out of Chaos: Man's New Dialogue with Nature*. New York: Bantam, 1984.

Rackham, Oliver. *Trees and Woodlands in the British Landscape: The Complete History of Britain's Trees, Woods and Hedgerows*. London: Weidenfeld and Nicolson, 1995.

Radford, Tim. "Futurology: A Pastime with a Rosy Future?" *Guardian* (11 May 1995), Online Section: 8-9.

—. "A Clean Pair of Wheels." *Guardian* (28 November 1996), Online Section: 5.

—. "Star Wars Scientists Find Water in Crater on the Moon." *Guardian* (4 December 1996): 1.

Rapport, David J. "What Constitutes Ecosystem Health?" *Perspectives in Biology and Medicine* 33 (1989): 120-32.

—. "Ecosystem Health: More than a Metaphor?" *Environmental Values* 4 (1995): 287-309.

Reed, Peter. "Man Apart: An Alternative to the Self-Realization Approach." *Environmental Ethics* 11 (1989): 53-69.

Regan, Tom. "The Nature and Possibility of an Environmental Ethic." *Environmetal Ethics* 3 (1981): 19-34.

Regis, Edward. *Nano!* London and New York: Bantam Press, 1995.

Report to the Intergovernmental Panel on Climatic Change. Bracknell, UK: Meteorological Office, 1990.

Rifkin, Jeremy. *The Biotech Century*. London: Gollancz, 1998.

Rindos, David. *The Origins of Agriculture: An Evolutionary Perspective*. Orlando and London: Academic Press, Inc., 1984.

Robbins, Jim. "Do Not Feed the Bears." *Natural History* (January 1984).

Roberts, H. F. *Plant Hybridization Before Mendel*. Princeton: Princeton University Press, 1929.

Robinson, Kim Stanley. *Red Mars*. New York: Bantam Books, 1993.

—. *Green Mars*. New York: Bantam Books, 1994.

—. *Blue Mars*. New York: Bantam Books, 1996.

Rolston, III, Holmes. "The Preservation of Natural Value in the Solar System." In *Beyond Spaceship Earth: Environmental Ethics and the Solar System*, edited by Eugene C. Hargrove. San Francisco: Sierra Club Books, 1986.

—. *Environmental Ethics: Duties to and Values in the Natural World*. Philadelphia, Temple University Press, 1988.

—. "The Wilderness Idea Reaffirmed." *Environmental Professional* 13 (1991): 370-77.

—. *Conserving Natural Values*. New York: Columbia University Press, 1994.

—. "Nature for Real: Is Nature a Social Construct?" In *The Philosophy of the Environment*, edited by T. D. J. Chappell. Edinburgh: Edinburgh University, 1997.

Rothenberg, David. *Hand's End: Technology and the Limits of Nature*. Berkeley, Los Angeles and London: University of California Press, 1993.

Routley, Richard. "Is There a Need for a New, an Environmental Ethic?" In *Proceedings of XVth World Congress of Philosophy*. Bucharest: World Congress of Philosophy, 1973.

Routley, Richard, and Val Routley. "Against the Inevitability of Human Chauvinism." In *Ethics and Problems of the 21st Century*, edited by K. E. Goodpaster and K. M. Sayre. Notre Dame and London: University of Notre Dame Press, 1979.

Routley, Val. "Critical Notice of John Passmore's *Man's Responsibility for Nature*." *Australasian Journal of Philosophy* 53 (1975): 171-85.

Roy, D. J., B. E. Wynne, and R. W. Old, eds. *Bioscience⇌Society*. Chichester and New York: John Wiley and Sons, 1991.

Ryder, Richard. "Animal Genetic Engineering and Human Progress." In *Animal Genetic Engineering: Of Pigs, Oncomice and Men*, edited by Peter Wheale and Ruth McNally, 1-16. London: Pluto Press, 1995.

Sagan, Carl. "The Search for Extraterrestrial Life." Scientific American 271 (1994): 70-77.

Sagan, Dorian, and Lynn Margulis. "Gaia and Philosophy." In *On Nature*, edited by Leroy S. Rouner. Notre Dame, Indiana: University of Notre Dame Press, 1984.

Sagoff, Mark. "Process or Product? Ethical Priorities in Environmental Management." *Environmental Ethics* 8 (1986): 121-38.

—. "Zuckerman's Dilemma: A Plea for Environmental Ethics." *Hastings Center Report* 21 (September-October 1991): 32-40.

Scheffer, V. B. "The Olympic Goat Controversy: A Perspective." *Conservation Biology* 7 (1993): 196-99.

Science. "Engineering a Small World: From Atomic Manipulation to Microfabrication." 254 (1991): 1300-304.

Scitovsky, Tibor. *The Joyless Economy: An Inquiry into Human Satisfaction and Consumer Dissatisfaction*. London: Oxford University Press, 1976.

Searle, John. "What is an Intentional State?" *Mind* 87 (1979): 74-92.

—. "Intentionality and the Use of Language." In *Meaning and Use*, edited by A. Margulit. Dordrecht, Holland: D. Reidel Publishing Co., 1979.

—. *The Construction of Social Reality*. London: Allen Lane, The Penguin Press, 1995.

Shapere, Dudley. *Galileo: A Philosophical Study*. Chicago and London: Chicago University Press, 1974.

Shapiro, K. J. "The Death of the Animal: Ontological Vulnerability." *Between the Species* 5 (1989): 183-94.

Shiva, Vandana. *Staying Alive: Women, Ecology and Development*. London: Zed Press, 1990.

Shrader-Frechette, Kristin, and Laura Westra, eds. *Technology and Values*. Lanham, Maryland, and Oxford: Rowman and Littlefield Publishers, Inc., 1997.

Simmonds, Norman W. *Principles of Crop Improvement*. New York: Longman, 1979.

Singer, Peter. "Not for Humans Only: The Place of Nonhumans in Environmental Issues." In *Ethics and Problems of the 21st Century*, edited by K. E. Goodpaster and K. M. Sayre. Notre Dame and London: University of Notre Dame Press, 1979.

Slote, Michael. "Virtue Ethics." In *Three Methods of Ethics: A Debate*, by Marcia W. Baron, Philip Pettit and Michael Slote. Oxford and Malden, Massachussetts: Blackwell, 1997.

Sober, Elliott, and Richard Lewontin. "Artifact, Cause and Genic Selection." *Philosophy of Science* 49 (1982): 157-80.

Solter, Davor. "Dolly *Is* a Clone — and No Longer Alone." *Nature* 394 (23 July 1998): 315-16.

Soper, Kate. *What Is Nature? Culture, Politics and the Non-Human.* Oxford, UK, and Cambridge, USA: Blackwell, 1995.

Soulé, Michael E. "The Social Siege of Nature." In *Reinventing Nature? Response to Postmodern Deconstruction*, edited by Michael E. Soulé and Gary Lease. Washington D.C. and California: Island Press, 1995.

Soulé, Michael E., and Gary Lease, eds. *Reinventing Nature? Response to Postmodern Deconstruction*. Washington D.C. and California: Island Press, 1995.

Sprigge, Timothy. "Respect for the Non-Human." In *The Philosophy of the Envionment*, edited by T. D. J. Chappell. Edinburgh: Edinburgh University Press, 1997.

Staudenmaier, John M., S. J. *Technology's Storytellers: Reweaving the Human Fabric*. Massachusetts and London: The Society for the History of Technology and the MIT Press, 1985.

Stevenson, Peter. "Patenting of Transgenic Animals: A Welfare/Rights Perspective." In *Animal Genetic Engineering: Of Pigs, Oncomice and Men*, edited by Peter Wheale and Ruth McNally, 156-61. London: Pluto Press,

1995.

Stone, Christopher D. *Earth and Other Ethics: The Case for Moral Pluralism*. New York: Harper and Row Publishers, 1988.

Sylvan, Richard. *In Defence of Deep Environmental Ethics: Holding the Temple Against Local Depradation*. Canberra: Philosophy Department, Research School of Social Sciences, Australian National University, 1990.

—. *Against the Main Stream: Critical Environmental Essays*. Canberra: Department of Philosophy and Law, Research School of Social Sciences, Australian National University, 1994.

Sylvan, Richard, and David Bennett. *Greening Ethics*. Isle of Harris: The White Horse Press, 1994.

Sylvester, E. J., and L. C. Klotz. *The Gene Age: Genetic Engineering and the Next Industrial Revolution*. New York: Charles Scribner's Sons, 1983.

Takacs, David. *The Idea of Biodiversity: Philosophies of Paradise*. Baltimore and London: The Johns Hopkins University Press, 1996.

Taylor, Paul. *Respect for Nature*. Princeton: Princeton University Press, 1986.

Tiles, Mary, and Hans Oberdiek. *Living in a Technological Culture: Human Tools and Human Values*. London and New York: Routledge, 1995.

Torrance, John, ed. *The Concept of Nature: The Herbert Spencer Lectures*. Oxford: Clarendon Press, 1992.

Tylecote, Andrew. "Ecology, Technology and the Next Long Wave Upswing." In *Environment, Technology and Economic Growth: The Challenge to Sustainable Development*, edited by Andrew Tylecote and Jan van der Straaten. Cheltenham, UK, and Northampton, MA, USA: Edward Elgar, 1997.

Varela, F. J. *Principles of Biological Autonomy*. New York and Oxford: North Holland, 1979.

Verhoog, H. "The Concept of Intrinsic Value and Transgenic Animals." *Journal of Agricultural and Environmental Ethics* 5 (1992): 147-60.

Vidal, John. "Eco Soundings." *Guardian* (8 February 1995), Society Section: 4.

Vidal, John, and John Carvel. "Lambs to the Gene Market." *Guardian* (13 November 1994): 25.

Vilkka, Leena. *The Intrinsic Value of Nature*. Amsterdam and Atlanta, Georgia: Editions Rodopi B.V., 1997.

Vogel, Steven. *Against Nature: The Concept of Nature in Critical Theory*. Albany: State University of New York Press, 1996.

Weinberg, Steven. "Life in the Universe." *Scientific American* 271 (1994): 22-27.

Weisheipl, James A., O. P. "Aristotle's Concept of Nature: Avicenna and Aquinas." In *Approaches to Nature in the Middle Ages*, edited by Lawrence D. Roberts. New York: State University of New York at Binghamton, 1982.

Westfall, Richard S. "The Scientific Revolution of the Seventeenth Century: The Construction of a New World View." In *The Concept of Nature: The Herbert Spencer Lectures*, edited by John Torrance. Oxford: Clarendon Press, 1992.

Wheale, Peter, and Ruth McNally. *Genetic Engineering: Catastrophe or Utopia?* Brighton and New York: Harvester, Wheatsheaf and St. Martin's Press, 1988.

—. eds. The Bio-Revolution: Cornucopia or Pandora's Box? London: Pluto Press, 1990.

—. eds. *Animal Genetic Engineering: Of Pigs, Oncomice and Men*. London: Pluto Press, 1995.

Wick, Warner. "Aristotelianism." In *The Encyclopedia of Philosophy*, vol. 1, edited by Paul Edwards. London: Collier-Macmillan, 1967.

Wilkerson, T. E. *Natural Kinds*. Aldershot: Avebury Press, 1995.

Williams, Raymond. "Ideas of Nature." In *Problems in Materialism and Culture*. London: Verso, 1980.

Wilmut, I., A. E. S. Schnieke, J. McWhir, A. J. Kind, and K. H. S. Campbell. "Viable Offspring Derived from Fetal and Adult Mammalian Cells." *Nature* (27 February 1997): 810-13.

Wilson, E. O. Address to the American Association for the Advancement of Science. As reported in *Guardian* (18 February 1995): 8.

—. *The Diversity of Life*. London: Penguin Books, 1992.

Wilson, E. O., and F. M. Peter, eds. *Biodiversity*. Washington, D.C.: National Academy Press, 1988.

Woolley, Benjamin. *Virtual Worlds: A Journey in Hype and Hyperreality*. Oxford, UK, and Cambridge, USA: Blackwell, 1992.

Wright, H. E., and D. G. Frey, eds. *The Quaternary of the United States*. Princeton: Princeton University Press, 1965.

Wright, Nick. "Lucy, the Cat Killer, Released." *Guardian* (24 August 1995): 7.

Xenos, Nicolas. *Scarcity and Modernity*. London: Routledge, 1989.

Yoxen, Edward. *The Gene Business: Who Should Control Biotechnology?* London: Pan Books, 1983.

Zimmerman, Michael. *Heidegger's Confrontation with Modernity: Technology, Politics, Art*. Bloomington: Indiana University Press, 1990.

Zubrin, Robert, and Richard Wagner. *The Case for Mars: The Plan to Settle the Red Planet and Why We Must*. New York: Simon and Schuster, 1996.

Index

aesthetics, Hegelian, 120-21; objective, 121-122, 211n38; subjective, 120-22, 171, 179, 211n38; Romanticism and, 224. *See also* nature, as work of art

Agazzi, Evandro, 82, 98n4

Alford, C. Fred, 130, 155n40

Allaby, Michael, and James Lovelock, 90, 104n25

Allen, Garland E., 51, 71, 76n9, 80n46

Amherst, Alicia, 93, 105n31

Anaxagoras, 134, 156n40

anthropocentric, 51, 75n6, 111-12, 116, 130, 150n18, 155n40, 171-72, 177, 185-187, 195, 206n20, 207n25, 210 n35, 225, 233, 235-237, 242-249. *See also* anthropocentrism, nonanthropocentric

anthropocentrism, 5, 22, 32- 33, 39- 40, 48n78, 119, 122, 128, 130, 132, 136, 139-40, 145, 156n54, 157n69, 162-63, 172, 181, 185-187, 195, 200, 215nn56-57, 216-17n68, 224, 230-32, 238n10, 242-44, 247-48, 249, 249-50n5, 252n27; aggressive, 32, 39-40, 48n78, 185-187, 215n58; axiological, 195, 217n 75, 248, 252 n27; existential, 195, 217n75, 248, 252n27; less aggressive (passive), 32, 39-40, 46n56, 145, 157n69, 185-86, 215n58. *See also* human chauvinsim, nonanthropocentrism

anthropogenic, 51, 75n6, 82, 86- 87, 89-90, 98n3, 100-103n13, 103n23, 103-4n24,104n25 160-61, 165, 170-172, 175, 177, 186, 194-196, 199, 200, 206n20, 210n35, 213n46, 227, 231, 237, 242-43, 246-248, 249-50nn4-5, 252n25; philosophical significance of the distinction between nonanthropogenic and, 194-201, 206n20. *See also* nonanthropogenic

Aquinas, Thomas, 11-13, 24, 32, 34, 35, 47n58; Thomism and, 11-13, 42n6

Arendt, Hannah, 125, 152-53nn35-36

Aristotelian(s), 8, 12-16, 19-23, 26-28, 31-32, 42n9, 44-45n36, 124, 152n34, 224, 227; Aristotelianism and, 11, 12, 16, 19, 26, 41n2, 42n9, 145. *See also* Aristotle

Aristotle, 2, 4, 5, 11-14, 16, 19-23, 26, 32-40, 42nn7-8, 43n25, 45n37, 46 nn55-56, 47n60, 47nn63-66, 48nn71-73, 48nn74-75, 48n6, 50- 51, 53, 74n5, 83, 85, 117, 124, 134, 145,

152n34, 156n50, 157n69, 163-64, 185, 201, 215n58, 219n89, 224, 226; biology/philosophy of biology and, 12, 42n8; relationship between external and intrinsic/immanent teleology and, 164.
Aristotle's four causes, 2, 5, 14, 19-20, 27, 31-32, 36-37, 40, 48n71, 50, 82, 145, 224, 226-27; efficient, 5, 14, 19, 27, 36-37, 40, 48n71, 50, 53-54, 82, 85, 94-95, 145, 161, 224, 226-27; final, 5,14, 19, 27, 31-32, 33, 35, 36-38, 40, 46n55, 48n71, 50, 53-54, 82, 85, 95, 145, 224, 226-27, 236; formal, 5, 14, 19, 27, 31-36, 40, 48n71, 50, 53-54, 82, 85, 91, 95, 145, 224, 226-27; material, 5, 9n2, 14, 36, 40, 50, 52-54, 82, 85, 96, 145, 224, 226-27. *See also* artefacts
Arnhart, Larry, 33, 47n60
artefact(s), 2, 4, 6-7, 9n2, 21, 36-39, 41n2, 49-56, 59, 73-74nn1-3, 74-75n6, 82-86, 90-92, 94-97, 102-3n16, 110, 112, 114, 118-19,121-22, 126, 128, 130-31, 133-48, 151n25, 155n46, 155-56n50, 166, 179-80, 182, 191, 194, 197, 200-201, 205n10, 214n47, 216nn64-65, 217n69, 219n89, 223, 226-27, 229-32, 236-39, 247; Aristotle's four causes and abiotic/non-biotic, 49-54, 74-75n6, 86, 96, 226-27, 230, 236-37; definition of, 47-48n70. *See also* Aristotle's four causes, biotic artefacts, dyadism (ontological), material embodiment of human intentionality, ontogical categories, transforming the natural to become the artefactual
artefacticity (degree/level of), 2-3, 6, 49, 52-54, 84-86, 91, 93-96, 97-98n3, 112, 118, 121, 134, 223, 226-28. *See also* deep(er) control/manipulation of nature, deep(er) scientific theories and discoveries, deep(er) technologies
artefactual, 1-9, 36, 37, 47-48n70, 49-51, 53-54, 56-57, 73-74nn1-3,74-75n6, 81-87, 90, 93, 96, 97n1, 99-100n7, 102-3n16, 104n25, 107, 113-115, 118, 119, 121, 129, 131-32, 134-35, 138-40, 146, 148, 159-60, 173, 178, 180-82, 184, 188-93, 195, 201-3, 211n38, 212n41, 216nn64-65, 221n93, 223-29, 231, 237, 247. *See also* the natural and the artefactual, ontological categories (distinctions)
artefactual kinds, 4, 83, 84, 93, 99n7, 118, 173, 188, 224-25. *See also* natural kinds, naturenk, transforming the natural to become the artefactual
'artificial,' two different senses of, 50-51. *See also* artefactual
Ashoori, R. C., 83, 100n7
Asymmetry/Autonomy/No External Teleology Theses, 173-76, 194, 208n28. *See also* nature, abiotic; valuable 'by itself;' independent value.
Atkins, F. W., 58, 78n26
Attenborough, David, 170, 206n17
Attfield, Robin, 166, 167, 172, 173, 179, 204-5n9, 207n24, 211n38, 230, 238n6, 243, 250n7
autopoiesis, 140-41, 143-44, 146-48, 157nn59-60, 157-58nn63-73. *See also* machines, autopoietic
axiological, 2, 7-9, 73-74n3, 81, 114, 159-60, 179-80, 192-94, 200, 205n10, 212n39, 217n70, 226, 228, 248, 252n27. *See also* secondary characteristics
axiological similarities and ontological difference, 8, 160, 181, 190-94. *See also* axiology and ontology, primary characteristic, secondary characteristics
axiology and ontology, 159-60, 181, 189-94, 213n45, 217n60. *See also* axiology-led environmental ethic, ontology-led environmental philosophy
axiology-led environmental ethic, 190-91. *See also* ontology-led environmental philosophy, secondary characteristics
Ayala, Francisco J., 197, 218, n82

Bacon, Francis, 11, 29-31, 64, 41n2,44
Ball, Philip, 96, 105n33
Barnes, Jonathan, 12, 23, 42n7, 45n37
Barnet, J., and C. Morse, 117, 150-51nn21-22
Benton, Ted, 119, 122, 151n24, 152n30, 232-37, 238nn11-12, 239n18
Bergson, Henri, 125, 133-135, 152n35, 155n45, 155nn48-49. *See also homo faber*
bios praktikos, 124
bios theoretikos, 5, 40, 48n78, 124-25, 145, 157n69. *See also* anthropocentrism, less aggressive (passive); Aristotle
biotechnology, 1-4, 93n, 31, 33, 47n62, 51-54, 68, 74, 76n9n12, 81, 93, 107-23, 129, 133-34, 138-40, 145, 148, 1494n, 156n50, 157n58, 158n74, 159, 173, 188, 216n64, 223, 226; definition of, 139; radical threat to biotic nature and, 112-14. *See also* nature-replacing technology, molecular (DNA) genetics, valuable 'for itself,' valuable 'by itself'
biotic artefact(s), 51-52, 73-74n3, 86, 95-96, 110, 112, 133, 136-37, 145, 148, 166, 179, 182, 191, 213-14n47, 223, 227, 230, 237. *See also* artefacticity, degree of; biotechnology; language of machines; Mendelian genetics, hybridization technology and; transforming organisms to become (biotic) machines; valuable 'for itself;' valuable 'by itself'
Birch, Paul, 90, 104n25. *See also* terraformation
Birch, Thomas H., 173, 208n27
Bird, Elizabeth Ann R., 246, 251n16
Borgmann, Albert, 93, 105n28
Bowler, Peter J., 52. 75, 76n8
Brennan, Andrew, 73n1, 173, 208n27
Buber, Martin, and Rudolph Otto, 214n48
Bud, Robert, 51, 74n4, 139, 157n57
Bunge, Mario, 55, 64-67, 77n18, 79nn37-38, 79n40, 79n42
Callicott, J. Baird, 7, 82, 84, 86, 87, 97-98n3, 100n8, 101-2n13, 163, 164, 175, 179, 184-189, 204n5n7, 209-10n33, 211n38, 215n55, 215n57, 215nn61-62, 225, 231, 238n10, 246, 251n23
Callicott, J. Baird, and Karen Mumford, 189, 216n67
Callicott, J. Baird, and Fernando J. R. da Rocha, 188, 215n62, 216n67, 246, 251n23
Capra, Fritjof, 140-41, 157n62
Carnot, Sadi, 58
Carruthers, Peter, 166, 205n10
Caspar, Max, 15, 42-43nn14-15
Cherfas, Jeremy, 52, 76n9, 113, 150n9
Collingwood, R. G., 13, 18, 34, 36, 42nn10-11, 43n22, 47n64
Commoner, Barry, 107, 108, 149n1
Comte, Auguste, 22-29, 44n35, 45nn40-42, 46n50
conservationism and preservationism 232. *See also* human chauvinism, Sole Value Assumption and; instrumentalism; resource conservation; resource preservation
Costanza, Robert, 188, 215n62
Crick, Francis H. C., 52, 76n9
critical school of technology assessment, 115-16
Cronon, William, 241, 245, 246, 249n2, 250nn14-15, 251nn17-22
Cullen, James, 182, 214n49
culture (human)/nature, 4, 82, 87, 98n4, 102n13, 213-14n47; 216n67, 246, 251n23. *See also* Deep Ecology, dualism, dyadism (ontological)

D'Silva, Joyce, 150n10, 179, 211n38
Davion, Victoria, 244, 250n9
Dawkins, Richard, 71, 80n45
deconstructionism, 102n13, 105n28, 241-42, 246-47, 249n3, 251n16n23. *See also* postmodernism
deep(er) control/manipulation of nature, 31, 225. *See also* deep(er) scientific theories and discoveries, goals of the new science, deep(er) technologies

Deep Ecology, 146, 180-81, 185, 212n42, 214n48, 215n52

deep(er) scientific theories and discoveries, 1, 2, 6, 9, 51, 53, 56, 58, 62, 66-67, 69-71, 73n2, 75n7, 83, 85-86, 99n7, 112, 179, 181, 201, 203, 221n93, 223. *See also* deep(er) control/manipulation of nature, deep-(er) technologies

deep (er) technologies, 6, 9n3, 51, 53, 56, 62, 66-67, 69, 85-86, 112, 118, 179, 181, 201, 221n93, 223. *See also* deep(er) control/manipulation of nature, deep(er) scientific theories and discoveries

deep value, 181, 237. *See also* independent value, trajectory and independent value

DeGrazia, David, 166, 197, 205n11, 218n83

Dennett, Daniel, 169

Derrida, Jacques, 246, 251n16

Descartes, René, 5, 11, 22, 28, 29, 31, 33, 41n2, 46n51, 64, 161, 163, 185, 203-4n4, 224. *See also* dualism, human consciousness, modern science

Diamond v. Chakrabarty, 137. *See also* patenting of modified organisms

Dijksterhuis, E. J., 23, 44- 45n36, 45n39

DNA, 52, 53, 69, 76n9n12, 79n44, 110, 134, 137, 139, 150n8, 156n50. *See also* biotechnology, molecular (DNA) genetics

Doeleman, J. A., 109, 110, 112, 149n6

Dolly the cloned sheep, 110, 149-50n8. *See also* deep(er) control/ manipulation of nature, deep(er) scientific theories and discoveries, deep(er) technologies

Drake, Stillman, 12, 15-17, 41- 42n4, 43nn17-19

Drexler, K. Eric, 52, 76n10, 115, 116, 133, 136, 137-39, 150n16, 150n20, 155n47, 156n53, 223, 228n1. *See also* nanotechnology (molecular)

Dryzek, John, 108, 149n3

dualism, 4, 7, 11, 41n2, 82, 84, 97-98n3, 100n8, 160, 180-182, 184, 189, 202, 212n43, 214n49, 215n52, 217n68, 224-226, 246-248, 251n23; hyperseparation and, 97n3, 100n8, 181, 213n44. *See also* dyadism (ontological); Plumwood, Val

duties: distinction between direct and indirect, 250n8; direct, 32, 21n848, 243-44, 250n8; indirect, 215n56, 243-44, 250n8. *See also* anthropocentrism

dyadism (ontological), 4, 7, 82, 84, 98n4, 100n8, 160, 180-182, 184, 202, 214n48, 226, 247-48. *See also* dualism, hyperseparation and; independent value; nature, as 'the (ontological) Other'

Earthbound environmental philosophy, 9, 173-77

Easterbrook , Gregg, 3, 9-10n4, 149n5, 195, 199-201, 217n76, 218n87. *See also* ecorealism

Eckersley, Robyn, 124, 140, 141, 146, 148, 152n32, 157n59, 232, 233, 235, 238n11, 239

ecocentrism, 86, 101n13, 166, 172, 185, 204n8, 215n57

ecofeminine and ecofeminist, distinction between, 244, 250n9

ecorealism, 3, 9-10n4, 109, 149n5, 199, 200. *See also* Easterbrook, Gregg; anthropogenic and nonanthropogenic, philosophical significance of the distinction between

ecosystem health, 160, 184, 186-189, 215n62, 216n63, 216n65, 216n67, 225. *See also* primary characteristic, secondary characteristics

egomania, 202-3, 220-21nn90-91. *See also* narcissism

Elliot, Robert, 73-74n3, 178, 179, 211n37, 211-12n38, 229, 237-38n2

enframing, 233; standing reserve and, 131, 233, 235. *See also* Heidegger, Martin

Engels, Friedrich, 184, 215n52. *See also* Marx, Karl

environmental ethics, 2, 160, 172, 189,

190, 218n84, 229
environmental philosophy, 2-4, 6-9, 57, 73n2, 107, 112, 114, 124, 140-41, 148, 152n32, 159-161, 163, 164, 173, 178, 179, 181, 184, 187, 189-191, 193, 195, 212n39, 217n68, 218n84, 223, 225, 227
environmental ethics and environmental philosophy, distinction between, 189-93, 213n45, 217n70. *See also* axiology-led environmental ethic, ontology-led environmental philosophy
episteme, 14, 19-20, 123. *See also techne*
epistemological goals (respective) of science and technology, 6, 65-68
Ereky, Karl, 139
ethics: consequentialist, 201-2, 219-20n89; deontic, 201-2, 219-20n89; virtue (the aretaic), 201, 202, 219-20n89, 220-21n91. *See also* narcissism
Evans, Edward P., 182, 197, 213n46, 218n83
Evernden, Neil, 214n48
exbiotic, 50-51, 75n6, 84, 134, 142, 155n49, 178, 211n36, 226. *See also* nature, abiotic
extant technology, 3, 7, 107-8, 114, 116, 119, 151n25, 159, 232, 235-36, 239nn16-17. *See also* future technology, history of technology
external teleology, 5, 33, 39-40, 144-45, 157n69, 163-64, 173-74, 176, 194; simplified, 39-40, 144. *See also* anthropocentrism, less aggressive (passive); Aristotle; extrinsic/imposed teleology; intrinsic/immanent teleology; teleology, new; teleology, old
extrinsic/imposed teleology, 2, 5-6, 36, 39-40, 47n59, 47-48n70, 49, 54, 85, 94-95, 145, 147, 205n10. *See also* Aristotle's four causes; artefact(s); teleology, new; teleology, old

fabrication, 2, 4, 7, 56, 76n10, 124-129, 132-135, 145, 148, 152n35, 155n49, 189, 193, 194, 203. *See also homo faber*; humanization of nature
Feinberg, Joel, 166, 167, 204n9
Ferré, Frederick, 11, 41n1, 56, 77n22, 82, 97n2, 109, 188
Ferry, Luc, 182, 197, 213n46, 218n83
Feuerbach, Ludwig, 127, 184
Feynman, Richard, 52, 76n10
Fichte, Johann Gottlieb, 127-29, 153n40
Fisher, Anthony C., and Frederick M. Peterson, 91, 105nn26-27
Fox, Michael, 148, 158n74
Fox, Warwick, 140, 141, 146-48, 157n59
Frankel, Charles, 82, 98n4
Frey, Raymond G., 73n2, 87, 166, 205n10
future technology, 1, 3, 7, 9n3, 52, 76n10, 81, 97n1, 100n7, 107, 112, 114, 121, 129, 138-39, 149n4, 150n8, 151n25, 155n46, 156n53, 159, 173, 179, 218n87, 221n93, 223; radical threat to the natural and, 114-23. *See also* biotechnology, radical threat to biotic nature and; microcomputer technology; extant technology, nanotechnology (molecular); terraformation

Galileo, Galilei, 5, 11-20, 22, 25, 27, 33, 41n2,41-42n4, 43nn17-19, 59, 60, 64, 130, 161, 224. *See also* modern science
garden, 93-95, 105n31, 121-22, 125; of Eden, 116-17, 119, 153n36, 244
Gare, Arran E., 140, 157n59, 189, 216n67, 241, 247, 249n2, 251n24
genetics. *See* Mendelian genetics, molecular (DNA) genetics
germ-line therapy, 72, 133
Geras, Norman, 125, 153n36, 155n44
goals of the new science, 2, 5, 22, 28-33, 39-40, 46n49, 49, 56, 62-63, 70, 73n3, 85, 123-25, 134, 166, 181, 203, 225, 233; freedom and self-realization and, 7, 56, 73, 77n15, 125-26, 152n35, 225; material well-

being and, 2, 56, 62, 132, 225, 235, 238n12. *See also* deep(er) control/ manipulation of nature, deep(er) scientific theories and discoveries, deep(er) technologies
Goodin, Robert, 229, 238n2
Gotthelf, Allan, 12, 32, 42n8, 46n55
Gould, Stephen Jay, 90, 103n24
Grant, Edward, 11, 41n2, 56, 77n22
Guardian, 90, 104n25, 140, 172, 182, 188
Guha, Ramachandran, and Juan Martinez-Alier, 244, 250n11

Habermas, J., 130, 155n40, 233, 235, 247
Halfpenny, Peter, 22, 44n34
Hankinson, R. J., 12, 42n8
Hannah, Lee, 87, 102-3n16
Hare, R. M., 190, 217n71
Hargrove, Eugene, 173, 208n27, 231, 233, 236, 238n9
Harlan, Jack R., 52, 75n8
Hayles, N. Katherine, 246, 250n16
Hegel, G. W. F., 120-21, 126-128, 153n39
Heidegger, Martin, 43-44n32, 63-64, 78n33, 79n36, 131, 233, 235. *See also* modern science, as Theoretical Technology (Heideggerian inversion)
Hilpinen, Rosso, 37, 47-48n70, 86, 102n14
history of science, 15, 43n20, 43-44n32, 57, 60, 83, 99n7; philosophy of science and, 64, 223
history of technology, 6, 43-44nn27-32, 47n69, 49, 54-60, 136; Phase IA and B, 57- 60, 62, 64-65, 67-69, 71, 77-78nn24-25, 136; Phase IIA and B, 22, 43-44n32, 57, 60-62, 64-65, 67-72, 136, 224. *See also* history of science; modern technology; technology, 'found;' technology, premodern
Hobbes, Thomas, 11, 12, 19, 22, 23-31, 33, 41n2, 43n16, 44nn33-34, 45 nn52-53. *See also* positivism
homo faber, 2, 4, 7-8, 11, 41n2, 54, 56, 77n16, 123, 125-35, 148, 152-53n36, 160, 181, 189, 193-94, 197, 200-203, 218n82, 225. *See also* artefactual, humanization of nature, narcissism
homo ludens, 7, 125, 134
human chauvinism, 33, 46n51, 47n61, 163, 184, 198, 218n84; Greater Value Assumption and, 29, 46n51, 185-86; Sole Value Assumption and, 29, 46n51, 166, 185, 187, 215n57. *See also* anthropocentrism
human consciousness, 8, 128, 130, 134-35, 161-166, 170, 177, 185, 193, 196, 201, 203-4n4, 206n17, 210n35, 215n57, 217-18n78, 224, 228, 230-31, 234, 239, 242-43, 247; uniqueness of, 164, 196, 21718n78. *See also* Humean projectivism, sole locus of intrinsic value, sole source of all values
human design, 5, 9, 50-51, 53, 81, 91, 96, 98n4, 130, 138, 155-56n50, 180, 211n38, 231. *See also* Aristotle's four causes, artefacts
human intentional structure , 37, 82, 85, 118, 122, 126, 181, 201, 229. *See also* artefact, material embodiment of human intentionality
human intention(s), 8, 9, 38-39, 47-48n70, 56, 83-84, 86, 90, 92-93, 135, 140, 146-48, 168, 178, 191, 208n30, 211n38, 224, 229, 247
human interests, conflict with nonhuman interests and, 46n51, 129, 153-55n40, 185-87, 241; the social construction of nature and, 241-249. *See also* human chauvinsim, Greater Value Assumption and
human intervention, 50-51, 53, 75-76n8, 76nn9-10, 76n12, 82, 86- 87, 91-95, 98n4, 101-2n13, 102-3n16, 171, 191, 197, 199. *See also* anthropogenic, $nature_{fa}$, $nature_{hi}$
human labor, 7, 126, 128, 130-32. *See also* labor theory of value
humanization of nature, 1, 7, 97n3, 118, 123, 128-132, 134, 138, 189, 201-2,

227. *See also homo faber*; narcissism, naturalization of humans, transforming the natural to become the artefactual
human/nonhuman, 4, 82, 98n4, 160, 184, 225, 246-47. *See also* culture (human)/nature, Deep Ecology, dualism, dyadism (ontological)
Hume, David, 5, 18, 23, 43n23, 45n39, 128, 161-162, 203n2, 203-4n4, 224, 234, 239. *See also* Humean projectivism, means/end rationality, sole locus of intrinsic value, sole source of all values
Humean/Cartesian theses, 128, 164
Humean projectivism, 4, 8, 119, 123, 136, 151n26, 156n54, 161-66, 174, 177, 196, 210n35, 213n46, 224, 230, 231, 238n10, 239, 243, 249-50n5. *See also* human consciousness, Hume, sole locus of intrinsic value, sole source of all values
hybridization technology. *See* Mendelian genetics

Idealism, 127, 129, 132, 155n40. *See also* Fichte, Hegal, Marx
Ihde, Don, 57, 64, 65, 77n25, 78 79n35
independent and autonomous (of humans), 93, 119, 141, 146, 175, 177-79, 198, 226, 228, 229, 237. *See also* Kant, Immanuel, autonomy and; independent value
independent value , 2, 6, 8-9, 36, 81-83, 86, 90-93, 105n29, 118, 141, 146, 160, 164, 166, 168, 171, 175, 177-179, 189, 193, 198, 203n1, 204n8, 205n10, 206n19, 208n30, 210n35, 217n69, 226, 228-31. *See also* axiology and ontology, deep value, ontology-led environmental philosophy, primary characteristic, trajectory and independent value
industrial civilization, 1, 9, 108, 118, 126, 142, 237, 239. *See also* goals of the new science, history of technology
instrumentalism, 39-40, 65, 85, 119-20, 122, 140, 145, 148, 159, 181, 185, 187, 194, 213n44, 224, 231-233, 235, 238-39n12, 243, 250n8; Instrumental Value Assumption and, 185-86, 194, 215n59. *See also* anthropocentrism, conservationism and preservationism, instrumental value, intrinsic value
instrumental rationality, 233, 235-36. *See also* Habermas, J., means/end rationality
instrumental value, 2, 22, 29, 32, 39, 40, 46n51, 124, 128, 130, 135, 139-40, 145, 159, 161, 163-64, 167, 172, 174, 176, 185, 186, 194, 227, 230, 232, 233, 235, 239, 243, 250n8. *See also* intrinsic value
intentionality, 37-39, 166, 169, 197, 201, 203, 225
intentional stance, 38, 48n73, 161, 164, 166-72, 177-78, 204n6, 205-6n12. *See also* Plumwood, Val
interests and intrinsic value, 4, 29, 38, 48n73, 141, 161, 164, 166-168, 178, 193, 204n6, 205n10, 242
intrinsic/immanent teleology, 2, 5, 33, 36, 39-40, 85, 145-46, 157n69, 164, 169, 173-74, 205n10. *See also* external teleology; extrinsic/imposed teleology; teleology, new; teleology, old; *telos/tele*; valuable 'for itself'
intrinsic value, 3-4, 8, 22, 29, 32, 39, 40, 46n51, 111-12, 119, 123-24, 128, 130, 136, 140-41, 145, 148, 156n54, 158n74, 160-61, 163-168, 181, 185, 187, 203-4n4, 212n40, 224, 238n10, 231, 243; abiotic nature and, 161, 164, 166, 167, 172-77, 204n7, 205n10, 207-08n25; secondary characteristics and, 176-79, 190-92, 212n39, 228; two senses of, 8, 164-66. *See also* independent value, primary characteristic, valuable 'for itself,' valuable 'in itself'

Jantsch, Erich, 140, 157n60
Janzen, Daniel, 113
Johnson, Lawrence E., 166, 172, 173,

204n8, 208n26
Jonas, Hans, 63, 64, 78n34
Judson, H. F., 52, 76n9

Kant, Immanuel, 32, 33, 185, 46n57, 84, 215n56; Kantian, 63, 186, 193, 209n32, 219n89; autonomy and, 93, 209n32
Katz, Eric, 50, 73n2, 178, 188, 211n37, 216n66, 238n2
Kepler, Johannes, 5, 15, 42-43nn14-15. *See also* modern science
Keynes, John Maynard, 59
Kloppenburg, Jack Ralph, Jr., 52, 76 nn8-9
Kolakowski, Lezek, 22, 44n34
Kuhn, Thomas S., 51, 79n39
Kullman, Wolfgang, 32, 46n55
Krimsky, Sheldon, 52, 76n9, 110, 113, 115, 149n7, 150n10, 150nn16-18.

labor theory of value, 129, 130-32. *See also* Locke, John; Marx, Karl
language of machines, 7, 11, 41n2, 135, 141. *See also* Maturana and Varela
Latour, Bruno, 246, 251n16
Lee, Keekok, 11, 18, 22, 23, 29-31, 41n2, 43nn23-24, 44nn33-35, 45n39, 46nn52-53, 53, 77n12, 87, 108, 121, 126, 130, 149n2, 152n28, 153n37, 155n41, 163, 164, 184, 204n5, 215n52, 229, 231, 233, 237n1, 238-39n12, 239nn20-21
Leggett, Jeremy, 87, 103n18
Leiss, William, 119, 151n25, 203, 221-22n94, 245, 250nn13-14
Lewis, C. S., 84- 85, 115, 100n11
Lloyd, G. E. R., 12, 42n8, 82, 97n2
Locke, John, 12, 17, 129-132, 155nn41-42, 224. *See also* Scientific Naturalism (metaphysics of)
logic, extensional, 168, 205-6n12; intensional, 168-69, 205-6n12
Lovelock, James, 82, 90, 98n3, 104n25

machine(s), 11, 41n2, 52, 55, 56, 58, 60, 93, 134-140, 142-144, 146-47, 155-56n50, 156n52; 203; 221-22; allopoietic, 143-44, 146-47; autopoietic, 144, 146-47, 179; (MV) machines and, 147-48, 157-58nn63-73; transforming organisms to become, 11, 138-48, 227. *See also* language of machines, Maturana and Varela
Maclachlan, James, 15, 43n17
McKibben, Bill, 6, 86-93, 100-101n13, 102n15, 103nn19-22. *See also* nature, end of
McLaughlin, Andrew, 130, 155n40
McNally, Ruth, 113, 150n10
McNeilly, F. S., 23, 45n39
Magner, Lois N. A., 52, 76n8
Manzini, Ezio, 96, 105n33
Marcuse, Herbert, 129, 130, 153-55n40
Marietta, Don E., 177, 210n35
Mars, 3, 85, 90, 104n25, 172, 176, 180, 206-07n22. *See also* Asymmetry/Autonomy/No External Teleology Theses; independent value; intrinsic value, abiotic nature and; solar system; terraformation; trajectory and independent value; valuable 'by itself;' Venus
Marx, Karl, 31, 124-127, 129-134, 152n30, 152n32, 152-53n36, 155 n40; Engels and, 184, 215n51, 215 nn53-54; Marxism and, 77n23, 129, 151n24, 152-53n36, 244-45, 251n24
material embodiment of human intentionality, 2, 47-48n70, 118, 166, 195, 197, 201, 205n10, 221n93, 225. *See also* artefact, human intentional structure, the natural and the arteactual, ontological categories (distinctions)
materialism (ontology of), 5, 11, 22, 25-27, 29-30, 33, 36, 39-41, 41n2, 65, 70, 127, 129, 132. *See also* mechanism (philosophyof), Scientific Naturalism (metaphysics of)
Mathews, Freya, 140, 180, 157n59, 212n42
matter, first, 4, 34, 37, 53, 74n6, 83; second , 4, 34, 36-37, 50, 74n6, 83, 92, 99-100n7, 117-18, 129. *See also*

Ashoori, nature$_{f}$, nature$_{nk}$
Maturana and Varela, 7, 137, 140-148, 157n59, 157-58nn63-73, 179, 227. *See also* language of machines; machines, (MV) machines and; transforming organisms to become (biotic) machines
Mayr, Ernst, 71, 80n45, 173, 175, 208-09n31
means/end rationality, 162, 203n3, 233-235, 237, 239. *See also* Hume, Humean projectivism, instrumental rationality
mechanism (philosophy of), 5, 11, 14-19, 22, 25-27, 30, 33, 36, 39-40, 41n2, 145. *See also* materialism (ontology of), Scientific Naturalism (metaphysics of)
Mendelian genetics, 52-54, 65, 70-71, 75-76n8, 145, 226; hybridization technology and, 52-54, 75-76n8, 76 n12, 226. *See also* biotechnology, molecular (DNA) genetics
Merchant, Carolyn, 245-46, 250n12, 251n20
microcomputer technology, 3, 52, 81, 93, 110, 133, 138, 173, 223. *See also* biotechnology, future technology, nanotechnology (molecular)
Mill, J. S., 46n52, 84- 85, 100n8n12
Mitcham, Carl, 20, 21, 36, 43-44nn27-32, 47n69, 54-56, 63, 64, 67, 77n13, 77n19, 77nn21-22, 78n32, 79n36, 121, 135, 152n29, 156n52
modernity, 2, 5, 7, 11-12, 20, 22, 28, 31-33, 39, 41nn1-2, 56, 63, 85, 123-126, 132, 134, 140, 145, 152 n35, 152-53n36, 155-56n50, 182, 189, 201, 213n46, 225, 228, 234-35, 237, 239, 246, 251n23. *See also* worldview, modern; new philosophy; new science
Modern Project of Science and Technology, 2, 63-64, 70, 224. *See also* goals of the new science, materialism (ontology of), modern science, modern technology, Scientific Naturalism (metaphysics of), transforming the natural to become the artefactual
modern science, 2, 3-6, 11, 22, 29, 31, 33, 36, 39- 40, 43-44n32, 46nn52-54, 49, 51-54, 56-57, 59-60, 62-71, 73, 78n25,78-79n35, 81, 83, 85, 93, 96, 99-100n7, 117-18, 120, 123, 125, 129, 151n25, 153n36, 161, 179, 181, 184, 189, 201, 203, 221n93, 223-24, 227-28, 248; as Theoretical Technology (Heideggerian inversion), 22, 43-44n32, 63-64. *See also* deep(er) scientific theories and discoveries, deep(er) technologies, goals of the new science, new science, philosophy of science and of technology
modern technology, 1-7, 20-22, 29, 31, 40, 42n12, 43n28, 43-44n32, 46n49, 50-54, 56-57, 59-60, 62-71, 75-76 nn8-9, 78-79n35, 81, 85, 93, 96, 99-100n7, 107, 109, 115, 117, 120, 125-29, 131-34, 143, 149nn4-8, 150-51 nn21-22, 152n35, 153-55n36, 159, 161, 173, 176, 179, 181, 184, 189, 200-201, 203, 221n93, 223-25, 227-28, 232; as applied science, 43-44 n32, 62, 64-65. *See also* biotechnology; deep(er) technologies; history of technology, (Phase IIA and B); extant technology; future technology; philosophy of science and of technology
molecular (DNA) genetics, 31, 33, 52-54, 68, 70, 72, 76n9, 139, 146, 148. *See also* biotechnology, Mendelian genetics
moral, agent, 182, 197-199, 213n46, 218n85, 251-52n25; responsibility and agency, 182, 197-98, 213n46, 218nn82-84, 221n91, 251-52n25; patient, 198-99, 247, 251-52n25; subject, 198, 218n84, 247-48, 251-52 n25. *See also* anthropogenic and non-anthropogenic, philosophical significance of the distinction between
moral considerability, 2, 3, 9, 73n2, 119, 141, 161, 166-69, 187, 178-79, 190-91, 194, 204n4, 207n25, 208n27,

212n39, 242-44, 252n25. *See also* independent value, intrinsic value
Morris, William, 237, 239n20
Mumford, Lewis, 52, 56, 57, 60- 61, 64, 75n8, 77n24, 78nn28-31, 79n36, 93, 105n31, 115, 150n16
Munro, D. H., 192, 217n73
mutely-enacted values, 8, 161, 165,177, 193, 210n35, 243, 249n5. *See also* Humean projectivism, recognized-articulated values, sole locus of intrinsic value, sole source of all values, valuable 'for itself'

Naess, Arne, 180, 182, 185, 212n42, 214n48
nanotechnology (molecular), 1-4, 9n3, 14, 31, 33, 42n12, 52, 76n10, 81, 83, 85, 92, 93, 99-100n7, 107, 110, 112, 114-119, 121, 123, 129, 133, 136, 138, 150n21, 151n25, 159, 173, 180, 183, 212n41, 223, 225. *See also* Drexler, K. Eric, future technology
narcissism, 2, 8, 160, 194-94, 201-2, 219-21nn89-91, 221nn93. *See also* *homo faber*, ontological elimination (impoverishment)
the natural and the artefactual, 2, 4, 6-7, 48n71, 50, 57, 73-74n3, 84, 100 n10, 119, 146, 159, 178, 181-82, 184, 189, 211n37, 216n67, 225, 229. *See also* artefactual, nature$_{fa}$, dyadism (ontological), transforming the natural to become the artefactual
naturalization of humanity, 97n3, 123, 132-134. *See also* humanization of nature
natural kinds, 3-5, 51-54, 70, 74-75n6, 83, 90, 93, 96, 99n7, 104n24, 117-23, 151n22, 178-79, 183-84, 223-25; supersession of, 120, 122. *See also* nature$_{nk}$; future technology, and radical threat to the natural; transforming the natural to become the artefactual
naturally-occurring beings (entities) or processes, 1, 4, 9, 32, 36-37, 39-40, 50, 52, 73n1, 83, 85, 92, 95, 97-98n3, 99n7, 117, 129, 136, 138-40, 146, 148, 160, 166, 178-80, 190-94, 200, 203, 205n10, 212nn40-41, 226-27, 230-31, 236, 239, 243. *See also* independent value, nature$_{fa}$, trajectory and independent value
Nature, 110, 149-150n8
nature$_{c}$, 83-84, 97-98n3, 99n5, 165, 183, 187, 189, 217n68, 241. *See also* nature (nonhuman), different senses of
nature$_{f}$, 74-76n6, 83-84, 184. *See also* nature (nonhuman), differrent senses of; first matter
nature$_{fa}$, 82, 84-88, 90, 93, 98n4, 101-2n13, 103n22, 117, 159, 179, 182, 184, 188-89, 193-94, 201, 241. *See also* nature (nonhuman), different senses of; the natural and the artefactual
nature$_{hi}$, 83-84, 102-3n16. *See also* nature (nonhuman), different senses of; human intervention
nature$_{nh}$, 82-86, 88-90, 93, 100-101n8, 101n13, 103n22, 103-4n24, 159, 179, 182, 189, 193, 201-2. *See also* nature (nonhuman), different senses of; culture (human)/nature; dualism; dyadism (ontological)
nature$_{nk}$, 74-75n6, 83-85, 90, 93, 99, 103-4n24, 117, 159, 179, 182, 188-89, 201. *See also* nature (nonhuman), different senses of; natural kinds; second matter; transforming the natural to become the artefactual
nature$_{p}$, 83-84, 86-90, 92, 98n3, 99n6, 101-2n13, 102-3n16, 103n22, 188. *See also* nature (nonhuman), diffferent senses of; solar system
nature (nonhuman), abiotic, 3, 6-9, 50, 114, 136, 160-61, 163, 164, 166-169, 172-174, 176, 204n7, 206n12, 206 n19, 207-8n25, 252n25; biotic, 7-8, 90, 103n16, 112-13, 118, 139, 141, 160, 161, 163, 166, 168, 172-73, 176-77, 189, 206n12, 206nn17-18, 206-7n22, 207n23, 207n25, 211n36, 217n68, 223, 226-27, 252n25; dif-

ferent senses of, 6, 9, 81-86, 98n4, 100-102nn8-13, 102-3n16, 103n22, 102-4n24, 187, 241; end of , 6, 86-91, 100-101n13; as 'the (ontological) Other,' 4, 7-8, 83, 101-2n13, 119, 140, 173, 180-81, 183, 187, 189, 193-94, 199, 203, 211-12n38, 221 n91, 221n93, 225-27, 246, 248; as social construct, 8-10, 65, 79n39, 82, 98n4, 177, 195, 210n35, 217n75, 241-49, 251; as work of art, 8, 162, 203, 229-37, 238nn4-5. *See also* dualism, dyadism (ontological)

nature being, morally and ontologically latent or void, 127-128, 228; morally void, 130, 132, 145, 161-62; ontologically void, 128-29, 153-55nn38-42. *See also* Idealism, Scientific Naturalism (metaphysics of)

nature-replacing technology, 3, 7, 109-13, 160; nature-saving technology and, 7, 109-13, 160. *See also* transforming the natural to become the artefactual; future technology, radical threat to the natural and

new philosophy, 5, 11, 12, 16, 19-20, 22, 23, 25, 27, 28, 31, 62, 161, 224. *See also* new science, Scientific Naturalism (metaphysics of)

new science, 5-6, 11-12, 15-19, 22-23, 25, 27-29, 43-44n32, 44-45n36, 59, 161. *See also* modern science, new philosophy

New Scientist, 72, 80n48, 84, 90, 100n7, 104n24, 109, 133, 149n4, 155n46, 170, 206n22

Newton, Issac, 5, 9n1, 11, 22, 41n2, 44n33, 59, 70

Nicholas of Autrecourt, 23, 45n39

Nolt, John, 201, 219n89

nonanthropocentric/nonanthropocentrism, 112, 116, 150n18, 177, 185-187, 195, 210n35, 215n57, 242-245, 247-249, 250n7. *See also* anthropocentric, anthropocentrism

nonanthropogenic, 82, 86, 87, 98n3, 101-2n13, 102-3n16, 160, 165, 194-196, 198-99, 200, 231, 242-43, 249, 249-250n5. *See also* anthropogenic; anthropogenic and nonanthropogenic, philosophical significance of the distinction between

Norton, Bryan, 86, 87, 101n13, 164, 172, 184-187, 189, 204n7, 206n21, 215n55, 215nn60-61, 217n68, 225

Nussbaum, Martha, 124, 152n34

Observer, 68, 72, 79n43, 80n49, 83, 90, 99n7, 104n25

Ockham, William (of), and John Duns Scotus, 23, 45n39

Olby, Robert C., 52, 76nn8-9

old philosophy and old science, 12-15. *See also* modern science, new philosophy, new science

ontological categories (distinctions), 2, 4, 7, 57, 73-74n3, 85, 95, 100n8, 119, 122-23, 139-40, 159, 177, 179-82, 184, 189, 191-94, 200, 203, 205 n10, 217n69, 225-26, 229, 231, 237, 247. *See also* dyadism (ontological)

ontological deflationism, 122, 194

ontological elimination (impoverishment), 2, 7-8, 73n3, 83, 85-86, 93, 107, 113-14, 117-20, 138, 140, 148, 160, 173, 180-81, 189, 199, 200-201, 220n89, 225-28, 247-248. *See also* *homo faber*, narcissism

ontological independence, 4, 8-9, 81, 92, 160, 177-80, 192-194, 199, 210n35, 220n89, 228, 237. *See also* the natural and the artefactual; primary characteristic; dyadism (ontological); nature (nonhuman), as 'the (ontological) Other'

ontology-led environmental philosophy, 2, 8-9, 73-74n3, 114, 160, 190-91, 193, 212n39, 217n70, 223. *See also* axiology-led environmental ethic, independent value, primary characteristic, trajectory and independent value

Pacey, Arnold, 56, 77n20

Passmore, John, 229-30, 232, 238nn3-

4, 244, 250n7
patenting of modified organisms, 136-37, 157n58
philosophy of science, 6, 12, 17, 22, 28, 42n12, 43nn23-24, 43-44n32, 49, 62, 64-66, 69, 71, 78-79n35, 223. *See also* goals of the new science, history of technology, modern science
philosophy of technology, 6, 20, 43-44nn27-32, 49, 55-57, 62, 64-67, 69,77n22, 77-78n21, 78n35, 82. *See also* goals of the new science, history of technology, modern science
philosophy of science and technology, 6, 43-44n32, 49, 62-69, 78-79n35. *See also* goals of the new science, history of technology, modern science
Plato, 16, 33-35, 123-24; Platonic, 152 n34; Platonism, 13
Plumwood, Val, 82, 97n3, 140, 157n59, 161, 168-173, 178, 181-82, 205-6 n12, 208n27, 212n43, 213n44, 243, 250n7; hyperseparation and, 97n3, 100n8, 181, 213n44
Popper, Karl, 29, 46n52, 66, 79n41; Popperian, 44n35
positivism. *See* Comte, Hobbes, new philosophy, new science
postmodernism, 82, 97n3, 140, 157n59, 184, 189, 215n52, 216-17nn67-68, 225, 241, 246-47, 249nn2-3, 251n16, 251n20, 251nn23-24. *See also* deconstructionism
pre-Mendelian techniques of breeding, 76n8
pre-modern science, 57. *See also* old philosophy and old science
Preus, Anthony, 12, 42n8
Prigogine, Ilya, and Isabelle Stenger, 140, 157n61
primary characteristic, 9, 81, 178-79, 192-93, 205n10, 211-12n38, 226. *See also* independent value; nature, as 'the (ontological) Other;' ontological categories (distinctions); ontology-led environmental philosophy; secondary characteristics; trajectory and independent value; valuable 'by itself'
Proctor, James, and Richard White, 246

qualities, primary, 17, 22, 70, 129-30, 161, 224; secondary, 12, 17-18, 70, 130, 224. *See also* Humean projectivism, materialism (ontology of), mechanism (philosophy of), Scientific Naturalism (metaphysics of), *res extensa*

Rackham, Oliver, 93, 105n31
Rapport, David J., 188, 215-16n62
realism: critical/metaphysical, 51, 65-66, 74n6, 79n39, 98-99n4, 251; instrumental, 64, 78-79n35; as pragmatism, 109, 149n5. *See also* ecorealism, Ihde
recognized-articulated values, 8, 161, 165, 177, 193, 210n35, 249-250n5. *See also* Humean projectivism, mutely-enacted values, sole source of all values, valuable 'in itself'
Reed, Peter, 182, 214n48
Regan, Tom, 185, 230, 238n6, 243, 250n7
Regis, Edward, 52, 69, 76n10, 79n44, 84, 100n7
res extensa, 129, 161, 224. *See also* mechanism (philosophy of), materialism (ontology of), Scientific Naturalism (metaphysics of)
resource conservation, 187, 231-32, 236. *See also* conservationism and preservationism, resource preservation
resource preservation, 187, 232. *See also* conservationism and preservationism, resource conservation
Rifkin, Jeremy, 33, 47n62
Rindos, David, 52, 76n8
Roberts, H. F., 52, 75n8
Robinson, Kim Stanley, 90, 104n25
Rolston, Holmes, 87, 102n13, 163, 173, 178, 182, 185, 197-99, 204n5, 208n27, 211n37, 213n47, 215n57, 218n84, 231, 238n10, 241, 249n1
Ross, David, 34, 47n64

Rothenburg, David, 7
Routley, Richard, 29, 33, 46n51, 47n61, 86, 96, 163. *See also* Sylvan, Richard
Routley, Richard, and Val Routley, 33, 47n61, 163
Routley, Val, 33, 47n61, 163, 230, 238n6. *See also* Plumwood, Val

Sagan, Carl, 172, 207n22
Sagan, Dorian, and Lynn Margulis, 82, 98n3
Sagoff, Mark, 113, 150nn11-14
Sartre, J.-P., 164, 204n7
Science, 52, 76n10, 83, 100n7, 90, 104n25, 223, 228n1
science as facts and science as advocacy, 248, 252n26.
Scientific American, 83, 90, 99n7, 104n24, 134, 155n50, 172, 206-7n22
scientific laws (laws of nature), 6, 11, 13, 18, 24, 28, 41n2, 51, 52, 63, 67, 69, 70, 72-73, 74-75n6, 114, 117, 122, 128, 150n21, 173, 182, 184. *See also* epistemological goals (respective) of science and technology, technological rules
Scientific Naturalism (metaphysics of), 2, 4-5, 7, 18, 43-44n32, 62, 82, 98n4, 117, 123, 129-30, 132, 155n40, 165, 183, 223. *See also* materialism (ontology of), mechanism (philosophy of), *res extensa*
Scitovsky, Tibor, 202, 221n92
Searle, John, 51, 74n6, 169, 206n13, 241, 246, 249n1, 251n19
secondary characteristics, 160, 166, 178-79, 189-90, 192-94, 205n10, 211n36, 211-12n38, 226. *See also* axiological, axiology-led environmental ethic, primary characteristic
Self-Realization, 140, 157n59, 182, 212nn42-43, 214n48. *See also* Deep Ecology
Shapere, Dudley, 12, 41n4
Shiva, Vandana, 117, 151n22
Simmonds, Norman W., 52, 76n8
Singer, Peter, 166, 185, 204n8, 230, 238n6, 243, 250n7
Slote, Michael, 201, 219n89.
Sober, Elliott, and Richard Lewontin, 71, 80n45
solar system, 1-2, 9n3, 87, 103n16, 160-61, 172-73, 176-77, 203, 208 n27, 221n93, 227. *See also* Asymmetry/Autonomy/ No External Teleology Theses; independent value; intrinsic value, abiotic nature and; Mars; terraformation; trajectory and independent value; valuable 'by itself;' Venus
sole locus of intrinsic value, 8, 29, 46n51, 128, 130, 136, 156n54, 163, 166, 181, 185, 204nn4-6, 224, 228, 231, 238n10. *See also* human chauvinism, Sole Value Assumption and; Humean projectivism; mutely-enacted values; recognized-articulated values; valuable 'by itself;' valuable 'for itself;' valuable 'in itself'
sole source of all values, 8, 22, 119, 123, 128, 130, 132, 156n54, 161-66, 185, 198, 204nn4-6, 210n35, 215n57, 224, 228, 231, 238n10, 249-50n5. *See also* human chauvinism, Sole Value Assumption and; Humean projectivism; mutely-enacted values; recognized-articulated values; valuable 'by itself;' valuable 'for itself;' valuable 'in itself'
Solow, Robert, 117, 151n22
Soper, Kate, 82, 83, 98-99nn4-5, 241, 249n2
Soulé, Michael E., 87, 90, 93, 102n13, 103n17, 241, 249n2
Sprigge, Timothy, 229, 238n4
St. Francis of Assisi, 244, 250n8
Staudenmaier, John M., 64, 79n35
Stone, Christopher, 173, 213n47
Sylvan, Richard, 50, 73n2. *See also* Routley, Richard; Routley, Richard, and Val Routley
Sylvester, E. J., and L. C. Klotz, 52, 76n9

Takacs, David. *See* science as facts and

science as advocacy
Taylor, Paul, 172, 185-87, 207n24, 215n61, 243, 250n7
techne, 14, 20-21, 123-24. *See also* *episteme*
techniques and technologies, distinction between, 54-56, 77n18
technological rules, 6, 21, 63, 67, 69, 71-73, 182, 242. *See also* deep(er) scientific theories and discoveries, deep(er) technologies, epistemological goals (respective) of science and technology, history of technology, philosophy of technology, scientific laws (laws of nature)
technology, definition and scope of, 54-56, 62, 69, 77n18; etymology of, 20-21, 43n27; 'found', 77n25; pre-modern, 6, 56. *See also* deep(er) control/manipulation of nature, deep-(er) technologies, history of technology, modern technology, philosophy of technology, *techne*
teleology, 14, 19, 27, 31-32, 36, 39- 40, 49, 73n3, 144, 157n69, 164, 169, 208n28, 208n30; new, 5-6, 32, 40; old, 32, 40. *See also* external teleology, extrinsic/imposed teleology, intrinsic/immanent teleology
teleomatic, 167-170, 173. *See also* nature (nonhuman), abiotic; teleonomic
teleonomic, 169, 173, 206n17. *See also* intrinsic/immanent teleology, *telos/tele*
telos/tele, 4-5, 8, 22, 33, 36, 39-40, 53-54, 92, 94- 96, 97, 117, 140, 145-148, 157, 164, 166, 169, 177, 182, 191, 205n10, 210n34, 214n47, 226-27, 236-37. *See also* intrinsic/immanent teleology
terraformation, 2-3, 9n3, 29, 46n50, 90, 104n25, 107, 114-15, 122, 203, 221n93. *See also* independent value; intrinsic value, abiotic nature and; Mars; solar system; trajectory and independent value; valuable 'by itself;' Venus
Tracy the sheep, 140, 157n58. *See also* biotechnology
trajectory and independent value, 8, 93, 105n30, 160-61, 177-180, 192-94, 210n34. *See also* independent value; nature, as 'the (ontology) Other;' ontology-led environmental philosophy; primary characteristic
transforming the natural to become the artefactual, 1-5, 7, 9n3, 49-50, 54, 73-74n3, 77n22, 81, 84-85, 90, 92, 95, 97-98n3, 103n22, 107, 113-15, 118, 129, 132-35, 139-40, 146-47, 155n40, 159, 166, 173, 176, 180-81, 188, 195, 201-3, 219-20n89, 221n93, 223-28, 231-32, 247. *See also* deep-(er) control/manipulation of nature, deep(er) scientific theories and discoveries, deep(er) technologies, goals of the new science, history of technology, *homo faber*, humanization of nature, modern science, modern technology, philosophy of technology
transforming organisms to become (biotic) machines. *See* machines
'true socialists', 184, 215n52. *See also* Deep Ecology, postmodernism
Tylecote, Andrew, 33, 47n62

Vaihinger, H., 29, 44n35, 46n52
valuable 'by itself,' 8, 10n5, 161, 167, 172, 174-76, 193, 204n7, 205n10, 208n28n30, 210n34, 227; the 'No External Teleology Thesis' and, 174, 208n28n30. *See also* independent value, trajectory and independent value
valuable 'for itself,' 8, 10n5, 39, 136, 139, 159, 161, 164-67, 172; 174, 204n7, 208n28, 210n34, 227. *See also* intrinsic/immanent teleology; intrinsic value, two senses of; biotechnology, radical threat to biotic nature and; valuable 'by itself'
valuable 'in itself,' 8, 10n5, 141, 164-66, 181, 204n7. *See also* intrinsic value, two senses of; valuable 'for

itself;' Sartre
Venus, 90, 104n25, 122. *See also* Asymmetry/ Autonomy/ No External Teleology Theses; independent value; intrinsic value, abiotic nature and; Mars; solar system; terraformation; trajectory and independent value; valuable 'by itself'
Verhoog, H., 148, 158n74
Vilkka, Leena, 177, 210n35
Vogel, Steven, 241, 243, 247-48, 249-50n5

Weinberg, Steven, 134, 155-56n50
Weisheipl, James A., 34, 35, 47n63
Westfall, Richard S., 17, 43n20
Wick, Warner, 13, 42n9
Wilkerson, T. E., 51, 74-75n6
Wilmut, Ian. *See* Dolly the cloned sheep. *See also* biotechnology, future technology
Wilson, E. O., 90, 103n23, 172, 196, 206n21
Woolley, Benjamin, 116, 120, 150n19, 151-52n27
worldview, pre-modern, 1, 5, 11-12, 14-15, 21-22, 32, 36, 39-40, 54, 135, 213n46, 251n23; modern, 1, 2, 5, 11, 14-16, 19, 27, 36, 40-42, 49, 70, 97n3, 140, 148, 234, 246, 251n23; post-modern, 82, 97n3, 140, 157n59, 184, 189, 215n52, 216-17nn67-68, 225, 241, 246-47, 249nn2-3, 251n16, 251n20, 251nn23-24. *See also* teleology, mechanism (philosophy of), materialism (ontology of)
Wright, H. E., and D. G. Frey, 87, 102n16

Xenos, Nicolas, 124, 152n33

Yoxen, Edward, 52, 76n9

Zimmerman, Michael, 63, 78n33
Zubrin, Robert, and Richard Wagner, 90, 104n25

About the Author

Keekok Lee received in philosophy the degree of B.A. from the University of Singapore, of B.Phil from the University of Oxford, and Ph.D. from the University of Manchester .

Lee taught philosophy for a short while at the University of Singapore before joining the University of Manchester; but, after many years, she will be leaving it shortly, out of despair.

In very broad terms, it could be said that Lee's long standing research interest lies in attempting a philosophical assessment of modernity, in particular, of two of its distinctive features, namely, its notion of the legal-rational state and its conception of science, scientific methodology together with its accompanying aspiration to control and manipulate nature *via* the technology its science induces. This preoccupation, in turn, means that she is involved not only with moral, social, political, legal and environmental philosophy, but also with the philosophy and history of science as well as technology. Of late, she has been much involved with problems in environmental philosophy.

Apart from numerous articles on aspects of those subjects just mentioned, she is the author of *A New Basis for Moral Philosophy* (Routledge and Kegan Paul, 1985), *The Positivist Science of Law* (Avebury, 1989), *The Legal-Rational State: A Comparison of Hobbes, Bentham and Kelsen* (Avebury, 1990) and *Social Philosophy and Ecological Scarcity* (Routledge, 1989).